INTENT ON THE PALEOLITHIC

sidestone press

INTENT ON THE PALEOLITHIC

Papers in honour of Prof. Dr. Wil Roebroeks

edited by
G.L. DUSSELDORP , W. CHU, C.C. BAKELS AND M.A. SORESSI

ANALECTA
PRAEHISTORICA
LEIDENSIA 55

Published by Sidestone Press, Leiden
www.sidestone.com
E-mail: info@sidestone.nl
Phone: (+31)(0)71-7370131

Series: Analecta Praehistorica Leidensia
Series editors: Victor Klinkenberg, Carol van Driel-Murray, Richard Jansen, Nathalie Ø. Brusgaard

Lay-out & cover design: Sidestone Press

Cover illustration: (front) Elephants in Madikwe Game Reserve, South Africa (Udo Kieslich, stock.adobe.com); (back) Elephants at a water hole (2630ben, stock.adobe.com); flint tool (46/106-11) from Site G, Maastricht-Belvédère, wear traces indicate it may have been used to cut elephant or rhinoceros.

ISBN 978-94-6426-382-4 (softcover)
ISBN 978-94-6426-383-1 (hardcover)
ISBN 978-94-6426-384-8 (PDF e-book)

ISSN 0169-7447 (Print)
ISSN 2665-9573 (Online)

DOI: 10.59641/i75887lu

Universiteit Leiden

Contents

In Dusseldorp, G.L., Chu, W., Bakels, C.C. and M.A. Soressi (eds) 2025. Intent on the Paleolithic. Papers in honour of Prof. Dr. Wil Roebroeks. *Analecta Praehistorica Leidensia* 55. Leiden: Sidestone Press, pp. 7-8.

Editorial

Gerrit L. Dusseldorp, Wei Chu,
Corrie C. Bakels, Marie Soressi

Egalité

How many scholars, how many laureates of the prestigious Spinoza prize[1] have a record as member of a punk band? How many have students that named their pet dog after him, meant as a tribute to the animal lover Wil Roebroeks is? Taking out students to watch red deer and wild boar in the forest during courses on the palaeolithic? It reflects the egalitarian, relaxed atmosphere Wil fostered in combination with his professional relationships. The latter can be characterized by a critical examination of claims, a commitment to fostering debate by posing questions rather than providing answers, and a fearless integration of data and theories from various relevant fields at the international level.

Many scales to comprehend a fragmented archaeological record

Wil's entrance on the Palaeolithic stage started with his excavations at Maastricht-Belvédère (Roebroeks 1988). With a group of archaeologists, several Limburg quarry walls had been scoured for Palaeolithic artefacts. The identification of the Belvédère quarry as a promising location led to extensive excavations with several high-density lithic scatters documented in detail. The conditions often required grit, demonstrated by the photos of Wil and team excavating in the snow on an island in the ever-deepening loess quarry (familiar to generations of Leiden students attending his classes).

The work at Belvédère showcases a recurring theme of Wil's work: an earnest attempt to reconcile the different scales of information to understand past human behaviour. By meticulously documenting flint artifacts crafted at least 250,000 years ago along the banks of an oxbow lake in the Meuse Valley and implementing an extensive refitting program, the study of the site complex provided "snapshots" of Neanderthal behaviour. Yet, Roebroeks also provided a landscape-scale view of human behaviour at this interglacial riverside by, for example, excavating off-site locations with the same care as the classical sites. These revealed a low-density "veil of stones" covering the landscape whose composition differed in important respects from those from the high-density scatter "sites" where tools were produced, while they were used and discarded; sometimes at the end of their use-life in the "veil of stones". However, Roebroeks also recognized the significance of the broader perspective. Expanding on raw material provenance studies, he developed a

Gerrit L. Dusseldorp
Faculty of Archaeology, Leiden University, Leiden, The Netherlands; Palaeo-Research Institute, University of Johannesburg, Johannesburg, South-Africa
g.l.dusseldorp@arch.leidenuniv.nl

Wei Chu
Faculty of Archaeology, Leiden University, Leiden, The Netherlands.
w.chu@arch.leidenuniv.nl

Corrie C. Bakels
Faculty of Archaeology, Leiden University, Leiden, The Netherlands
c.c.bakels@arch.leidenuniv.nl

Marie Soressi
Faculty of Archaeology, Leiden University, Leiden, The Netherlands
m.a.soressi@arch.leidenuniv.nl

1 Wil Roebroeks was awarded the Spinoza Prize (often referred to as the Dutch Nobel Prize) in 2007. The Dutch Organization of Scientific Research hailed him as the most outstanding Dutch archaeologist, both nationally and internationally.

scenario on transport distance and planning depth spanning human history (Roebroeks *et al.* 1988). This exemplifies an enduring strength: connecting meticulous archaeological analysis with the wider continental picture.

The Palaeolithic communities of Europe

The Netherlands is currently not particularly rich in Palaeolithic sites and it seems that when the United Netherlands were broken up in 1830, the Belgians got all the good sites. By the 1980s the Dutch Palaeolithic community was still small and reckoning with the aftermath of a forgery scandal in the northern Netherlands. To Roebroeks, one must imagine, tackling the Palaeolithic on a European scale seemed like the only logical way forward and the only scale that his students could even consider. Wil's emphasis on scrutinizing the evidence for Europe's earliest occupation appears as a logical progression. Through a workshop on the topic at Tautavel, two more themes become apparent. In terms of content, one can equivocally say that Roebroeks is not averse to taking controversial (but well-argued) standpoints such as that in the 1990s there was no good evidence for human presence in Europe prior to 500,000 years ago. The workshop and subsequent collaborations demonstrate an intermediary of sometimes diametrically opposed personalities.

Wrong for all the right reasons

Not one to shy away from academic debate on these frequently controversial theses, Roebroeks also is acquiescent when new data falsify his hypotheses. With the discovery of the site Pakefield with a small assemblage of artefacts laid down ~700,000 years ago (but see Chu 2016), it is Roebroeks himself who announces in the accompanying "News & Views" that the short chronology for Europe represents a now flawed hypothesis. (Although earlier there had been a strategic retreat on the short chronology to 40^{0}N (cf. Roebroeks & Van Kolfschoten 1994; Roebroeks & Dennell 1996)).

La méthode Roebroeks can perhaps be characterized as based on highly critical scrutiny of any claim, placing high value on debate, but also as a holistic approach that considers data and theories from any and all relevant fields. His students will recall especially his praise of the work of Lewis Binford, who was "wrong for all the right reasons". This is a higher honour under the méthode Roebroeks, than getting it right. Binford was thus duly awarded an honorary doctorate by Leiden University. This really is a rare and august honour, which was bestowed on the eponymous New Archaeologist just one year after Nelson Mandela received the same honour. More importantly perhaps, it allowed his students to attend lectures and crucially have drinks in the local bar with the American guest of honour.

A role model: Inspirational teacher and non-authoritarian scholar

In a business rife with hierarchy, Wil has (to our experience) always been approachable, provided opportunities, great and small to his students and other Palaeolithic enthusiasts. In short he has always acted as a connector between themes and between people.

Always one to ensure friends and colleagues got the recognition they deserved, Wil shied away from momentous honours such as the Spinoza Prize, and this Festschrift, dedicated to the hooggeleerde Roebroeks, is likely to make him recoil, but really, is less for him than it is for those he will continue to inspire in the future. The content of the volume is testament of some of his legacy and surely there will be more in here to debate!

References

Chu, W. 2016. *Fluvial processes in the Pleistocene of Northern Europe*. Oxford. (BAR International Series 2797).

Dennell, R. and W. Roebroeks. 1996. The earliest colonization of Europe: the short chronology revisited. *Antiquity* 70: 535-542.

Roebroeks, W. 1988. From Find Scatters to Early Hominid Behaviour. A Study of Middle Palaeolithic Riverside Settlements at Maastricht Belvédère (The Netherlands). *Analecta Praehistorica Leidensia* 21.

Roebroeks, W. and Th. van Kolfschoten. 1994. The earliest occupation of Europe: a short chronology. *Antiquity* 68: 489-503.

Roebroeks, W., J. Kolen and E Rensink. 1988. Planning depth, anticipation and the organization of middle Palaeolithic technology: the "archaic natives" meet Eve's descendants. *Helinium* 28: 17-34.

In Dusseldorp, G.L., Chu, W., Bakels, C.C. and M.A. Soressi (eds) 2025. Intent on the Paleolithic. Papers in honour of Prof. Dr. Wil Roebroeks. *Analecta Praehistorica Leidensia* 55. Leiden: Sidestone Press, pp. 9-10.

Preface

Jean-Jacques Hublin

In the shadows of prehistory, where the origins of present-day humankind lie hidden, Wil Roebroeks has illuminated paths that have deepened our understanding of the complex journey of human evolution. This volume, a Festschrift in his honour, stands as a testament to the broad impact of his work, bringing together contributions from colleagues and students inspired by his research, methodology, and profound insights into the human past.

Human evolution has been a long and complex process of interaction between biology and culture, unfolding in a world vastly different from our own. How can we imagine humans who were neither the people of today nor even akin to recent hunter-gatherers? The many subfields of palaeoanthropology strive to reconstruct human presence in these lost worlds. Unfortunately, we can only grasp certain aspects of these mysterious territories. Today, we can reconstruct past landscapes and climates with some accuracy. Yet, the people who inhabited these environments are often known to us only through their material production, notably stone tools, nearly indestructible evidence that has accumulated on the Earth's surface. Raw materials inform us about past mobility. In well-preserved assemblages, we can analyse human subsistence with some details and more broadly the way hominins interacted with their environments. Palaeontology seeks to reconstruct the anatomy and appearance of ancient hominins, and palaeogenetics has recently added other elements to our understanding of their biology and behaviour. However, both rely on a limited evidence whose representativeness can often be questioned. Finally, except in some rare cases, dwellings remain difficult to reconstruct. Social structures, reproductive strategies, and more generally interactions between individuals are mostly domains of speculation, and sometimes of pure storytelling. Especially for the earliest periods, the daily lives of deep-past humans remain elusive. Beyond analyzing tools, faunal remains, or describing fossil hominins when we have them, we must therefore turn to numerous related sciences, for example cultural anthropology, human ecology, or evolutionary psychology, in order to form an image of them.

Throughout his career, Wil Roebroeks has eschewed specialization, consistently exploring diverse avenues that broaden our grasp of Palaeolithic societies in all their complexity. He quickly demonstrated his expertise in the field of prehistory and critically questioned how Palaeolithic sites can be interpreted. But his intellect was also constantly engaged by a variety of very broad subjects, beyond the narrow fields of traditional prehistory. Wil Roebroeks is a generalist able to seize the wider picture of human condition in the deep past, and one of the few palaeoanthropologists I know able to creatively deal with subjects as varied as the oldest traces of humans outside the African continent, the mechanisms of innovation among hunter-gatherer societies, or the many effects of fire production and use. Often challenging conventional wisdom, Wil Roebroeks has revealed

Jean-Jacques Hublin
Collège de France, France
jean-jacques.hublin@college-de-france.fr

new perspectives about our ancestors and relatives. Personally, and undoubtedly it is the case for many of his colleagues, it is this originality and openness associated with a great intellectual rigor that make Wil Roebroeks' work so interesting and engaging.

A significant portion of this volume focuses on the elemental force of fire, a subject central to Roebroeks' research. The contributions delve into how early humans harnessed fire for cooking, protection, and landscape management, fundamentally altering their environment and themselves in the process. The studies on Neandertal and Denisovan genomes provide fascinating insights into how early humans adapted biologically to the use of fire, highlighting the evolutionary advantages conferred by these adaptations. The exploration of hearths and huts within the context of hospitality and social structures also revisits themes central to Roebroeks' own inquiries – how our ancestors harnessed fire and constructed spaces not just for survival, but for the foundational social practices that underpin humanity.

Another theme that emerges strongly in this volume is the construction of early human environments and the social structures they reveal. Not only on the scale of a hut but also on the scale of a landscape. Like the milieu reconstruction around the site of Maastricht Belvédère, the exploration of Eemian hornbeam forests of Neumarkt Nord, their burning by archaic humans, and the impact of large mammals like the straight-tusked elephant on these environments further enrich our understanding of landscapes the early humans called home.

This Festschrift also showcases the importance of material culture and subsistence strategies in understanding the behaviours of ancient hominins, in particular of Neandertals, that have been the focus of many works by Wil Roebroeks. Today, one can see the complexity of ancient hominin behaviours, not only in stone-tool making technologies, and Wil Roebroeks largely contributed to this advance also.

Reflecting Roebroeks' interdisciplinary approach, the volume encompasses a wide range of theoretical explorations. They span from agent-based modelling of fire-driven hunting strategies to the settlement patterns of early humans in north-west Europe. These contributions demonstrate the power of combining archaeological evidence with insights from genetics, geography, and cultural anthropology to construct a more nuanced picture of human evolution.

This Festschrift is a journey through time, guided by the insights and discoveries of one of the most respected figures in the field, yet it is more than a collection of essays; it is a celebration of the spirit of inquiry that Wil Roebroeks has instilled in his colleagues. It reflects his enduring curiosity, his rigorous scientific approach, and his generous mentorship, which has influenced generations of archaeologists. By presenting this volume, we not only acknowledge the past achievements of a distinguished career but also the ongoing influence of his work on future research in Palaeolithic archaeology.

In Dusseldorp, G.L., Chu, W., Bakels, C.C. and M.A. Soressi (eds) 2025. Intent on the Paleolithic. Papers in honour of Prof. Dr. Wil Roebroeks. *Analecta Praehistorica Leidensia* 55. Leiden: Sidestone Press, pp. 11-28.

Four decades of Middle Palaeolithic research at Maastricht-Belvédère (Limburg, The Netherlands): current state of knowledge and perspectives

Dimitri De Loecker

Abstract

At the commercially exploited Maastricht-Belvédère quarry (the Netherlands) a series of well-preserved late Middle Pleistocene find distributions has been studied for more than four decades. The early Middle Palaeolithic relics were excavated from fine-grained fluvial sediments. The local deposits contained fully temperate faunas and could probably be attributed to the late Middle Pleistocene (MIS 7, possibly MIS 9). The material evidence generally seems to reflect very mobile, episodic, and short-term Neanderthal occupations of parts of the river Meuse (Maas) valley, at least 250,000 years ago. Yet, some data suggest that considerable time, energy and skill was invested at certain localities, in order to execute specific technology or food related maintenance and production tasks.

This article gives a brief summary of the local Pleistocene archaeology, focusing on the high residential mobility character of Early Neanderthals. Data from more recent Maastricht-Belvedere studies is concisely integrated to illustrate the complexity of the artefact distributions. It will be argued that we are not dealing here with a 'hit-and-run' record only. Moreover, the fossil remnants of numerous but episodic and inter-related Neanderthal visits eventually created a so-called palimpsest taskscape. Finally some arguments will be mentioned to initiate an appropriate up-to date dating programme at Maastricht-Belvédère.

Keywords: Maastricht-Belvédère (the Netherlands); early Middle Palaeolithic;MIS 7-9; Early Neanderthals; lithic technologies; well-equipped foragers; mobile – episodic – short-term occupation; time – energy – skill investment; dating program.

Introduction

Since the dawn of Palaeolithic archaeology at the beginning of the 19th century, northwest European researchers were predominantly concerned with large lithic surface collections and diagnostic stray finds (*e.g.* Bosinski 1967; Daniel 1972; Ulrix-Closset 1975; Roe 1981). In Dutch southern Limburg these secondary context (surface) assemblages were generally located in the higher landscapes (outside the river valleys) and the investigations were mainly concentrated on typo-/technological interpretations (*e.g.* Stapert 1979; Roebroeks 1980; Kolen *et al.* 1999; Rensink 2005; Verpoorte *et al.* 2016). In fact, well into the 1970s, strenuous efforts were made to answer geological questions, solve dating problems, and study artefacts using culture-historical frameworks. Similarities and differences,

Dimitri De Loecker
Faculty of Archaeology, Leiden University, Leiden, The Netherlands
dimitri.loecker@pandora.be

within and between assemblages and sites, were systematically analysed. Excavations and analysis were dominated by the prospection and sampling of single stratigraphical columns. Detailed typo-chronologies of distinctive Palaeolithic cultures were built, which at the time were believed to fully represent former ethnic groups (Bordes 1972).

In the mid-1960s many archaeologists became disenchanted with this culture-historical branch of Palaeolithic Archaeology. They argued that archaeology was a social science like anthropology, and it should therefore explain the past in terms of social and economic systems, not just simply describe them according to typological schemes (Binford and Binford 1968). Through deductive reasoning from ethnoarchaeological examples, hypotheses and models were constructed to explain observed patterns and changes. These models were tested on the basis of solid evidence. Experts in Palaeolithic research now needed multi-dimensional information on archaeological levels to better reconstruct and understand the behaviour of prehistoric hunters and gatherers. Spatial data on the horizontal distribution of artefacts, within the given levels, was required to reconstruct activity areas. This meant that during the 1970s and 1980s, new excavation methods, involving more precise documentation of the finds, were introduced in Palaeolithic archaeology. Additionally, elaborate statistical computer analyses were initiated and new analytical sources were explored for lithic and spatial analysis. These included experimental flint knapping, pioneered by Crabtree (1972) and Bordes (1961); refitting analysis, triggered by the innovative use of the method at Pincevent (France; Leroi-Gourhan and Brézillon 1966), and use-wear analysis developed by Semenov (1964) and Keeley (1980). The combined ethnoarchaeological and experimental approach, based on a functionalist approach to artefact studies and on the distribution of tools and waste, provided opportunities to answer new questions relating to subsistence strategies and the spatial patterning of activity areas (*e.g.* Cahen *et al.* 1979; Van Noten *et al.* 1980). Moreover, since archaeologists became conscious of the fact that (early) human behaviour is spatially continuous, archaeological interpretations went beyond the site boundaries (whatever that means).

Sites were now explained as the remains of temporary camps, consisting of spatial accumulations of specific activity areas. The wider variation of settlements within a specific region and during a specific period could then be seen as indicative of the land-use system practiced by Palaeolithic groups. These reconstructions of the socio-economic organisation of hunters and gatherers were analysed in relation to former environmental conditions to judge their adaptive success (Binford 1980). By that time, most researchers of the Middle Palaeolithic were in search of absolute dates, stratigraphical contexts, well-preserved artefact distributions and palaeoenvironmental evidence. Thus, the emphasis shifted systematically towards the excavation of stratigraphically sealed and well-preserved settlement remains in sediment receiving areas, such as the loess covered plateaus and river valleys with fine-grained fluvial deposits in northwestern Europe (Bosinski 1966; Tuffreau *et al.* 1985; Roebroeks 1988; Tuffreau and Sommé 1988; Bosinski *et al.* 1995). Moreover, sealed artefact scatters were expected to represent the thin time slices necessary for reconstructing early human on-site activities.

It was against the background of this theoretical debate (descriptive/culture-historical archaeology *versus* explanatory/processual archaeology) and armed with new Palaeolithic research questions and methodological innovations, that Leiden-based archaeologist Wil Roebroeks instituted a long-standing excavation program at Maastricht-Belvédère in the Netherlands. Here, in the hilly riverine setting of southern Limburg a snapshot of the landscape's history of at least 250,000 years ago was investigated from the early 1980s onwards. At the former Belvédère loess and gravel pit, extensive fieldwork has focused on an interdisciplinary study of flint distributions in primary geological context, which occasionally were associated with faunal remains.

The late Middle Pleistocene (Marine Isotope Stage [MIS] 7, see below) Maastricht-Belvédère-evidence is generally seen as reflecting highly mobile behaviour with episodic and short-term occupations of parts of the Meuse (Dutch: Maas) valley. Four decades after the first excavations the Maastricht-Belvédère record still provides new insights. A recent re-evaluation of the data suggests that considerable time, energy and skill was invested in the production, maintenance and use of toolkits at certain Belvédère localities, in order to execute specific food-related activities (De Loecker and Roebroeks 2012; De Loecker 2014). Some of the performed activities may, therefore, have gone beyond the limits of opportunistic '15-minute' flint knapping episodes or short butchering

Site	Area dug (m^2)	Total number of artefacts	Density per square metre						
			Total number of artefacts	**Chips <30 mm**	**Flaked artefacts ≥30 mm**	**Flakes**	**Blade-like flakes**	**Chunks**	**Burned arte-facts**
A	5	80[2]	16.0/6.8	9.4	6.4	6.0	0.2	0.2	0.2
B	20	6	0.3	-	0.3	0.25	0.05	-	-
C	264	3,067	11.61	10.11[3]	1.5[3]	1.48[3]	-[4]	0.071[4]	0.5[4]
D	-	11	-	-	-	-	-	-	-
F	42	1,177	28.0	24.28	3.69	3.19	0.14	0.35	0.35
G	50[1]	75	1.5/1.22	1.06/0.89	0.44/0.36	0.4/0.32	0.02/0.016	0.02/0.016	-
H	54	270	5.0	3.33	1.66	1.55	0.055	0.055	0.018
K	370	10,912	29.49	20.96	8.27	8.01	0.17	0.091	1.66
N	765	450	0.58	0.47	0.11	0.11	0.0013	-	0.0013
July '90	7	15	2.14	1.14	1.0	1.0	-	-	-
L	-	8	-	-	-	-	-	-	-
M	-	44	-	-	-	-	-	-	-
O	-	10	-	-	-	-	-	-	-
Site N, Level X	-	29	-	-	-	-	-	-	-
Section finds	-	67	-	-	-	-	-	-	-

Site	Area dug (m^2)	Total number of artefacts	Density per square metre					
			Cores	**Core Trimming Elements *sensu sticto***	**Tools**	**Tools *sensu stricto***	**Long Sharpening Flakes**	**Transversal Sharpening Flakes**
A	5	80[2]	0.2	0.4	0.4	0.2	-	0.2
B	20	6	-	-	-	-	-	-
C	264	3,067	0.015[3]	0.05[4]	0.087	0.018	-	-
D	-	11	-	-	-	-	-	-
F	42	1,177	0.047	0.12	0.19	0.071	-	-
G	50[1]	75	-	-	0.16/0.13	0.06/0.049	-	0.02/0.016
H	54	270	-	-	0.19	0.074	-	-
K	370	10,912	0.25	0.27	0.37	0.3	0.0027	0.0027
N	765	450	0.0013	0.0026	0.033	0.016	-	-
July '90	7	15	-	-	0.14	0.14	-	-
L	-	8	-	-	-	-	-	-
M	-	44	-	-	-	-	-	-
O	-	10	-	-	-	-	-	-
Site N, Level X	-	29	-	-	-	-	-	-
Section finds	-	67	-	-	-	-	-	-

Table 1. Maastricht-Belvédère. A comparison (basic count and densities per m^2) of the Unit IV primary context sites and section/ test pit assemblages. For details the reader is referred to De Loecker (2006). Table after De Loecker (2006: 236). Remarks: 1 The excavated Site G area together with the test pit has a surface of 61 m^2; 2 within the excavated Site A area only 34 artefacts were found; 3 Site C figures after Roebroeks (1988; n= 3,067); 4 Site C figures after Schlanger's sample (1994; n= 1,438).

events (*cf.* Roebroeks and Tuffreau 1999; McCrone 2000; Hallos 2005).

The objective of this article is twofold. First, a brief summary, mainly based on De Loecker and Roebroeks (2012), of the early Middle Palaeolithic record at Maastricht-Belvédère is given. The emphasis is on the high residential mobility (*cf.* Binford, 1980) of early Neanderthals. Data from recent Maastricht-Belvédère studies are briefly integrated to illustrate the complexity of the artefact distributions. Furthermore, a number of arguments are cited which demonstrate that technologically well-equipped hunter-gatherers did not only use the Maas Valley (*i.e.* Maastricht-Belvédère) for 'hit-and-run' extractions of raw materials. Subsequently, some perspectives for future research will be presented as well. For a suitable understanding of the available data, we will start with a brief outline of the Maastricht-Belvédère setting and its geology.

The Maastricht-Belvédère framework

The former Belvédère loess and gravel quarry is located on the left bank of the river Meuse. Approximately one km north-northwest of the Dutch town of Maastricht (50° 52' 09.40"N, 5°40' 27.33"E),

Area		Number of artefacts					Ratio			
Site	**Area dug (m^2)**	**All tools and fragments**	**Tools and fragments *sensu stricto***	**Cores**	**Flakes and chips**	**Total**	**Tool: waste**	**Tool *sensu stricto*:waste**	**Core: waste**	**Core: tool**
A	5	2	1	1	77	80[2]	1:39	1:78	1:77	1:2
B	20	-	-	-	6	6	-	-	-	-
C	264	23	5	4	3,040	3,067[3]	1:132	1:609	1:760	1:6
D	-	-	-	1	10	11	-	-	1:10	-
F	42	8	3	2	1.167	1,177	1:146	1:389	1:584	1:4
G	50[1]	8	3	-	67	75	1:8	1:22	-	-
H	54	10	4	-	260	270	1:26	1:65	-	-
K	370	137	111	91	10,684	10,912	1:79	1;97	1:117	1:2
N	765	26	12	1	423	450	1:16	1:35	1:423	1:26
July '90	7	1	1	-	14	15	1:14	1:14	-	-
L	-	-	-	-	8	8	-	-	-	-
M	-	3	2	-	41	44	1:14	1:21	-	-
O	-	-	-	-	10	10	-	-	-	-
Site N, Level X	-	-	-	-	29	29	-	-	-	-
Section finds	-	4	3	1	62	67	1:16	1:21	1:62	1:4

Table 2. Maastricht-Belvédère A comparison (basic count and ratios) of the Unit IV primary context sites and section/test pit assemblages. For details the reader is referred to De Loecker (2006). Table after De Loecker (2006: 239). Remarks: 1 The excavated Site G area together with the test pit has a surface of 61 m^2; 2 Within the excavated Site A area only 34 artefacts were found; 3 Site C figures after Roebroeks (1988; n= 3,067).

the locale is situated on the northern border of the northwest European loess-belt. The research of the early Middle Palaeolithic sites of the locale, that started in the early 1980s, focused on an interdisciplinary study of flint distributions, which occasionally were associated with faunal remains (Roebroeks 1988; De Loecker 2006). These distributions, in the form of dense find concentrations, thin artefact spreads and more or less isolated artefacts (*i.e.* scatters and patches, *cf.* Isaac 1981), were generally well preserved in fine-grained deposits of the Middle Pleistocene Meuse. High rates of sedimentation of fine-grained fluvial sediments created artefact assemblages of a very limited time-depth (in Pleistocene terms) and enabled extensive refitting, spatial analysis, and behavioural interpretations beyond the site-level.

A total of eight archaeological sites, together with a series of test pits (in total 1577 m^2) were manually excavated between 1981 and 1990 (Table 1 and 2; Roebroeks 1988; Roebroeks *et al.* 1992; De Loecker 2006). The Middle Pleistocene river deposits yielded a full interglacial vertebrate fauna with 26 species (van Kolfschoten 1993) and a rich mollusc fauna containing more than 70 land and freshwater species (Meijer 1985). Both the terrace and loess stratigraphy, as well as the mammal and mollusc biostratigraphical evidence, indicate an age prior to MIS 6 (van Kolfschoten and Roebroeks 1985; Vandenberghe *et al.* 1993; van Kolfschoten *et al.* 1993). Thermoluminescence (TL) dating of heated flint artefacts produced an age of 250 ± 20 ka (Huxtable 1993), and electron spin resonance (ESR) dating of shells yielded an age of 220 ± 40 ka, both corresponding to MIS 7 (van Kolfschoten *et al.* 1993). However, amino acid racemization (AAR) dating of *Corbicula* shells as well as biostratigraphically important elements of the mollusc fauna suggest an earlier, MIS 9, age for the 'Belvédère-interglacial' and its associated archaeology (Meijer and Cleveringa 2009). The most important archaeological levels at Belvédère were excavated in the upper part of a series of fine-grained sediments (Unit IV in the local stratigraphy) deposited by a meandering river system during MIS 7 or MIS 9. These fine-grained late Middle Pleistocene river deposits were subsequently covered by a thick sequence of Saalian and Weichselian silt loams consisting of both reworked and primary loesses.

Detailed geological and biostratigraphical research indicates that during this 'Belvédère interglacial' the meandering riverbed of the Meuse was situated in a broad valley that was formed in the previous cold stages. Along the riverbed was a mosaic of relatively high, sandy levees, low-lying backswamps, oxbow lakes, and gentle slopes towards the plateau edges. The molluscs and mammal remains (Meijer 1985; Duistermaat 1993; van Kolfschoten 1993) indicate that the flood plain was characterised by a forested environment. The low and wet parts were dominated by marshes with dense aquatic vegetation and were surrounded by alder forests with ash trees. On the higher parts in and around the stream valley, alder forests gradually gave way to open woodland. The latter consisted of deciduous forest and open areas with grasses and herbs. This variation of biotopes attracted a large number of animal species (van Kolfschoten 1993) of which the remains were recovered in association with early Middle Palaeolithic lithic artefact configurations. For more details the reader is referred to van Kolfschoten and Roebroeks (1985), Roebroeks (1988) and Vandenberghe *et al.* (1993).

Characterization of an early Middle Palaeolithic technological landscape

Excavations at the Maastricht-Belvédère quarry showed that the fossil surfaces of the interglacial valley bottom were littered with tools, production waste (almost exclusively flakes) and animal bone fragments. This continuous 'veil' of finds (or 'Veil of Stones', *cf.* Roebroeks *et al.* 1992) is the result of a number of well-preserved early Neanderthal 'on-site' activities and probably covered large parts of the original stream valley landscape.

Mainly through the work of Isaac (1981) and his colleagues in Koobi Fora (East Turkana, Kenya), it became clear that the dense concentrations of finds (high-density patches, commonly called "sites") are usually present against a background of continuous distributions of artefacts and bones (low-density scatters). Both these low-density scatters and high-density patches should be studied as integral parts of the forager landscape. Isaac's (1981) so-called scatters-between-the-patches project proposed a hierarchy of levels for structuring and understanding the spatial configurations, with their quantitative and typo-/technological artefact variations, across palaeo-landscapes. He suggested that there may have been significant functional (behavioural) differences between the dense patches and the more diffuse scatters surrounding them (*e.g.* Isaac and Harris 1978; Isaac 1978, 1981; Isaac and Crader 1981; Isaac *et al.* 1981; Stern 1991). In fact, both high and low artefact distributions were treated as integral parts of a single mobile system (see also Foley 1981a and b). By doing so, the solitary site focus was surmounted.

In an attempt to explain the spatial artefact continuum at Maastricht-Belvédère, Isaac's scatters-between-the-patches model was used as a heuristic device. One of the results of this was in the large-scale excavation of a very thin scatter of artefacts and bones (*i.e.* Site N). Here, an area of 765 m^2 of river deposits was carefully excavated for the explicit purpose of studying a low-density background scatter. The obtained information was subsequently used to interpret the vast and continuous artefact distribution, with its low and high densities. A detailed lithic inquiry showed that, apart from differences in find densities, the most important discrepancies between the various Belvédère sites can mainly be found in fine typological and technological variations (differences in percentages and ratios). At the same time, an in-depth refitting analysis proved to be of fundamental importance for the study of these subtle, but important, differences (De Loecker *et al.* 2003; De Loecker 2006).

In this way, at Maastricht-Belvédère, an effort was made to exceed the individual site-level by integrating all the available data into an off-site approach (*cf.* Foley 1981a and b; Isaac 1981). The excavated find distributions were treated as part of a spatially continuous subsistence system, in which different activities were performed in different parts of the landscape. At least at Belvédère it seemed legitimate and justified to compare the different lithic scatters and patches because:

- All available data comes from the same fine-grained (Unit IV) sediments.
- The archaeological material is probably contemporaneous (in Pleistocene terms, that means in terms of hundreds of years), as it was discarded during a relatively short phase within the same warm-temperate interglacial period.
- The find distributions were documented in a relatively small area of approximately six hectares. This could suggest that the sites were formed under the 'same' (micro) environmental conditions. For example, there is no evidence for significant changes in lithic raw material provenance.
- The spatially continuous archaeological patterning at Belvédère is basically characterized by two types of find distributions. The first type concerns dense and clearly delimited concentrations, such as Sites C, F, H and K, which mainly consist of compact clusters of waste from flint working activities (*i.e.* the patches). High rates of dorsal/ventral refits dominate (Roebroeks 1988; De Loecker 2006). The second type of artefact distribution mainly consists of widely spread flakes, tools and faunal remains (*i.e.* Site G

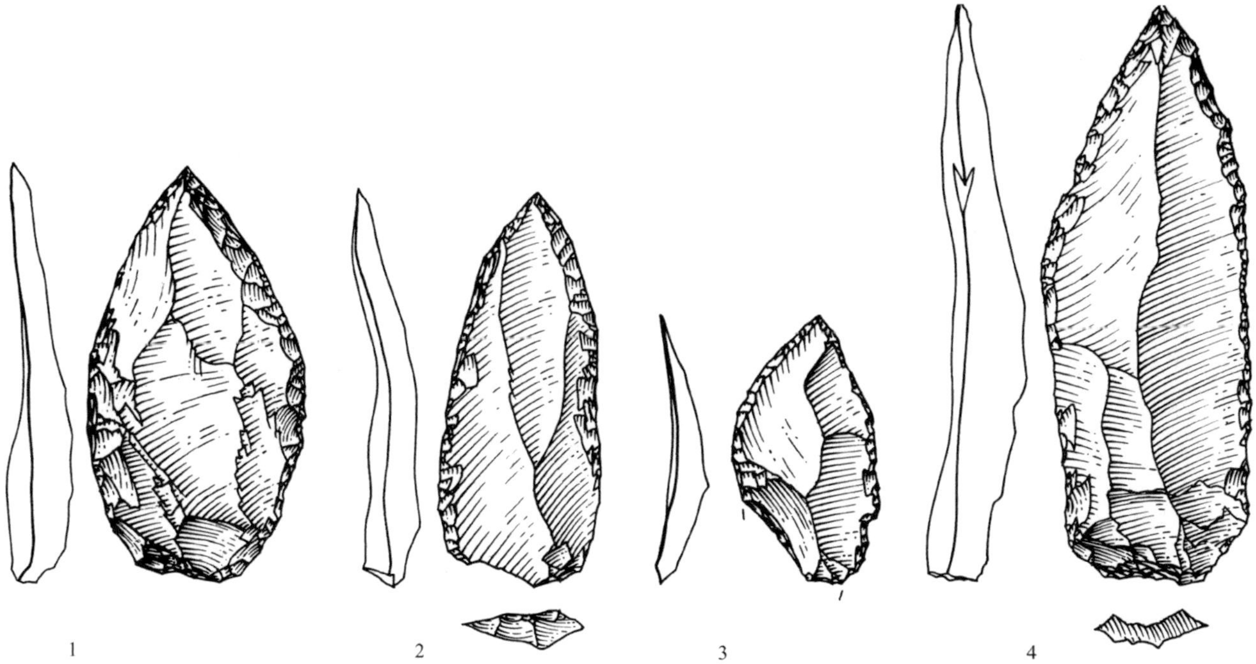

Figure 1. Maastricht-Belvédère Site K. Four of the 111 retouched tools *sensu stricto* (Mousterian points/convergent side scrapers). These tools show a rather triangular morphology and could not be incorporated into one of the many refit sequences from Site K. Maximal dimension of the tools: 1. = 86 mm, 2. = 81 mm, 3. = 55 mm, 4. = 120 mm. Figures after De Loecker (2006: 507).

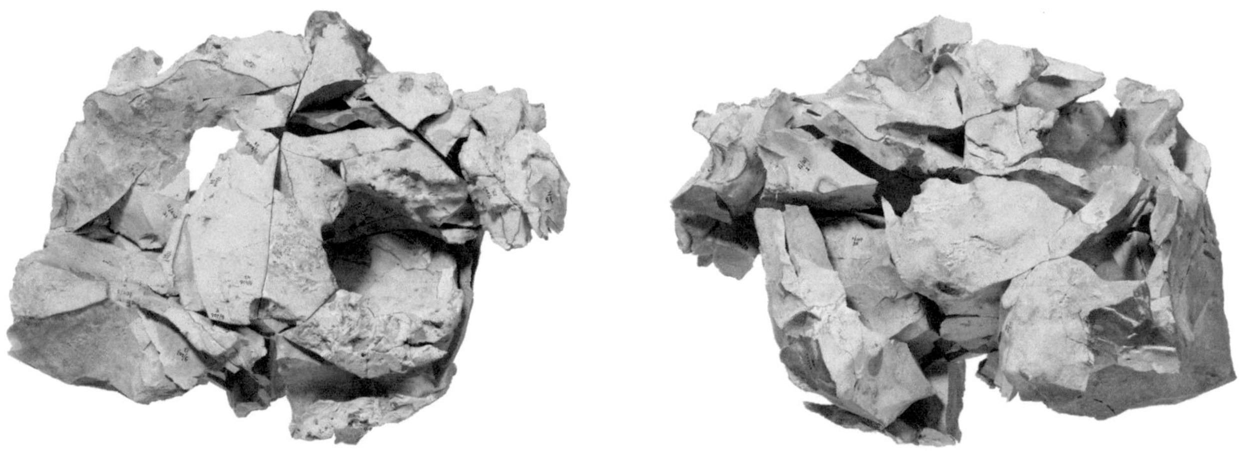

Figure 2. Maastricht-Belvédère Site K. Refitted composition I shown from two sides (length = 253 mm, width = 330 mm, thickness = 285 mm). The nodule weighs 9.286 kg (15.4% of the total weight of all conjoined Site K artefacts) and consists of 160 artefacts. It represents nine separately reduced parts or cores. (Figure after De Loecker 2006: 41).

and N). Densities within these scatters are low and dorsal/ventral refits are relatively scarce (Roebroeks *et al.* 1992).

Tools seem to play a more prominent role in the scatters than in the patches. However, the immense Site K flint cluster seems to be an exception as, by Belvédère standards, relatively numerous tools *sensu stricto* were recovered (n = 111, Fig. 1). Moreover, refitting showed that none of these tools could be integrated in one of the many reconstructed reduction sequences (flint nodules). Most of the Site K tools *sensu stricto* are therefore interpreted as well-prepared (Levallois *sensu stricto*) products, which entered the location as ready-made products. This seems to be confirmed by the used raw materials: fine-grained types of flint which do not occur locally. In addition, Site K also stands out because of its large number cores (n = 91), all of which were produced on-site. The many cortex remains on the cores indicate that they were probably only intended for *ad hoc* use (De Loecker 2006).

During the Site K analysis, a total number of 1,828 artefacts (16.8% of 10,912 artefacts) could be refitted, in other words 60.4 kg (61.7 %) of the total weight of the flint assemblage (97.8 kg). The obtained compositions, together with high percentages of cortex, indicate that several large flint nodules were brought to the location without any, or at most a minimal, form of core preparation, decortication or testing of the flint (Fig. 2). Inside the excavated area, the raw material nodules were primarily split up into smaller units (*i.e.* rough manageable cores) and stripped of their cortex. Intra-site spatial patterning show that parts or cores were subsequently moved to other zones within the excavated area. There they were further reduced and eventually discarded (De Loecker 2006). Although large flakes were used occasionally as cores (*i.e.* flaked-flakes), hidden natural (frost) fissures in the flint played a significant role in splitting up the raw material nodules. Together with a number of technological flaking failures (*i.e.* hinges, steps, face battering and stacked steps on the cores; *cf.* Shelley 1990) this could indicate a non-selective choice of raw material or a shortage of better-quality flint.

Just as for Sites F and H, the Site K cores and flakes are characterized by large dimensions and few, but large, dorsal negatives. These assemblages show little attention for core preparation. As a rule, the negatives of previous flakes in the reduction sequence are used as a striking platform. Refitting furthermore indicates that usually long 'uninterrupted' series of flakes were produced by means of single-sided and/or alternating bifacial disc/discoid core reduction techniques (Boëda 1993). This strategy involved a constant rotating of the cores to maintain good striking angles. As a result of this method, the Site K cores can be described as the result of a continuous, sometimes radial, production of flakes. Yet such a reduction technique makes it difficult, or even impossible, to interpret certain groups of flakes as waste or as a desired end-product. It can be concluded that the above-mentioned high-density patches (*i.e.* Sites K, F and H) are mainly characterized by a wasteful and expedient *ad hoc* flint technology.

A completely different type of high-density patch was excavated at Site C (Roebroeks 1988). The spatial pattern shows a configuration of a number of 'smaller' related find clusters which consist of production waste and are located next to each other. The Site C flint assemblage is mainly characterized by smaller artefact sizes and by a sophisticated technique in which cores were carefully prepared. 'Classic' Levallois flakes and products of a *débitage Levallois recurrent* (Boëda 1986, 1993, 1994), manufactured on fine-grained flint, occur regularly. Refitting indicates that several cores (and tools and larger flakes), entered the location in different stages of reduction. These finds were further reduced within the site boundaries. A spatial study of the refitted artefacts shows that out of the *ad hoc* produced *débitage*, certain elements were moved from one zone to another within the excavated surface. There they were discarded (*i.e.* small, thin, exhausted cores) and the processing of another raw material unit (core) was initiated in the same spot. Next, this new 'reduction sequence' was moved to a 'third' activity area, where the core was further reduced and discarded. At this location a new series of flakes, from a new core, was produced once again, etc. Some carefully prepared elements from these *chaînes opératoires (i.e.* cores and flakes) were eventually discarded outside the excavated area (Fig. 3). All this is in contrast to Sites K, F and H, where most of the reduction sequences started and ended within the site boundaries.

Generally, at Sites C and K we are dealing with two different, but related, flint knapping strategies. On the one hand, there are a number of carefully prepared (*Levallois recurrent*) cores, flakes and tools, which were made on non-local fine-grained types of flint. These artefacts were produced at other places in the palaeo-landscape and subsequently carried to the location in reduced

Figure 3. Maastricht-Belvédère Site C, Raw Material Unit 5. This refitted composition has a maximal dimension of 128 mm, consists of 162 elements and shows the flaking debris of a flat disc(oidal)-like core. The actual core was not recovered within the excavation boundaries but must have been nearly exhausted. The initial flaking and decortication occurred outside the excavated area. About 10% of the elements were burned. (Figure after Roebroeks 1988: 53).

forms or as 'desired' ('ready-made') end-products. The earliest phases of these reduction sequences are therefore missing within the excavated area. At the location the imported cores were further reduced and parts of the *chaînes opératoires*, or individual artefacts, were discarded on the spot. Eventually, some carefully prepared cores and flakes were taken to other locations for future use (Roebroeks 1988; Roebroeks *et al.* 1992). On the other hand, there are artefacts produced by a disc/discoid core reduction strategy (Boëda 1993). These show minimal attention for core preparation and are predominantly manufactured on more coarse-grained and locally available flint types (*i.e.* large flint nodules at Site K). These products were probably intended for *ad hoc* use only.

Between the patches of debris a thin, uninterrupted, scatter of widely spread artefacts was found. This low density find distribution was present in the fine-grained river sediments, throughout the whole Belvédère quarry. The vast and diffuse find distributions do not show any clustered occurrence and parts were excavated at Sites G and N (Roebroeks 1988; Roebroeks *et al.* 1992). The average artefact densities are generally low and range between 0.58 (Site N) and 1.22 (Site G) artefacts per m^2 (Tab. 1). Within the scatters the highest percentages, by far which were recorded, were of tools. The tools encountered mainly consist of damaged and exhausted examples. Strikingly, a number of large, sometimes fragmented, flakes were recovered from these scatters as well. These flakes have a long cutting edge, a clear back and could not be refitted. An example is a backed knife on an *éclat débordant* (*cf.* Beyries and Boëda 1983; Fig. 4) which was excavated at Site G. The tools and large flakes at Sites G and N probably reflect the discarded components of a mobile toolkit, while the few dorsal/ventral refits represent parts of spatially fragmented reduction and retouching sequences (modification of cores, flakes and tools).

More than half of the refitted artefacts consists of broken flakes. At Site N a sequence of five small flakes could be refitted (dorsal/ventral, Fig. 5). This unit represents a short-lived activity which involved the refreshment of a core edge. The core itself has been taken from the location to another activity area in the landscape. Some very small resharpening flakes (*cf.* Cornford 1986), which could not be refitted, indicate that scraper-like objects have been brought to and resharpened within the excavated parts of the scatter. After freshening up the working edges, the tools were taken to other places (*cf.* Sites A and G, Fig. 6).

An exhausted double convex/concave scraper, excavated at Site N, also shows traces of resharpening. A negative of a Transverse Sharpening Flake (Cornford 1986) indicates that the scraper was already resharpened at a location outside the excavated surface before it was discarded at Site N (Fig. 7). Flakes recovered from the low-density scatters are on average larger than those from the high density patches. They exhibit higher numbers of dorsal negatives and show very little cortex. The striking surfaces and dorsal patterns are frequently prepared and the faceting indices are among the highest at Belvédère. Moreover, the used raw materials show considerable heterogeneity, which is also evident from the limited refitting results (dorsal/ventral refits). Altogether, the continuous scatter of isolated or small sets of artefacts at Maastricht-Belvédère seems to represent a series of unrelated activities.

To conclude, the low-density scatters, compared to the high-density patches, clearly show spatial, typological and technological idiosyncracies (Table 1 and 2; see also De Loecker 2006) that correspond to differences in processed raw materials (flint types). In its entirety, the uninterrupted 'veil of stones and bones' (*cf.* Roebroeks *et al.* 1992), with its low and high artefact densities, represents a very mobile and flexible use of lithic artefacts in a riverside landscape, at least 250,000 years ago.

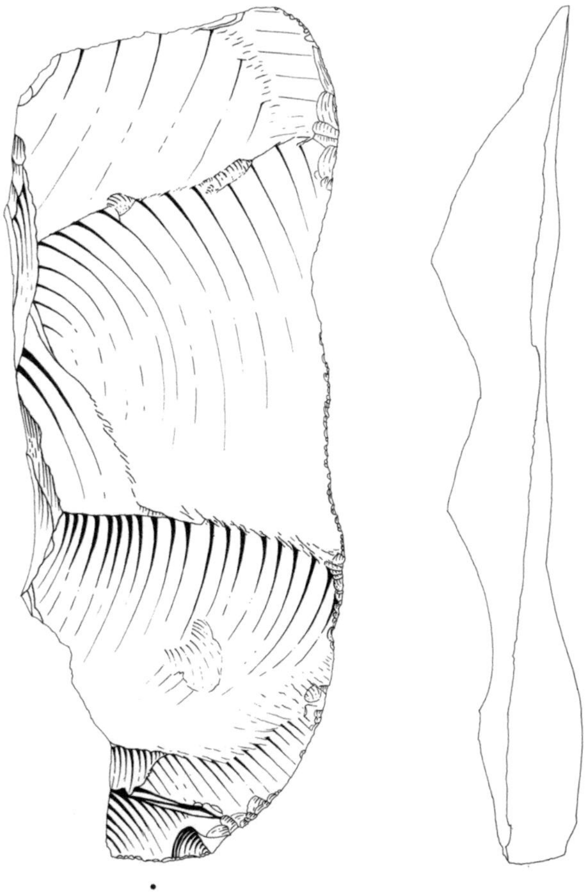

Figure 4. Maastricht-Belvédère Site G. Large flake with a long and sharp cutting edge. The flake has been produced of the lateral side of a prepared core. This so-called *éclat débordant* (Beyries and Boëda 1983) has a triangular cross-section and exhibits a clear back (*i.e.* the side of a core). The tool resembles a back knife and could not be refitted to the rest of the Site G lithic material. Maximal dimension of the tool: 170 mm (Figure after Roebroeks 1988: 71).

Time, energy and skill investment

In addition to flint working, there were most probably also other (more time-consuming) activities carried out at the various Maastricht-Belvédère locations. Due to a number of analytical limitations, it is, however, not possible to determine the exact nature of these actions. For example, only a limited amount of bone material was recovered from the excavated locations. On the one hand, at certain locations (*e.g.* Site K), no animal remains were present. On the other hand, their absence could be the result of a local decalcification of the sediments. Post-depositional processes are also responsible for the absence of unambiguous microscopic signs of use. In that case, artefacts that do show use-wear traces are too heavily weathered to determine the exact nature of the traces (van Gijn 1988, 1989). However, there are some Belvédère micro-wear results that are indicative for the activities carried out, such as food (meat) extraction at Site C. At the low-density Site G scatter, the remains of a young rhino were found near a large backed knife on an *éclat débordant* (Fig. 4). Micro-wear analyses independently showed that this knife was used for dissecting a thick-skinned animal like an elephant or rhino (van Gijn 1988).

The European Middle Palaeolithic dataset indicates that Neanderthals were constantly transporting lithic materials through the landscape. By doing so, they could anticipate the future need for 'cutting edges' (*e.g.* Geneste 1985; Roebroeks *et al.* 1988; Féblot-Augustins 1993, 1999; Gamble 1999). The elaborate lithic analyses at Maastricht-Belvédère suggest that a wide range of morphologically different artefacts like scrapers, large flakes and prepared cores,

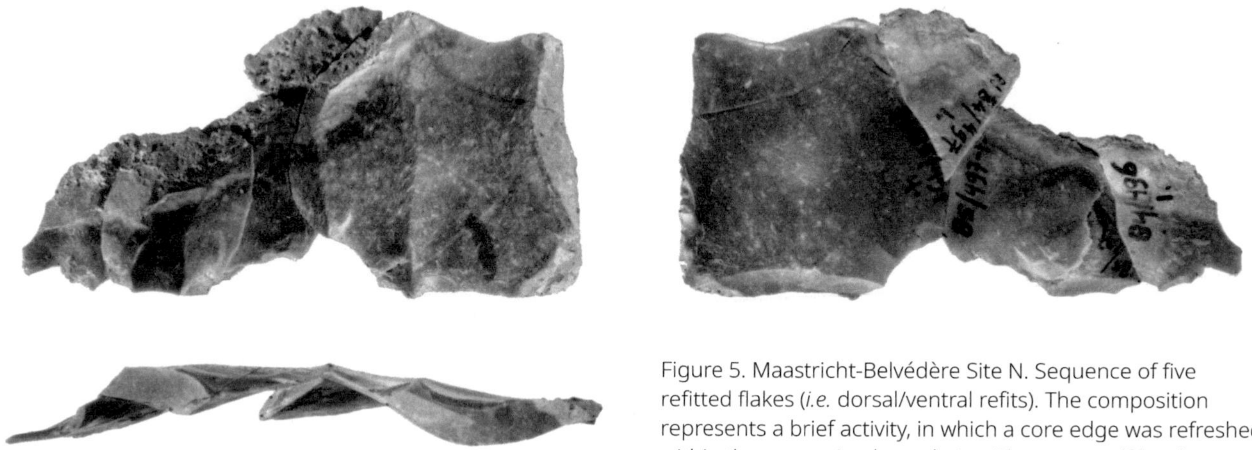

Figure 5. Maastricht-Belvédère Site N. Sequence of five refitted flakes (*i.e.* dorsal/ventral refits). The composition represents a brief activity, in which a core edge was refreshed within the excavation boundaries. The core itself has been taken to the a next activity location, somewhere outside the excavation area. Maximal dimension of the composition: 73 mm. Figure after Roebroeks *et al.* (1992: 13).

Figure 6. Maastricht-Belvédère. Two small Transverse Sharpening Flake (*cf.* Cornford, 1986) which could not be refitted. These refreshing flakes indicate that scraper-like objects briefly entered the excavated areas. After freshening up the working edges the tools were taken back to other places in the palaeo-landscape. 1. Site A, maximal dimension: 19 mm. 2. Site G, maximal dimension: 23 mm. Figures after De Loecker 2006: 192, 208.

usually made of 'non-local' raw materials, were part of the mobile toolkit (Roebroeks *et al.* 1992; De Loecker 2006). It is difficult to gain reliable insight into the initial choices and selection of these raw materials, which were used for the production of certain tools, cores and/or tasks. In addition to contextual factors such as quality, quantity and accessibility, considerations in the planning of future activities also played an important role in the procurement of raw materials. Each type of raw material needs a separate evaluation, or even better, another 'conceptual scheme' to evaluate its suitability for specific activities.

As mentioned earlier, there are striking qualitative differences between the raw materials used for the *ad hoc* produced artefacts and the ones used for the transported and prepared Levallois (-like) objects (respectively termed 'coarse-grained' *versus* 'fine-grained' flint). Because the Levallois technique has a complicated *chaine opératoire*, it seems plausible that considerable time and energy were invested in the procurement and selecting of suitable raw materials for these products. Generally, the older Pleistocene gravels of the Meuse river (Unit III in the local stratigraphy) contain a variety of different types of flint. The aforementioned fine-grained flint also occurs regularly in these gravels. From this point of view all Belvédère artefacts which were recovered can therefore be described as 'local'. Moreover, at all (Unit IV) sites the artefacts have a fluvially rolled cortex. This indicates that the raw materials (*i.e.* large flint nodules at Site K) were probably collected from nearby riverbeds. They must have been eroded from chalk walls which were located a few kilometres south of Maastricht (Roebroeks 1988). According to palaeo-ecological reconstructions (Meijer 1985; Duistermaat 1987), the (Unit IV) Belvédère sites were located approximately 100 to 200 meters away from the main river flow. This means that within a radius of at least 100 to 200 meters around Site K, no gravels (raw material sources) were present. This may also mean that after extracting the raw material, energy and time were invested in transporting the large and heavy flint nodules to Site K over some distance. A total weight of at least 97.8 kg was probably moved at least 100 to 200 meters (De Loecker 2006). The fact that these large flint nodules have been brought to the location without significant preparation, decortication or testing may indicate an opportunistic form of raw material use. This strategy ran parallel to, or was complementary to, well-planned activities which are suggested by the transported products of prepared core technologies and curated tools (*e.g.* scrapers). The Site K activity zone was apparently of major importance to the early Middle Palaeolithic hunter-gatherers, as a considerable amount of energy and time was invested in transporting large and heavy flint nodules. This could also suggest that a longer period of time was spent at these locations.

The analysis of a number of specific find categories indicates that some activities at Maastricht-Belvédère lasted longer than the short-lived and opportunistic episodes of flint working or carcass dissecting. This particularly holds for hematite or 'red ochre' remains (Roebroeks *et al.* 2012) and burnt artefacts (Stapert 1990; 2007). Especially at Sites C and K large amounts of heated artefacts were recovered. These occurred spatially *between* the dense clusters of *débitage*, tools and cores. Visible hearth structures were not documented. However, several analytical studies (Stapert 1990; 2007) resulted in the identification of a number of possible 'phantom' fireplaces (Alperson-Afil *et al.* 2007). This could possibly mean that at the Site C and K patches, the maintenance and production-related tasks were associated with (derived) anthropogenic hearths.

During the Site C and Site F excavations a number of reddish hematite specks were identified (Arps 1988; Fig. 8). Roebroeks *et al.* (2012) concluded that the red

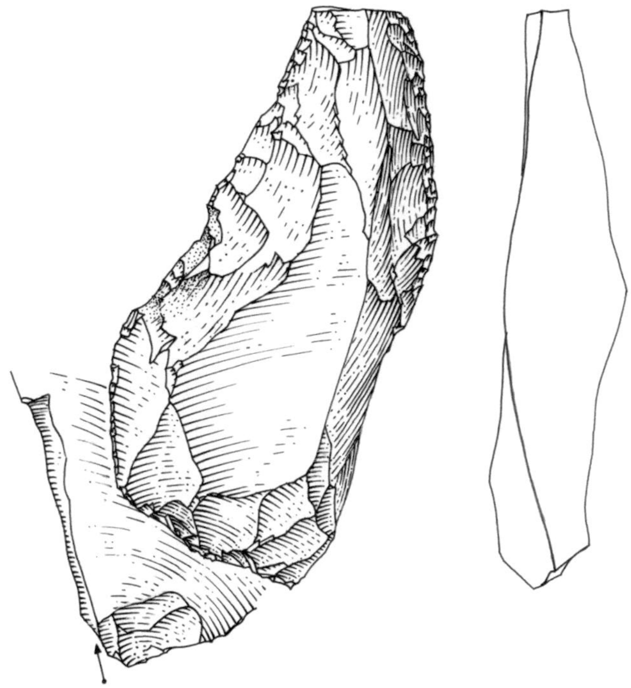

Figure 7. Maastricht-Belvédère Site N. Exhausted double convex/concave scraper, which shows traces of resharpening behaviour. Somewhere in the palaeo-landscape a 'Long Sharpening Flake' was removed to resharpen the cutting edge. Subsequently, the scraper was transported to the excavated area and was discarded. Maximal dimension of the scraper: 130 mm. Figure after Roebroeks *et al.* (1992: 8).

ochre stains did not belong to the natural sedimentary environment. They had entered the matrix after the deposition of the sediments. The nearest natural sources of hematite are situated at least 40 km from Maastricht: in the German Eifel region and in the Belgian Ardennes (Liège-Dinant-Namur vicinity).

Overall, the combined evidence of on-site observations, studies on raw material provenance, and the character of the red concentrates, justifies the inference that the hematite fragments at Sites C and F were deliberately transported to these locations. A number of experiments conducted by Roebroeks *et al.* (2012) suggests that the small hematite concentrations may originally have entered the sites as a liquid solution. During use a few drops of this ochre-rich liquid was spilled onto the palaeo-surface. Strikingly, these hematite droplets were found at the patchy locations (*i.e.* Site C) where possible anthropogenic hearths were present and

Figure 8. Maastricht-Belvédère Site C. Two hematite ('red ochre' iron oxide) concentrates, which were submitted for further analysis. (Left) Concretion Dz23-16 against the background of the fine-grained fluvial deposit from which it was recovered. Desiccation of the matrix caused breakage of the concentrate. (Right) Concretion Bv-894 which is ca. 2 mm thick. Figures after Roebroeks *et al.* 2012: 1890.

where flint and food processing seem to have taken place as well.

Discussion

From the early 1980s onwards, the large-scale archaeological excavations and multidisciplinary research at Maastricht-Belvédère have grown into a flagship of Middle Palaeolithic investigations in Europe. A number of well-preserved open-air findspots have been excavated in the quarry. These recorded a series of non-related activities (separated in time and space) of early Neanderthals in a river valley landscape during various foraging trips. Typological, technological, refitting and spatial analyses allowed detailed interpretations on the nature of Late Middle Pleistocene human behaviour (Roebroeks 1988; De Loecker 2006).

The find distributions are the result of a series of episodic foraging-related visits to the valley floor of the river Meuse. These activities resulted in widely spread find scatters that also included a series of more compact artefact concentrations. The so-called low-density scatters and high-density patches show clear quantitative and qualitative variations. The early Neanderthals that once foraged in these landscapes were clearly technologically well-equipped. Ready-made tools, well-prepared flakes, flint nodules and other materials such as red ochre concentrates were imported from other locations to this part of the river valley and found at Sites C, G, K, and N. Lithic artefacts undoubtedly played a major role in the daily extraction and processing of animal food sources. The studies conducted also suggest that a considerable amount of time and energy was invested in the maintenance of used flint technologies. In addition, the burnt flint artefacts found in the high-density Site C and K patches indicate the presence of anthropogenic ('phantom') hearths. A variety of tasks were carried out at a short distance from these hearths, some of which may have been related to the manipulation and/or use of red ochre (Roebroeks *et al.* 2012). The diversity of vital resources in the Meuse Valley will have been large and included flint, drinking water, fuel for fireplaces and of course the necessary food supplies such as game.

The spatially continuous 'veil' of finds can be seen as the fossil remnants of foraging trails that crisscrossed the landscape. Each of the abovementioned toolkit-components requires an individual consideration, investment of time and knowledge of the raw materials. Besides the well-prepared tools, cores and flakes, which were often manufactured at other locations and carried to the riverbanks, the early Neanderthals also made use of locally available raw materials. The large flint nodules that were collected a short distance from Site K were used opportunistically and *ad hoc* to produce 'situational gear' (Binford 1979; De Loecker 2006). The geographical range of their technologies, the skills to produce fire and the knowledge to manipulate hematite were all elementary in the daily life of early Neanderthals. This way of life, which was clearly more than just a 'hit-and-run' extraction of food and raw materials, made it possible to survive for hundreds of thousands of years in Pleistocene Europe. In that respect we can picture these early Neanderthals as highly experienced foragers who continuously used their skills and knowledge during daily food-related visits to the Meuse valley. The fossil remnants of their numerous but episodic and inter-related visits eventually created a so-called palimpsest taskscape (Ingold 1993).

To place the remains of this Middle Pleistocene taskscape and its environmental setting in a chronological framework, dating methods play a significant role. As mentioned before a series of absolute dates, corresponding to MIS 7 (Huxtable 1993; Vandenberghe *et al.* 1993), were obtained in the 1980s. However, more recent amino acid racemization (AAR) dating of *Corbicula* shells as well as elements of the mollusc fauna suggest an earlier, MIS 9, age for the 'Belvédère-interglacial' and its associated archaeology (see Meijer and Cleveringa 2009 for more details).

During the MIS 9 – MIS 6 period rapid behavioural changes and technological innovations took place in northwest Europe (Hérisson *et al.* 2016). Levallois core technologies appear from around 300 ka (late MIS 9/early MIS 8) onwards, and conventionally mark the Lower/Middle Palaeolithic transition (*cf.* Gamble 1999; White and Ashton 2003; Pettitt and White 2012). The earliest Levallois products appear within layers that are generally considered to be Lower Palaeolithic in character and only represent a small part of these assemblages. During the early Middle Palaeolithic (MIS 7) various Levallois technologies became increasingly dominant and geographically more widespread (White *et al.* 2006; Emery 2010; Scott and Ashton 2011). Only insignificant evidence of human occupation during the full Pleniglacial conditions of MIS 8 has been noted in the northern latitudes of Europe (*e.g.* Hublin and Roebroeks 2009; Antoine and Locht 2015; Hérisson *et al.* 2015; Locht *et al.* 2016).

The northwest European MIS 9-8 record shows only a restricted number of sites in primary geological context and especially the early Middle Palaeolithic site of Kesselt-Op de Schans (level 3; Limburg, Belgium) stands out in this respect (Hérisson *et al.* 2016). This open-air site is situated near the western border of Maastricht, about five kilometres southwest of the Belvédère quarry. The archaeological record from Kesselt-Op de Schans shows a variability in flint knapping strategies, combining Levallois core technologies with discoidal production systems. The transport of lithic items was also documented and substantiated by refitting. The assemblage suggests a technological behaviour which is typical for the Middle Palaeolithic (Van Baelen 2014; 2017). A correlation of the observed stratigraphic Kesselt sequence with the local chronostratigraphic framework, together with several optical stimulated luminescence (OSL) and thermoluminescence (TL) dates on loess deposits, shows that the main occupation horizon (level 3) can be assigned to the MIS 9 to MIS 8 transition (ca. 280,000 years ago). If this dating evidence is accurate, the Kesselt-Op de Schans record suggests that classical Middle Palaeolithic assemblages were already produced in the region at the end of MIS 9 or from the beginning of MIS 8 onwards. Speculatively, this could constitute a technological argument in favour of a MIS 9 age for the Maastricht-Belvédère scatters and patches. Like Kesselt-Op de Schans, the Maastricht-Belvédère record also shows a typical Middle Palaeolithic assemblage, where a prepared core technology (Levallois *sensu stricto* and *recurrent*) was combined with a disc/discoid core reduction strategy.

Epilogue: a new dating program

To move beyond this discussion, the initial Belvédère dating results await verification by initiating an appropriate up-to date dating programme. "It should also be borne in mind that due to methodological advances in the laboratories, present-day dates are in general more significant than those carried out years ago when it comes to such old chronologies." (Maroto *et al.* 2012, 23). Since the 1980s absolute dating methods (radiometric and non-radiometric) have been fine-tuned and new dating methods have been developed constantly. Moreover, huge advances have been made in constructing chronological frameworks over the last forty years. It is therefore a challenge to put a new dating program on the Maastricht-Belvédère agenda for the coming years. This call for better dating evidence could possibly push back the age of the 'Belvédère-interglacial' and its associated archaeology to MIS 9 (*cf.* Meijer and Cleveringa 2009). Moreover, it is expected that future dates will contribute to optimize the northwest European, or at least the regional, chronological framework. Since absolute dating options are still available or possible at Maastricht-Belvédère this new research may include:

- Thermoluminescence (TL) dating of heated flint artefacts which were carefully collected in the 1980s and properly stored in in light-resistant plastic containers (*e.g.* Sites C and K).
- Optical stimulated luminescence (OSL) dating of excavated and stored sediment blocks (mainly containing faunal remains), which were recovered from the excavated areas (*e.g.* Site C, K and N).

- Initiating a field campaign and implementing the sampling for (new) non-radiometric and/or radiometric dating methods. This means that we have to create a new west-east orientated section at the most southern extant profile-wall in the former Belvédère quarry. By means of a digging machine, the middle Pleistocene sediments must be uncovered and the chosen section must be cleaned and cut straight for sampling. The main objectives of this profile will be to document and date the local channel/gully deposits of Unit IV (Roebroeks 1988: 13-24), which represent three consecutive, but in energy decreasing, phases of rather identical meander development in the Belvédère sequence (*i.e.* Subunits IV-A, -B, -C; Vandenberghe 1993; Vandenberghe *et al.* 1993). The sampling exercise should, in particular, focus on the Subunit IV-B greyish-olive green silty clay zone (which is overlain by calcareous tufas), and the so-called 'mottled zone' of Subunit IV-C-ß. These two Subunits respectively yielded the archaeological assemblages of Sites B, C and G, and Sites A, D, F, H, K and N.

To conclude, new dating opportunities at Maastricht-Belvédère could possibly shed light on the organization of early Neanderthal technologies during MIS 9, or at least produce dates confirming a MIS 7 correlation. The consequences of an older date for the main archaeological levels at Belvédère could imply that fully evolved Middle Palaeolithic technologies, consisting of well-prepared Levallois technologies (*sensu stricto* and *recurrent*) were jointly used with disc/discoid reduction strategies, and were already applied in full swing during the so-called Lower/Middle Palaeolithic transition (late MIS 9/early MIS 8). This is a statement that can also be suggested for the Kesselt-Op de Schans archaeological record (Van Baelen 2014; 2017).

Acknowledgements

The author would like to thank Jan Pauptit (Leiden, the Netherlands) for most of the photographic work and Gilles Leroy (Amiens, France) for the drawings of the flint artefacts. Special thanks go to Nancy Lenaert (Antwerp, Belgium) and especially to Jan Kolen (Leiden, the Netherlands), who were kind enough to read and constructively comment on earlier versions of this manuscript.

I would like to dedicate this article to my mother Gilberte Cornelis (°30-04-1933 – †01-01-2020), who always believed in my dreams. I miss you more than words can say.

References

Alperson-Afil, N., D. Richter and N. Goren-Inbar. 2007. Phantom hearths and the use of fire at Gesher Benot Ya'aqov, Israel. *Paleoanthropology* 1: 1-15.

Antoine, P. and J.-L. Locht. 2015. Chronostratigraphie, paleoenvironnements et peuplements au Paleolithique moyen: les donnees du Nord de la France. *Memoire la Societe Prehistorique Française* 59: 11-23.

Arps, C.S.. 1988. Appendix III – The identification of haematite as the colouring agent in red ochre from the Middle Palaeolithic site C at Maastricht-Belvédère, The Netherlands, by means of x-ray diffraction analysis. In: W. Roebroeks (ed.): From Find Scatters to Early Hominid Behaviour. A Study of Middle Paleolithic Riverside Settlements at Maastricht-Belvédère (The Netherlands): 163-165. *Analecta Praehistorica Leidensia* 21.

Beyries, S. and E. Boëda. 1983. Étude technologique et traces d'utilisation des 'éclats débordants' de Corbehem (Pas-de-Calais). *Bulletin de la Societé Préhistorique Française* 80: 270-275.

Binford, L.R.. 1979. Organization and formation processes: looking at curated technologies. *Journal of Anthropological Research* 35: 255-273.

Binford, L.R.. 1980. Willow smoke and dogs' tails: hunter-gatherer settlement systems and archaeological site formation. *American Antiquity* 45: 4-20.

Binford, S.R. and L. Binford. 1968. *New Perspectives in Archaeology*. Chicago: Aldine Pub. Co.

Boëda, E. 1986. Le débitage de Biache-Saint-Vaast (Pas de Calais): première étude technologique. *Bulletin Association Française pour l'Étude du Quaternaire* 26: 209-218.

Boëda, É. 1993. Le débitage discoïde et le débitage Levallois récurrent centripète. *Bulletin de la Société Préhistorique Français* 90: 392-404.

Boëda, E. 1994. *Le Concept Levallois: Variabilité des Méthodes*. Paris (Monographie De CRA. 9).

Bordes, F. 1961. *Typologie du Paleolithique ancien et moyen*. Bordeaux.

Bordes, F. 1972. *A Tale of Two Caves*. New York, Evanston, San Francisco, London: Harper & Row,

Bosinski, G. 1966. Der paläolithische Fundplatz Rheindahlen, Ziegelei Dreesen-Westwand. *Bonner Jahrbücher* 166: 318-344.

Bosinski, G. 1967. *Die mittelpaläolithische Funde im westlichen Mitteleuropa*. Köln, (Fundamenta A/4).

Bosinski, G. 1995. The Palaeolithic and Mesolithic in Central Europe. In: W. Schirmer (ed.). *Quaternary field trips in Central Europe vol.* 2: 829-999. München.

Cahen, D., L.H. Keeley and F.L. Van Noten. 1979. Stone tools, toolkits and human behavior in prehistory. *Current Anthropology* 20: 661-683.

Cornford, J. M.. 1986. Specialized resharpening techniques and evidence of handedness. In: P. Callow and J. M. Cornford (eds.), *La Cotte de St-Brélade, Jersey. Excavations by C. M. B. McBurney 1961-1978*: 337-351. Norwich: Geo Books Norwich.

Crabtree, D.E.. 1972. *An Introduction to Flintworking.* Pocatello.

Daniel, G.. 1972. The Idea of Man's Antiquity. In: C.C. Lamberg-Karlovsky (ed.), *Readings from Scientific American: Old World Archaeology: foundations of Civilization*: 9-13. San Francisco, Scientific American.

De Loecker, D. 2006. Beyond the Site. The Saalian Archaeological Record at Maastricht-Belvédère (the Netherlands). *Analecta Praehistorica Leidensia* 35/36.

De Loecker, D. 2014. Maastricht-Belvédère: Interpretation of a Technological Paleo-Surface. In: C. Smith (ed.): *Encyclopedia of Global Archaeology*: 4575-4587. New York: Springer Science + Business Media.

De Loecker, D. and W. Roebroeks. 2012. Beyond '15-minutes': revisiting the late Middle Pleistocene archaeological record of Maastricht-Belvédère (The Netherlands). *Analecta Praehistorica Leidensia* 43/44: 349-367.

De Loecker, D., J. Kolen, W. Roebroeks and P. Hennekens. 2003. Refitter's Paradise: on the conjoining of artefacts at Maastricht-Belvédère (The Netherlands). In: N. Moloney and M.J. Shott (eds.), *Lithic Analysis at the Millennium*: 113-136. London.

Duistermaat, H. 1987. Voorlopig resultaat van een malacologische analyse van een profiel uit Site G van de Belvédère. Unpublished typescript, Faculty of Archaeology, Leiden University, Leiden.

Duistermaat, H.. 1993. The interglacial mollusc fauna from Maastricht-Belvédère Site G, *Mededelingen Rijks Geologische Dienst* 47: 61-67.

Emery, K.. 2010. *A re-examination of variability in handaxe form in the British Palaeolithic.* Unpublished PhD Thesis, University College London.

Féblot-Augustins, J.. 1993. Mobility Strategies in the Late Middle Palaeolithic of Central Europe and Western Europe: Elements of Stability and Variability. *Journal of Anthropological Archaeology* 12: 211-265.

Féblot-Augustins, J.. 1999. Raw Material Transport Patterns and Settlement Systems in the European Lower and Middle Palaeolithic: Continuity, change and variability. In: W. Roebroeks and C. Gamble (eds.), *The Middle Palaeolithic Occupation of Europe*: 193-214. Leiden: Leiden University Press.

Foley, R.. 1981a. Off-site archaeology: an alternative approach for the short-sited. In: I. Hodder, G. L.L. Isaac and N. Hammond (eds.), *Pattern of the Past: Studies in Honour of David Clarke*: 157-183. Cambridge: Cambridge University Press.

Foley, R. 1981b. A model of regional archaeological structure. *Proceedings of the Prehistoric Society* 47: 1-7.

Gamble, C.. 1999. *The Paleolithic societies of Europe.* Cambridge: Cambridge University Press.

Geneste, J.-M.. 1985. *Analyse lithique d'industries moustériennes du Périgord: Une approche technologique du comportement des groupes humains au Paléolithique moyen.* Thèse de Doctorat de l'Université de Bordeaux.

Gijn, A. L. van.1988. Appendix I – A functional analysis of the Belvédère flints. In: W. Roebroeks (ed.). From Find Scatters to Early Hominid Behaviour. A Study of Middle Palaeolithic Riverside Settlements at Maastricht-Belvédère (The Netherlands). *Analecta Praehistorica Leidensia* 21: 151-157.

Gijn, A. L. van. 1989. The wear and tear of flint, principles of functional analysis applied to Dutch neolithic assemblages. *Analecta Praehistorica Leidensia* 22.

Hallos, J.. 2005. '15 Minutes of Fame': exploring the temporal dimension of Middle Pleistocene lithic technology. *Journal of Human Evolution* 49: 155-179.

Hérisson, D., E. Goval,, P. Antoine, J.-J. Bahain, S. Balescu, Y. Chantreau, E. Claud, S. Coutard, N. Debenham, C. Font, J.-L. Locht, S. Pirson and P. Spagna. 2015. *Les occupations du Paleolithique inferieur et du Paleolithique moyen d'Etricourt-Manancourt, Canal Seine Nord Europe.* Rapport de fouille, SRA de Picardie.

Hérisson, D., M. Brenet, D. Cliquet, M.-H. Moncel, J. Richter, B. Scott, A. Van Baelen, K. Di Modica, D. De Loecker, N. Ashton, L. Bourguignon, A. Delagnes, J.-P. Faivre, M. Folgado-Lopez, J.-L. Locht, M. Pope, J.-P. Raynal, W. Roebroeks, C. Santagata, A. Turq and P. Van Peer. 2016. The emergence of the Middle Palaeolithic in north-western Europe and its southern fringes. *Quaternary International* 411 A: 233-283.

Hublin, J.-J. and W. Roebroeks. 2009. Ebb and flow or regional extinctions? On the character of Neandertal occupation of northern environments. *Comptes Rendus Palevol* 8: 503-509.

Huxtable, J.. 1993. Further thermoluminiscence dates for burnt flints from Maastricht-Belvédère and a finalised thermoluminiscence age for unit IV Middle Palaeolithic sites. *Mededelingen Rijks Geologische Dienst* 47: 41-44.

Ingold, T.. 1993. The temporality of the landscape. *World Archaeology* 25: 152-174.

Isaac, G.L.L.. 1978. Food sharing and human evolution: archaeological evidence from the Plio-Pleistocene of East-Africa. *Journal of Anthropological Research* 34: 311-325.

Isaac, G.L.L.. 1981. Stone age visiting cards; approaches to the study of early land use patterns. In: I. Hodder, G.L.L. Isaac and N. Hammond (eds.), *Pattern of the Past: Studies in Honour of David Clarke*: 131-155. Cambridge: Cambridge University Press.

Isaac, G.L.L. and J.W.K. Harris.. 1978. Archaeology. In: M.G. Leakey and R.E.F. Leakey (eds.), *Koobi Fora Research Project. Volume 1, The Fossil Hominids and an Introduction to their Context, 1968-1974*: 47-76. Oxford.

Isaac, G.L.L. and D. Crader. 1981. To what extent were early hominids carnivorous – an archaeological perspective. In: R.S.O. Harding and G. Teleki (eds.), *Omnivorous Primates: Gathering and Hunting in Human Evolution*: 37-103. New York.

Isaac, G.L.L., J.W.K. Harris and F. Marshall 1981: Small is informative: the application of the study of mini-sites and least-effort criteria in the interpretation of the early Pleistocene archaeological record at Koobi Fora, Kenya. In: J.D. Clark and G.L.L. Isaac (eds.), *Las Industrias más Antiguas. X Congreso de la Unión Internacional de Ciencias Prehistóricas y Protohistóricas, México*: 101-119. Paris.

Keeley, L.H.. 1980. *Experimental Determination of Stone Tool Uses: a Microwear Analysis*. Chicago.

Kolen, J., D. De Loecker, A. Groenendijk and J.-P. de Warrimont. 1999. Middle Palaeolithic surface scatters: how informative? A case study from southern Limburg (the Netherlands). In: W. Roebroeks and C. Gamble (eds.), *The Middle Palaeolithic occupation of Europe*: 177-191. Leiden: University of Leiden.

Kolfschoten, T. van.. 1993. The mammal fauna from the interglacial deposits at Maastricht-Belvédère. *Mededelingen Rijks Geologische Dienst* 47: 51-60.

Kolfschoten, T. van and W. Roebroeks 1985 (eds.), *Maastricht- Belvédère: stratigraphy, palaeoenvironment and archaeology of the Middle and Late Pleistocene deposits*. Haarlem: Rijks Geologische Dienst. (Mededelingen Rijks Geologische Dienst 39).

Kolfschoten, T. van, W. Roebroeks and J. Vandenberghe. 1993. The Middle and Late Pleistocene sedimentary and climate sequence at Maastricht-Belvédère: The type locality of the Belvédère interglacial. *Mededelingen Rijks Geologische Dienst* 47: 81-91.

Leroi-Gourhan, A and M. Brézillon. 1966. L'habitation magdalénienne n.1 de Pincevent près Montrau. *Gallia Préhistoire* IX: 263-385.

Locht, J.-L., D. Herisson, E. Goval, D. Cliquet, B. Huet, S. Coutard, P. Antoine and P. Feray. 2016. Timescales, space and culture during the Middle Palaeolithic in northwestern France. *Quaternary International* 411 A: 129-148.

Maroto, J. , M. Vaquero, Á. Arrizabalaga, J. Baena, E. Baquedano, J. Jordá, R. Julià, R. Montes, J. Van Der Plicht, P. Rasines and R. Wood. 2012. Current issues in late Middle Palaeolithic chronology: New assessments from Northern Iberia. *Quaternary International* 247: 15-25.

McCrone, J.. 2000. Fired up: theory on when prehistoric man first used fire. *New Scientist* 166: 30-35.

Meijer, T.. 1985. The pre-Weichselian non-marine molluscan fauna from Maastricht-Belvédère (Southern Limburg, the Netherlands). *Mededelingen Rijks Geologische Dienst* 39: 75-103.

Meijer, T. and P. Cleveringa. 2009. Aminostratigraphy of Middle and Late Pleistocene deposits in The Netherlands, with notes on published data from the southern part of the North Sea Basin. *Global Planet Change* 68: 326-345.

Pettitt, P. and M. White. 2012. *The British Palaeolithic. Human Societies at the Edge of the Pleistocene World*. London: Routledge.

Rensink, E.. 2005. Het Midden-Paleolithicum in Zuid-Nederland. In: J. Deeben, E. Drenth, M.A.F. van Oorsouw and L.B.M. Verhart (red.), *De Steentijd van Nederland*: 119-141. Zutphen: Stichting Archeologie. (Archeologie 11/12)

Roe, D. A.. 1981. *The Lower and Middle Palaeolithic Periods in Britain*. London.

Roebroeks, W.. 1980. De "Middenpaleolithische" vindplaats Sint Geertruid (L). Hypothesen voor nader onderzoek. *Archaeologische Berichten* 8: 7-37.

Roebroeks, W.. 1988. From Find Scatters to Early Hominid Behaviour. A Study of Middle Palaeolithic Riverside Settlements at MaastrichtBelvédère (The Netherlands). *Analecta Praehistorica Leidensia* 21.

Roebroeks, W. and A. Tuffreau. 1999. Palaeoenvironment and settlement patterns of the Northwest European Middle Palaeolithic. In: W. Roebroeks and C. Gamble (eds.), *The Middle Palaeolithic Occupation of Europe*: 121-138. Leiden: Leiden University Press.

Roebroeks, W., J. Kolen and E. Rensink. 1988. Planning depth, anticipation and the organization of Middle Palaeolithic technology: the 'archaic natives' meet Eve's descendants. *Helinium* 28: 17-34.

Roebroeks, W., D. De Loecker, P. Hennekens and M. van Ieperen. 1992. 'A veil of stones': on the interpretation of an Early Middle Palaeolithic low density scatter at Maastricht-Belvédère (The Netherlands). *Analecta Praehistorica Leidensia* 25: 1-16.

Roebroeks, W., M.J. Sier, T.K. Nielsen, D. De Loecker, J.M. Parés, C.E. Arps and H.J. Mücher. 2012. Use of red ochre by early Neandertals. *PNAS* 109: 1889-1894.

Schlanger, N.. 1994. *Flintknapping at the Belvédère: Archaeological, Technological and Psychological Investigations at the early Palaeolithic Site of Maastricht-Belvédère (Limburg, The Netherlands).* Unpublished PhD thesis. Cambridge: University of Cambridge.

Scott, B and N. Ashton. 2011. The Early Middle Palaeolithic: the European context. In: N. Ashton, S.G. Lewis and C. Stringer (eds.), *The Ancient Human Occupation of Britain*: 91-112. Amsterdam: Elsevier.

Semenov, S.A. 1964. *Prehistoric Technology.* London.

Shelley P.H.. 1990. Variation in Lithic Assemblages: An Experiment. *Journal of Field Archaeology* 17: 187-193.

Stapert, D.. 1979. The hand-axe from Venray (province of Limburg, the Netherlands), and several other finds in the vicinity. *Berichten van de Rijksdienst voor het Oudheidkundig Bodemonderzoek* 29: 115-131.

Stapert, D.. 1990. Middle Palaeolithic dwellings: fact or fiction? some applications of the ring and sector method. *Palaeohistoria* 32: 1-36.

Stapert, D.. 2007. Youngsters knapping flint near the campfire: an alternative view of Site K at Maastricht-Belvédère (The Netherlands). *Archäologisches Korrespondenzblatt* 37: 19-35.

Stern, N.. 1991. *The Scatters-Between-the-Patches: A Study of Early Hominid Land Use Patterns in the Turkana Basin, Kenya.* Unpublished PhD thesis Harvard University Cambridge.

Tuffreau, A. and J. Sommé (eds.). 1988. *Le gisement paléolithique moyen de BiacheSaintVaast (PasdeCalais) Vol. I, Stratigraphie, Environnement, Études archéologiques.* Paris.

Tuffreau, A., S. Révillion, J. Sommé, M.J. Aitken, J. Huxtable and A. LeroiGourhan. 1985. Le Gisement Paléolithique Moyen de Seclin (Nord, France). *Archäologisches Korrespondenzblatt* 15: 131138.

UlrixClosset, M.. 1975. *Le Paléolithique Moyen dans le Bassin Mosan en Belgique.* Wetteren.

Van Baelen, A.. 2014. *Kesselt-Op de Schans (Limburg, Belgium) and the Lower/Middle Palaeolithic transition in Northwestern Europe.* Leuven: KU Leuven.

Van Baelen, A.. 2017. *The Lower to Middle Palaeolithic Transition in Northwestern Europe. Evidence from Kesselt-Op de Schans.* Leuven: Leuven University Press.

Vandenberghe, J.. 1993. River terrace development and its relation to climate: the Saalian Caberg terrace of the Maas river near Maastricht (The Netherlands). *Mededelingen Rijks Geologische Dienst* 39: 19-24.

Vandenberghe, J., W. Roebroeks, and T. van Kolfschoten (eds.). 1993. *Maastricht-Belvédère: Stratigraphy, Palaeoenvironment and Archaeology of the Middle and Late Pleistocene Deposits, Part II.* Haarlem: Rijks Geologische Dienst. (Mededelingen Rijks Geologische Dienst 47).

Van Noten, F., D. Cahen and L. Keeley. 1980. A Paleolithic Campsite in Belgium. *Scientific American* 242 (4): 48-55.

Verpoorte, A., D. De Loecker, M.J.L.Th. Niekus and E. Rensink. 2016. The Middle Palaeolithic of the Netherlands – contexts and perspectives. *Quaternary International* 411: 149-162.

White, M. and N. Ashton. 2003. Lower Palaeolithic core technology and the origins of the Levallois method in north-western Europe. *Current Anthropology* 44: 598-609.

White, M.J., B. Scott and N. Ashton. 2006. The Early Middle Palaeolithic in Britain: archaeology, settlement history and human behaviour. *Journal of Quaternary Science* 21: 525-541.

In Dusseldorp, G.L., Chu, W., Bakels, C.C. and M.A. Soressi (eds) 2025. Intent on the Paleolithic. Papers in honour of Prof. Dr. Wil Roebroeks. *Analecta Praehistorica Leidensia* 55. Leiden: Sidestone Press, pp. 29-38.

Forty years after Maastricht-Belvédère: The story of a Middle Palaeolithic quarry site at Roeselare (Flanders)

Philip Van Peer, Ward Caes, Dave Geerts, Bram Peeters

Abstract

In this contribution we report the discovery of a Middle Palaeolithic site in a claypit at Roeselare, in the low hills adjacent to the southwestern Flemish Valley. The archaeological assemblage occurs in near-primary context in peaty overbank sediments that were deposited in a gully eroding into a former depression of Last Interglacial age. It most probably dates from the end of the Last Interglacial. OSL dating is under way. Both lithic and bone artefacts have been recovered next to abundant faunal and macrobotanical remains.

Keywords: Middle Palaeolithic; Roeselare; Last Interglacial; bone tools

Philip Van Peer
Centre for Archaeological Research of Landscapes, KULeuven, Leuven, Belgium
Philip.vanpeer@kuleuven.be

Ward Caes
Centre for Archaeological Research of Landscapes, KULeuven, Leuven, Belgium
Ward.caes@kuleuven.be

Dave Geerts
ArcheoWorks Division, KULeuven, Leuven, Belgium
Dave.geerts@kuleuven.be

Bram Peeters
ArcheoWorks Division, KULeuven, Leuven, Belgium
Bram.peeters1@kuleuven.be

Introduction

A few months ago, the *Lithos* palaeontological society reported the occurrence of lithic artefacts in a claypit at Roeselare, Belgium, which they had been monitoring for several years and which by then had delivered an impressive assemblage of Pleistocene fauna. Upon an inspection of the lithics it was immediately clear that that they had encountered an exceptional site. The artefacts were amazingly fresh Middle Palaeolithic specimens, suggesting a stratigraphic position in or at least close to primary context. Furthermore, there was at least one indubitable bone tool.

Alongside many well-preserved bones, this small assemblage of artefacts had been rescued from a small area cleared from the southern quarry face as the clay exploitation was pushing on. This is reminiscent of a situation some 40 years back, when Wil Roebroeks started excavating the Maastricht – Belvédère quarry, which was to become a landmark site. The future will tell if Roeselare will have the same fate but a volume in honour of Wil's career seems an appropriate place to tell a short story of this new quarry site. His exemplary research at Maastricht-Belvédère has made us all aware of the importance of deep quarries for Pleistocene archaeology in the Low Countries. At the same time, it has demonstrated the challenges involved in finding and excavating these sites. Notwithstanding the spectacular growth of preventive archaeology in Western Europe, the same challenges remain forty years on.

A short chronicle of events

Let us leave the Meuse Valley and move west into the Scheldt Basin, towards the region where the Flemish hills border the Flemish Valley (Fig. 1A). This is an area where Eocene clays are exploited for brick manufacture. The site to be discussed

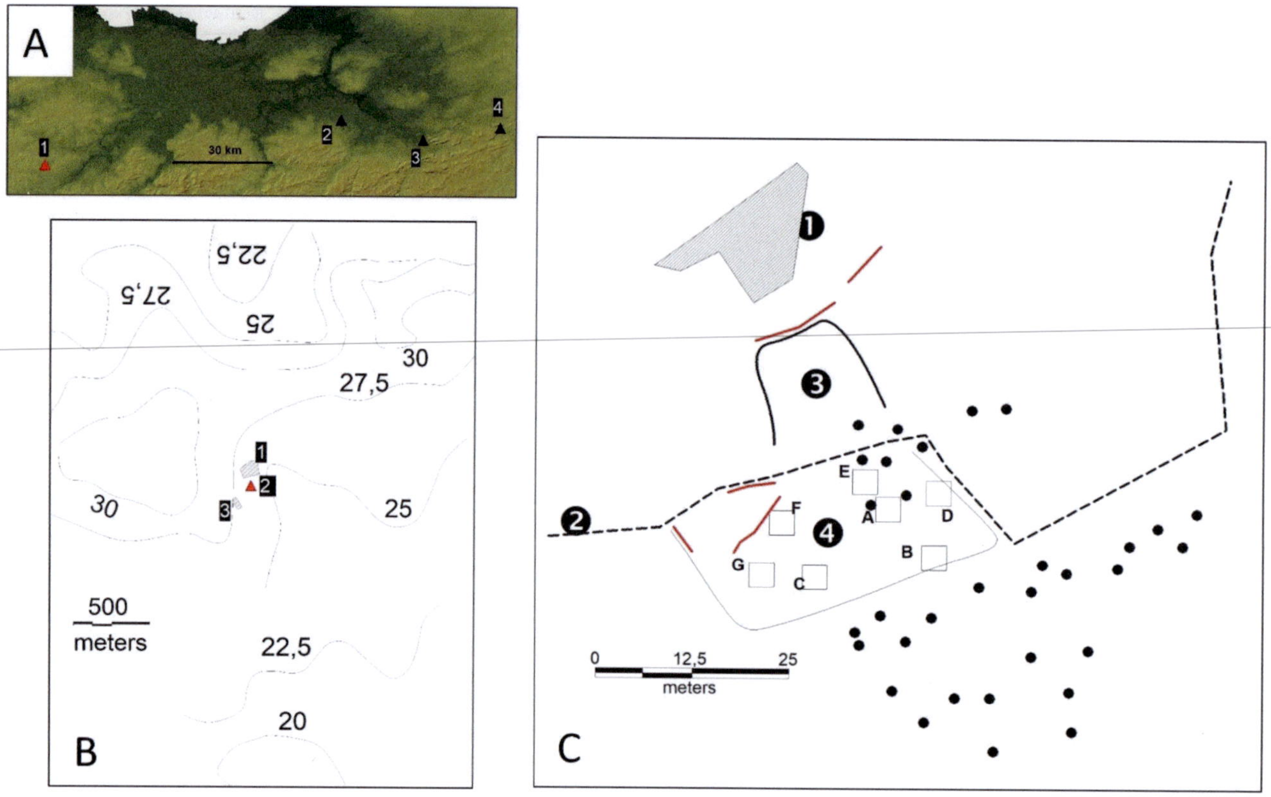

Figure 1. A: Geographic location of Roeselare site (1) and quarry sites in the eastern Flemish Valley (2: Zemst; 3: Rotselaar Torenter-Heide; 4: Schulen). B: Topographic map of the wider region of the site, with locations of the 3 quarry pits. Elevations in m above sea level. Adapted from topographic map sheet 28/4, National Geographic Institute, Belgium. C: Situation sketch of the southeast corner of the present claypit; 1: Area surveyed by the Lithos group; 2: Present limit of the quarry; 3: embankment standing in May 2023, recorded in the first stage of the rescue intervention and collapsed since; 4: Area cleared to ~4 m below surface for study in the second stage of the rescue intervention (A-G = profile pits). Black dots indicate position of mechanical drillings; red lines indicate recorded profiles. The southernmost profiles were exposed in erosional rills originating during rainstorms.

here is located in a quarry in the city of Roeselare. The first archaeological finds in the southern face of this pit were done just a few months ago but as a matter of fact, the site is not so new. More than 10 years ago, two reports had already been published on a few lithic and faunal finds in 2009. These came from a former quarry pit (Fig.1B, 2) located slightly to the northeast of the present one (Goderis and Hameeuw 2011; Hameeuw et al. 2012). They came from a stratigraphic context but at the time it seemed impossible to correlate the latter with any chronostratigraphic reference sequence and it was merely qualified as 'Pleistocene'. Yet, R. Paepe and R. Vanhoorne (1967) had published a beautifully detailed section which they had recorded in the very first exploitation at this location, just north of the 2009 pit (Fig. 1B, 1). They had proposed a chronostratigraphic interpretation based on solid pedostratigraphic, geomorphological, palynological, and radiocarbon evidence. As we shall see below, their recorded sequence is amazingly similar to what is observed in the present pit (i.e. the third at this location (Fig.1B, 3), some 200 m to the south of the one studied by Paepe and Vanhoorne) and undoubtedly to the 2009 section as well. This extremely valuable work really marks the beginning of the history of the Roeselare site.

The Lithos group was literally working in front of the quarry machinery and in the days after that very fortunate recovery, the exploitation had advanced in a southern direction over a distance of some 10 meters. It was around this time that the group reported the finds to the Centre for Archaeological Research of

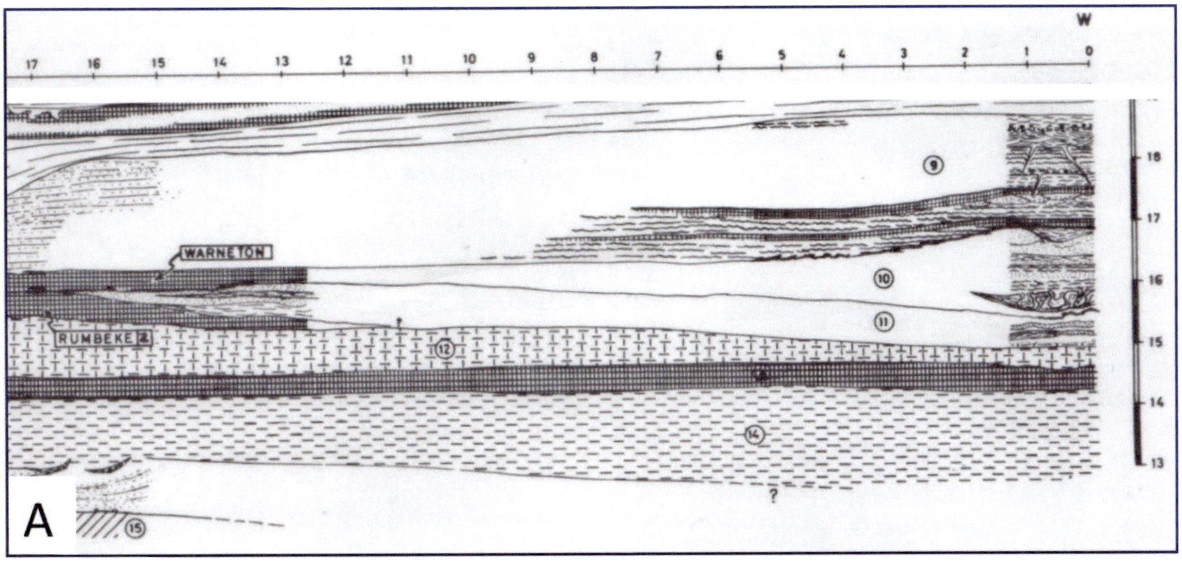

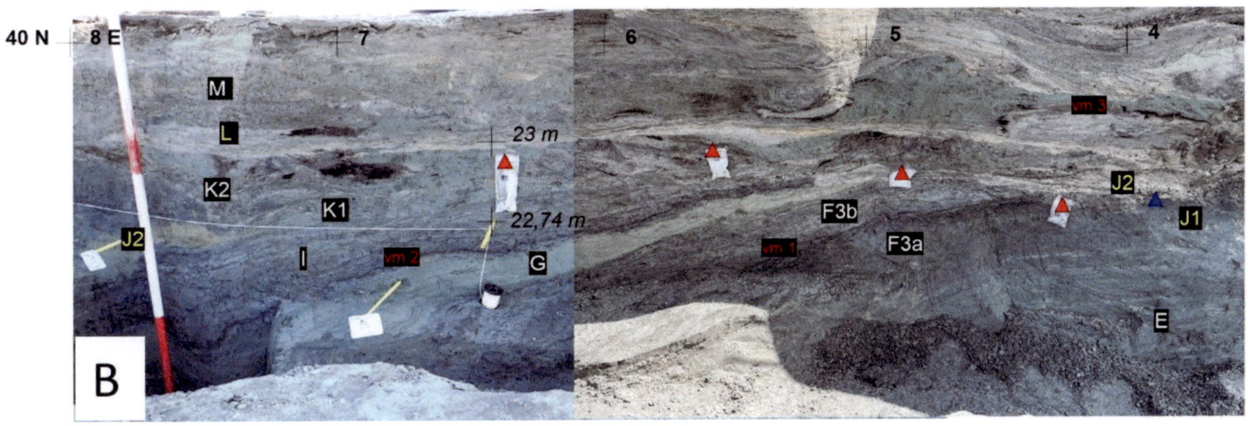

Figure 2. A: Basal part of the east to west profile of the southern face of the original quarry pit (Fig. 1B,1) at this location (reprinted from Paepe and Vanhoorne 1967, Fig. 12). For a full description of the stratigraphic units the reader is referred to the original source. B: Western part of our embankment profile with Layer ID's (E-M) indicated. Vm1-3 are thin peat layers, typically at the interface between larger stratigraphic units. Red triangles indicate faunal elements. Blue triangle is the findspot of a Mousterian point. Numbers (in m) on the X-axis refer to local grid coordinates.

Landscapes at Leuven and we visited the site together in the spring of 2023 to prepare a rescue intervention. We considered that such an operation should serve two goals: 1) to record the stratigraphic sequence closest to the original findspot and most importantly 2) to launch a deep mechanical coring program to survey for intact locations beyond the present quarry limits. The efficiency of the latter survey method had been demonstrated in the *Flemish Valley Survey Project* (Van Peer *et al.* 2017; Van Peer 2020) which had just come to an end and where we had found direct archaeological evidence at six new sites.

This operation took place in the first week of May 2023 (Fig. 1C). We studied a profile of a small embankment isolated from the southern quarry face with its surface at ~4.5 m below the surface. In the area south of the present quarry limit, we executed 30 geoprobe drillings to 6 m below surface, each delivering six sediment columns in 1 m plastic tubes with a diameter of 0.05 m. Unfortunately, the embankment collapsed soon after that as a consequence of exceptional rainfall in the spring of 2023. This resulted in another setback of the quarry front several meters to the south, well into the area where we had been drilling. During the initial rescue operation, we had not found any lithics whatsoever, but two artefacts were found soon after at the foot of a small, isolated ridge of a few m² that had withstood the collapse. They came from the same stratigraphic context as had been observed many times before: at the interface of a shallow gully with a laminated silty infilling, and green-blueish sandy clays.

This evidence, as well as the stratigraphic analysis of the drilling cores, urged us to continue the rescue operation. We selected a 600 m² area to the south of the collapsed profile (Fig. 1C, 4) adjacent to the present quarry limit for further monitoring. Here the top 4 m of sediments, well above the level at which the early Upper Pleistocene sequence was to be expected, were removed.

At this stage, we are arriving at the inevitable conclusion that an exceptional site had been encountered at Roeselare of which now only a small fraction remains. Dating from the end of the Last Interglacial, it is a riverside occupation in near-primary stratigraphic context with excellent preservation of fauna and with bone tools in its technological assemblage. Perhaps the most striking character of the site is its spatial layout and extent. As it stands, we have reason to believe that archaeological debris were scattered along the western bank of a shallow gully over a distance of at least 200 meters. Let us take a look at some of the supporting evidence.

General topographic and stratigraphic context

The three quarries are located on the gentle southern slope of a watershed ridge between the valley of the Leie to the south and the Mandel to the north, both tributaries to the Scheldt and the Flemish Valley (Fig. 1B). The present surface is at ~27-28 m asl, slightly higher than Middle Palaeolithic sites situated at the edge of the central and eastern Flemish Valley.

The original quarry, of which Paepe and Vanhoorne (1967: 22-23 and 87) recorded the profile of the southern wall, is situated some 200 m to the northeast of the present one. In Fig. 3, we reprise the western portion of this profile drawing. There are a few interesting observations that are relevant to the stratigraphic sequence to be treated below.

A first remarkable observation is that the top of the Eocene clay formation (Fig. 2A; Unit 15) appears be much lower in the northern quarry than in the present one, where it occurs at 22 up to 25 m above sea level. Unfortunately, Paepe and Vanhoorne seem to have made an error in levelling an altitude of 22 m for the top of their profile. This is clearly impossible according to the topographic map. But even adding 5 m, the top of the clay is only at 17 m. This observation immediately explains the presence of massive peat deposits at the base of the Quaternary sequence of both quarries: during Eemian times, a deep lake existed in this area.

In the east, we observe the presence of three thick peat deposits which Paepe and Vanhoorne identify from bottom to top as Rumbeke 1 (their unit 13), Rumbeke-2 and Warneton. Based on the pollen analysis of Rumbeke 2 and Warneton (cfr. *infra*), the former is established as Eemian in age and the latter as early Glacial. Although they do not report an analysis of Rumbeke 1, they consider it to be Eemian, together with the superimposed unit 12. In our research area, we have not observed the presence of the Rumbeke 1 peat nor unit 12.

Particularly important is their observation that towards the west, the two upper peats become separated by green silts. We have observed these very characteristic olive-green silts in our research area as well. In our profiles, they do clearly show an erosional contact with the underlying deposits, as the Paepe and Vanhoorne profile suggests as well. The latter furthermore seems to suggest that these green silts are continuous with a sequence of layers in the west (their unit 11). This, however, is an inferred continuity from the partial sections that they have actually recorded in detail. We shall see in a moment that in our sections, unit 11 is superimposed on the green silts.

The 2023 intervention

As described above, an embankment with the basal part of the Quaternary sequence preserved some 5 m below present surface and 10 m south of the area where lithic artefacts had been found, was still standing at the time of our first field intervention (Fig. 1C, 3). We cleared and recorded this profile over ~8 m of length and 1 m of depth (Fig. 2B). Even though it showed an intricate sequence of stratigraphic units due to gullying and cryoturbations, the general similarity to the 1967 sequence was obvious. In apparent contrast to the latter, however, our profile documented the presence of a few faunal remains and one lithic artefact.

The artefact was a frost-cracked fragment of a Mousterian point, situated at the base of a shallow erosional gully (J2) cutting into the underlying F3 sediment. The fill of the gully consists of coarse

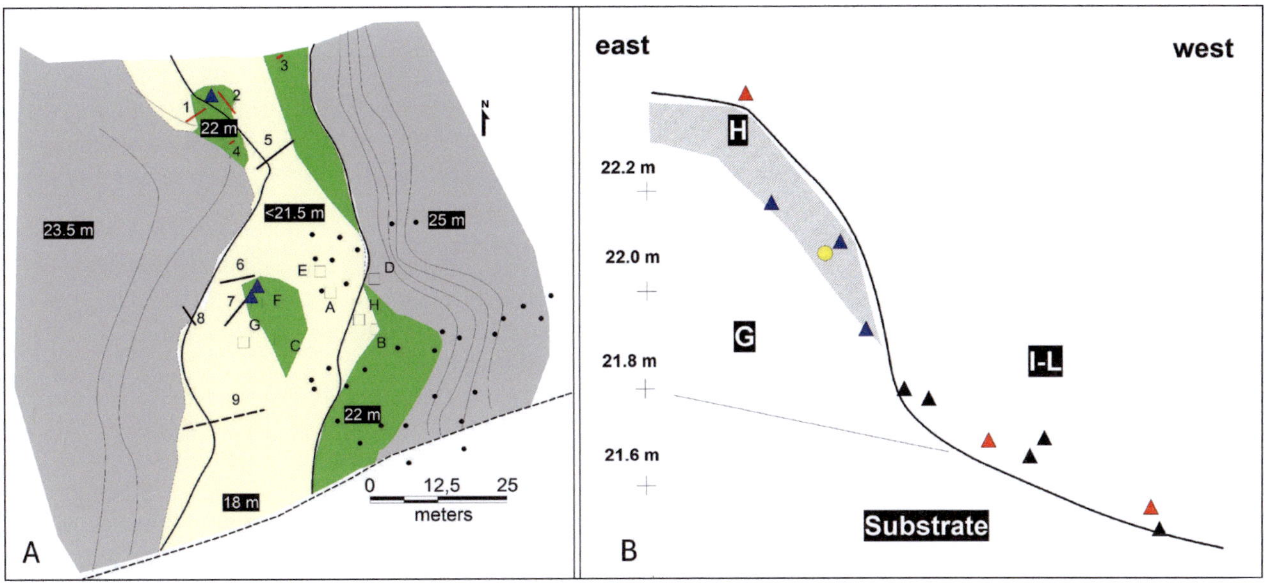

Figure 3. A: Simplified palaeotopographic plan of the 2023 research area as inferred from profiles (1-4 recorded by Lithos; 5-8 recorded during 2023 intervention; 10 recorded after rescue intervention), testpits (A-G), and drillings (black dots). Numbers in black boxes indicate top of Eocene clay in m asl. Thick black lines indicate channel incision in substrate deposits (grey). Early channel fill and bank deposits are preserved in green areas, eroded in yellow area. Blue triangles are artefacts in primary context. B: Vertical distribution of lithic artefacts in primary context (blue triangles), secondary context (black triangles), and bones (red triangles), plotted on a schematic representation of the stratigraphic succession in profile 7. Yellow dot indicates the position of a large fragment of tree wood.

yellowish sands with a few small gravels at the base. Faunal remains were present in several stratigraphic units but always below a characteristic continuous sandy layer (L) separating the basal sequence from overlying cryorturbated sediments topped by a peat deposit. The latter was only preserved in the profile further to the east, beyond the area shown in Fig. 2B. It is most probably to be correlated with the Warneton peat of Paepe and Vanhoorne. Our E through M sequence correlates with the one separating Rumbeke-2 and Warneton in Fig. 2A. Unit G comprises the characteristic olive-green silts with fine lamination. The overlying Layer I has a similar structure, but it has an important peat component. A manifest change in depositional regimes is evidenced starting with Layers J1/J2. Now we observe cross-bedded small gullies eroding into the fluvial sediments underneath.

Simultaneously, we executed a drilling program in an 800 m² area to the south of the present quarry, using a geoprobe mechanical coring technique. Informed by our experience with the Flemish Valley Survey Project (Van Peer *et al.* 2017), we sampled at an average density of 30 m² per core. This was sufficient to produce a general stratigraphic sequence and a reasonably adequate palaeotopography of the area concerned. At the same time, we used these sediment columns to search for direct archaeological evidence. All the cores were drilled to 6 m below surface, which in most cases was sufficient to reach either the top of the Eocene clays or the sands of unit 14 in the 1967 profile which may be of the Middle Pleistocene age (Paepe and Vanhoorne 1967: 22).

We identify this as the 'substrate' on which the Upper Pleistocene sequence has been accumulated. Its topography shows a steep gradient from east to west (Fig 3A). Whereas the top of the substrate is below 22 m asl in the west, it rises to 25 m in the east. The lowermost layers on top of this interface are the blueish clay of unit 12 and a massive black peat deposit (Rumbeke-2) which is encountered in some of the eastern cores up to 24 m asl. Our area, therefore, is located near a steep slope at the edge of an Eemian lake. In the northern coring area (see Fig. 1C) however, the Rumbeke-2 peat was absent and, directly on top of the substrate, we encountered a rather uniform sequence of laminated silts and sands sometimes with clay intercalations. These represent the infilling of a relatively recent erosion channel possibly dating from

MIS 4. In three cores in the southwest of the research area, Rumbeke-2 was absent as well, but the F3-I sequence seemed to be present at their bases (Fig. 3A). Core 60 delivered a lithic chip from this part of the section. This was a comforting find, suggesting that part of the site might still be preserved undamaged in the so far unexploited southern area.

The spatial distribution of the Rumbeke-2 peat and the F3-I sequence suggests that the latter occurs in a gully cut into the substrate and eroding the peat. In several profiles, we have clearly observed this erosional interface. It seems that we witnessed a climatological turnover during which the former Eemian lake fill was incised by meandering channels subsequently filled up with fine overbank sediments. The latter, however, have been observed in certain areas only. In most of the research area, they have been obliterated by the overlying J-L complex. This is a thick depositional sequence consisting of cross-bedded channel fills, heavily cryoturbated in some areas. In contrast to the earlier phase of incision, we now see a braided river at work creating a flat landscape of intersecting shallow gullies. The lithologies of the channel fills vary from greenish silts to coarse sands, often with a band of small gravels at the base.

As mentioned above, most of the lithic artefacts were found so far in these gully bases. Their spatial distribution, however, shows a clear association with the limits of the channel containing the fluvial deposits. This pattern strongly suggests that they were locally reworked from the latter and that the site still preserves its original layout. This hypothesis found confirmation in profile pit F (Fig.1C). This is one of the profiles where we observe the incision of the channel sediments by a gully (Fig. 3B). Three artefacts were found within layer H, the most recent channel fill. This bank is re-incised by a J-L gully at the base of which artefacts in secondary position occur.

This stratigraphic evidence corroborates the inference from the spatial distribution above. Our conclusion at the end of the field intervention, therefore, was that we had encountered the remnants of a Middle Palaeolithic site in near-primary stratigraphic context.

A few conclusions

How much of the site is preserved?

The clay exploitation at this location has been going on for more than 60 years. Whereas in Paepe and Vanhoorne's account of the original quarry there is no mention of archaeological evidence, artefacts were reported from the second pit (Fig.1B, 2) which was exploited during the first decade of this century (Goderis and Hameeuw 2011). The stratigraphic descriptions for these finds, a few lithics and one bone tool, are very limited but two profile pictures (Goderis an Hameeuw 2011: ill.8-9) allow us to recognize a stratigraphic context very similar to the one described above. The *laminated fill* identified in their Profile IV most probably corresponds to the F3-I complex; the artefacts were found in the overlying *loam, sand and clay* deposit.

Comparing the elevations of the base of the F3-I sequence in our research area with those in the 1967 profile (Unit 11 eroding into the Rumbeke-2 peat in fig. 2A), the channel can be inferred to stream down from the watershed area to the north. Hence, it is not surprising to encounter the same channel in the intermediate claypit. The finds here are most probably an extension of the site encountered in the present quarry. We can imagine a longitudinal distribution along the banks of a shallow river extending for at least 200 meters if not more.[1] This interpretation implies that the site has almost entirely been lost in the subsequent clay exploitations.

How old is the site?

The Eemian age of the basal Rumbeke-2 peat is confidently demonstrated by the palynological analysis (Paepe and Vanhoorne 1967: 61). The authors further note that

> *'In the two upper peaty layers, there is an increase of the anemophilous herbs, with Pinus becoming the dominating tree' (1976: 62).*

Plotting the elevations of these two samples on the recorded profile, it is clear that the uppermost sample comes from the 'Warneton' peat. The other one comes from the intermediate deposits between Rumbeke-2 and Warneton, our F3-I complex. They go on as follows:

> *'The comparison however with the Dutch diagrams allow us to make some suppositions. It is possible that the middle peat layer developed in the Amersfoort interstadial and the uppermost one in the following stadial' (1967: 62).*

1 We consider it not unlikely that archaeological evidence was also present in the original quarry.

It is impossible to say with which stratigraphic unit of the F3-I complex, the 'middle peat layer', should exactly be correlated. In any case, the palynological evidence suggests a chronostratigraphic position for an occupation at the end of the Eemian or the onset of the Last Glacial, possibly in the Herning Stadial (Behre 1989). This is entirely in line with the geomorphological evidence that indicates a period of incision at a time when the sea level was lowering, and the Flemish Valley was being reduced in size.

What is the function of the site?

Several lines of evidence lead to a preliminary interpretation of the Roeselare site as a large palimpsest of short hunting visits. First of all, there is the association of lithics with at least a part of a large faunal assemblage. The latter has not been systematically studied so far but a number of bones show indubitable butchering marks. The palaeotopographic situation of the site points in the same direction. It is located along the banks of a river flowing immediately below a talus at least 3 m above water level in the southeast of the research area. It is conceivable that an archaeological distribution would have been present on this talus but if so this must have been lost in the course of numerous erosional events during the Upper Pleistocene.

But there may be more to say about the site's function. As indicated above, there are at least two certain bone tools (Fig. 4) of the *lissoir* type (Soressi *et al.* 2013). A hypothesis coming from our previous work in the eastern part of the Flemish Valley (Van Peer et al. 2017; Van Peer 2020), is that Neanderthals were attracted to this marshy environment in order to exploit carcasses. As a matter of fact, the taphonomic analysis of lithics and bones at the site of Rotselaar-Toren ter Heide suggests that the latter were present at the site before the arrival of hominins. These carcasses would have accumulated along the edges of the valley according to the model proposed by A. Gautier (1985) and they may have been exploited as raw material sources. The evidence of Roeselare points in the same direction and this in particular is a research avenue to pursue in the future.

Roeselare and the Scheldt Basin Micoquian

The lithic assemblage retrieved from the present quarry is very small, comprising only some 20 artefacts awaiting detailed analysis. Among the retouched tools, Mousterian points are the best represented type. There is one handaxe preparation

Figure 4. Tip of bone lissoir (compare to Soressi *et al.* 2013, Fig.2).

or resharpening flake. The most common raw material is a fine-grained black flint of the Obourg variety. It was imported from the Haine Basin some 40 km to the south. Yet there was also use of local flint pebbles. From a taphonomic viewpoint, most of the artefacts are extremely fresh, confirming the idea that they are in a (near-) primary context. A few exceptions, however, show some rounding and patina development, suggesting that they have been exposed at the surface for a time and were slightly more reworked.

Even if small, the similar technological features of this assemblage as compared to others from Flemish Valley sites are obvious. Roeselare adds to the distinct cultural homogeneity of the early Glacial Middle Palaeolithic in the northern Scheldt Basin. Many of its traits are reminiscent of the Middle Pleistocene Micoquian in the east, for example, the presence of handaxes, large preferential Levallois flakes, and volumetric blade production. The Roeselare assemblage contains a few typical examples of such blades. We have proposed the name of *Scheldt Basin Micoquian* (SBM) to capture this particular variety of the European Middle Palaeolithic record (Van Peer, 2025). In terms of Neanderthals' settlement system, the record shows its quite typically fragmented character on the one hand (Turq *et al.* 2013), resulting in low density scatters in which perhaps even one single tool or a tool rejuvenation product may be the basic unit of site formation. On the other hand, these scatters may have spatial distributions much wider than we think.

The idea has been advanced for the Flemish Valley site of Rotselaar and the much better evidence at Roeselare points in the same direction.

Even if for the time being the Roeselare site is just a small, rescued remnant of an exceptionally well preserved early Last Glacial site, it delivers important new evidence for our understanding of Neanderthal history in northwestern Europe. It provides strong support for our recently proposed idea that the open-air sites of the early Upper Pleistocene in the northern Scheldt Basin all date from a narrow chronological window at the very onset of the Last Glacial. This is what the slowly accumulating chronostratigraphic evidence points to, for example from well documented sites as Remicourt on the Scheldt-Meuse watershed (Haesaerts *et al.* 1997). As sites from that age are practically non-existent in the neighbouring regions, the Scheldt Basin may have been a refuge area for Neanderthals adapting to new conditions in the aftermath of the Eemian upheaval.

In contrast to conventional and popularized belief (e.g. Wragg-Sykes 2020) these SBM Neanderthals were quite capable of exploiting the wet marshy lowlands adjacent to the North Sea coasts. The only reason why they were not there until now is our inability to find sites in these taphonomic conditions. If we succeed in overcoming the obstacles involved in surveying for deep buried palaeolandscapes, we may expect to open up a quite new window on Neanderthal ways of life. Until we do, quarry exploitations remain Pleistocene archaeologists' best survey instrument, just as it was forty years ago.

Epilogue

Wil Roebroeks once said that it is more important to find one flake in unknown territory then to excavate yet another flagship site. How right he was! This proclamation is particularly relevant to the Middle Palaeolithic record: that one flake may constitute the entire site. That is what Neanderthals do, at least before they went through a process of behavioural change ~70,000 years ago (Van Peer, 2025).

The anachronistic flavour to the story of discovery of the Roeselare site, after 30 years of preventive archaeology, teaches us a few important lessons. Obviously, those 30 years have not resulted in a *modus operandi* that would have saved this site. We are using the past tense because the intervention reported on above has simply exposed the margins of a site that was once there. The simple old truth is that the site would not have existed at all if it were not for the effort of the amateur archaeologists. Rather than comfortably relying on legislation implemented in stringent rules and civil duties, the profession should seek new ways to work with them and acknowledge their efforts.

One might do away with this case as the one exception that will always confirm the rule. The essential question then becomes the following: have present legislation and current practices resulted in a better recovery rate of quarry sites in the northern Scheldt Basin as compared to Maastricht-Belvédère times? The prudent yet worrying answer to that question is *'almost certainly not'*. Without a serious overhaul of current 'standard' field practices, discovery of quarry sites seems bound to remain a pure chance event of extremely low probability.

Actually, from a probabilistic perspective the number of quarries in and around the Flemish Valley that have delivered Pleistocene archaeological evidence would seem too large to be explained by chance. Does this mean that things are not so bad after all? Let us make no mistake and take this as a consolation or an excuse to carry on as before. This pleasing observation simply demonstrates that our conventional assumptions about the Palaeolithic record are wrong. In fact, it should be a loud plea to organise more effective prevention. The Palaeolithic record is a 'veil of stones' of much larger density and distribution than we have imagined (Roebroeks *et al.* 1992). The present sample of open-air sites in the lowlands of northwestern Europe is ever more unlikely to adequately represent that record and consequently, to produce confident knowledge of Neanderthal behaviour in these environments.

Quarries are bound to remain the lampposts in a dark Pleistocene world hidden under many meters of sediment fill. Will Roebroeks was among the first to turn on the light. Please, let us keep it on.

Acknowledgements

Many people must be credited for their merits in saving this site. First of all there are the members of Lithos society, Jan Vanheuverzwijn, Johan Vandamme, Yoeri Christiaens, Rudy Bruyneel, and Frans Bartier. Their line of action made scientific research possible at all and they assisted in the rescue operation. We thank Luc Vandermersch for granting access to the quarry and for making logistics available. Yannick Vanhove took part in the fieldwork. Peter Janssens (BP^2 company) facilitated the drilling operation. The latter was funded by the Flemish Heritage Agency.

References

Behre, K.-E.. 1989. Biostratigraphy of the Last Glacial period in Europe. *Quaternary Science Reviews* 8: 25-44.

Gautier, A.. 1985. Pleistocene zoogdieren van België en waar ze gevonden worden. *Lutra* 28: 121-123.

Goderis, J. and H. Hameeuw. 2011. Roeselare-Oekene: archeologisch onderzoek op vondsten uit het Paleolithicum. *West-Vlaamse Archeologica* 24: 40-49.

Haesaerts, P., H. Mestdagh and D. Bosquet. 1997. La séquence loessique de Remicourt (Hebaye, Belgique). *Notae Praehistoricae* 17: 45-52.

Hameeuw, H., A., G. J. Goderis and W. Hantson. 2012. Some preliminary notes on the Late Pleistocene contexts and Middle Palaeolithic finds in Oekene (West Flanders, B). *Notae Praehistoricae* 32: 141-145.

Paepe, R. and R. Vanhoorne. 1967. *The Stratigraphy and Palaeobotany of the Late Pleistocene in Belgium.* Toelicht. Verhand. Geologische Kaart en Mijnkaart van België, 8. Brussel: Aardkundige Dienst van België.

Roebroeks, W., D. De Loecker, P. Hennekens and M. van Ieperen. 1992. "A veil of stones": on the interpretation of an early Middie Palaeolithic low density scatter at Maastricht-Belvédère (The Netherlands). *Analecta Praehistorica Leidensia* 25: 1-16.

Soressi, M. , S.P. McPherron, M. Lenoir, T. Dogandžić, P. Goldberg, Z. Jacobs, T.E. Steel, S. Talamo and J.-P. Texien. 2013. Neandertals made the first specialized bone tools in Europe. *PNAS* 110: 14186-90.

Turq, A., W. Roebroeks, L. Bourgignon and J.-P. Faivre. 2013. The fragmented character of Middle Palaeolithic stone tool technology. *Journal of Human Evolution* 65: 641-655.

Van Peer, P.. 2020. Het Schulensbroek (Demervallei, BE): een uniek paleolandschappelijk en archeologisch archief. *Notae Praehistoricae* 40: 129-144.

Van Peer, P., M. Willems, P. Degryse, P. Claeys, and E. Marinova. 2017. A reconstruction of the stratigraphic position of a former Middle Palaeolithic surface site at Rotselaar – Toren ter Heide (Flemish Valley, Belgium) using mechanical sounding and geochemical fingerprinting. *Journal of Archaeological Science: Reports*, 16: 380-390.

Van Peer, P.. 2025. On Mousterian, Micoquian, and the origin of new Neanderthals in Western Europe ~70,000 years ago. In *Sitting on the Fence: Negotiating Archaeology, Anthropology and Philosophy. Festschrift for Prof. Dr Raymond H.A. Corbey in celebration of his 70th birthday. Edited by S.T. Hussain and G.L. Dusseldorp*. Analecta Praehistorica Leidensia 54, 151-161.

Wragg-Sykes, R.. 2020. *Kindred. Neanderthal Life, Love, Death and Art.* London: Bloomsbury.

In Dusseldorp, G.L., Chu, W., Bakels, C.C. and M.A. Soressi (eds) 2025. Intent on the Paleolithic. Papers in honour of Prof. Dr. Wil Roebroeks. *Analecta Praehistorica Leidensia* 55. Leiden: Sidestone Press, pp. 39-56.

Avocational archaeology and the Middle Palaeolithic of Dutch and Belgian Limburg

Phil Glauberman, Yannick Raczynski-Henk, Luc Amkreutz

Abstract

The careers of Wil Roebroeks as well as a number of his students are closely linked with the avocational archaeology community of (predominantly) the Limburg area of the Netherlands and Belgium. This connection is important through the exchange and dissemination of knowledge, in the symbiotic relationship between local enthusiasts and professionals and in the rooting of research in the region through the inclusion of, and cooperation with locals. This paper presents the exciting results of some of these exchanges which shaped the careers of Roebroeks as well as the authors, and it offers recommendations for improvements of the inclusion of avocational archaeologists in the process of researching the rich, yet largely unknown archaeological record of the Dutch and Belgian Middle Palaeolithic.

Keywords: Middle Palaeolithic; Neanderthals; surface artefact collections; avocational archaeology; citizen science

Phil Glauberman
The Catalan Institute of Human Paleoecology and Social Evolution (IPHES) and Universitat Rovira i Virgili, Tarragona, Spain; Department of Early Prehistory and Quaternary Ecology, University of Tübingen, Tübingen, Germany
phil.glauberman@gmail.com

Yannick Raczynski-Henk
Human Origins Institute, Leiden University, Leiden, The Netherlands
y.raczynski-henk@denoud-engroep.com

Luc Amkreutz
Rijksmuseum van Oudheden, Leiden; Faculty of Archaeology, Leiden University, Leiden, The Netherlands
l.amkreutz@rmo.nl

§ All authors contributed equally

Introduction

The avocational archaeology community in the Netherlands and Belgium assumes a prominent role in academic and professional Palaeolithic archaeology (De Vries *et al.* 2022). This community embodies much more than artefact collectors. They actively contributed to the construction of a theoretical framework for the Palaeolithic settlement of the region by proposing models about Neanderthal land use, subsistence, local/regional geology, and site formation. These theories have been disseminated through the journal *Archeologie in Limburg* and others. Some of their models have been tested by professional archaeologists, including Wil Roebroeks and the current authors, during research projects in Limburg in the last decades.

The sites at the Maastricht-Belvédère quarry were excavated and published by W. Roebroeks and some of these avocational archaeologists, and the site would not have been excavated in the first place without them. From the early 1980s onwards, they studied the profiles of the quarry, looking for soils and artefacts, which led to the discovery of the many sites in the Belvédère, but subsequently also the nearby sites of Veldwezelt-Hezerwater, Kesselt op de Schans and others (De Warrimont 2007). Avocational archaeologists have been instrumental in discovering nearly all of the excavated Palaeolithic sites in Limburg.

While the Belvédère site is famous for its fine-grained fluvial sediments and behavioural resolution, W. Roebroeks has always maintained an interest in the

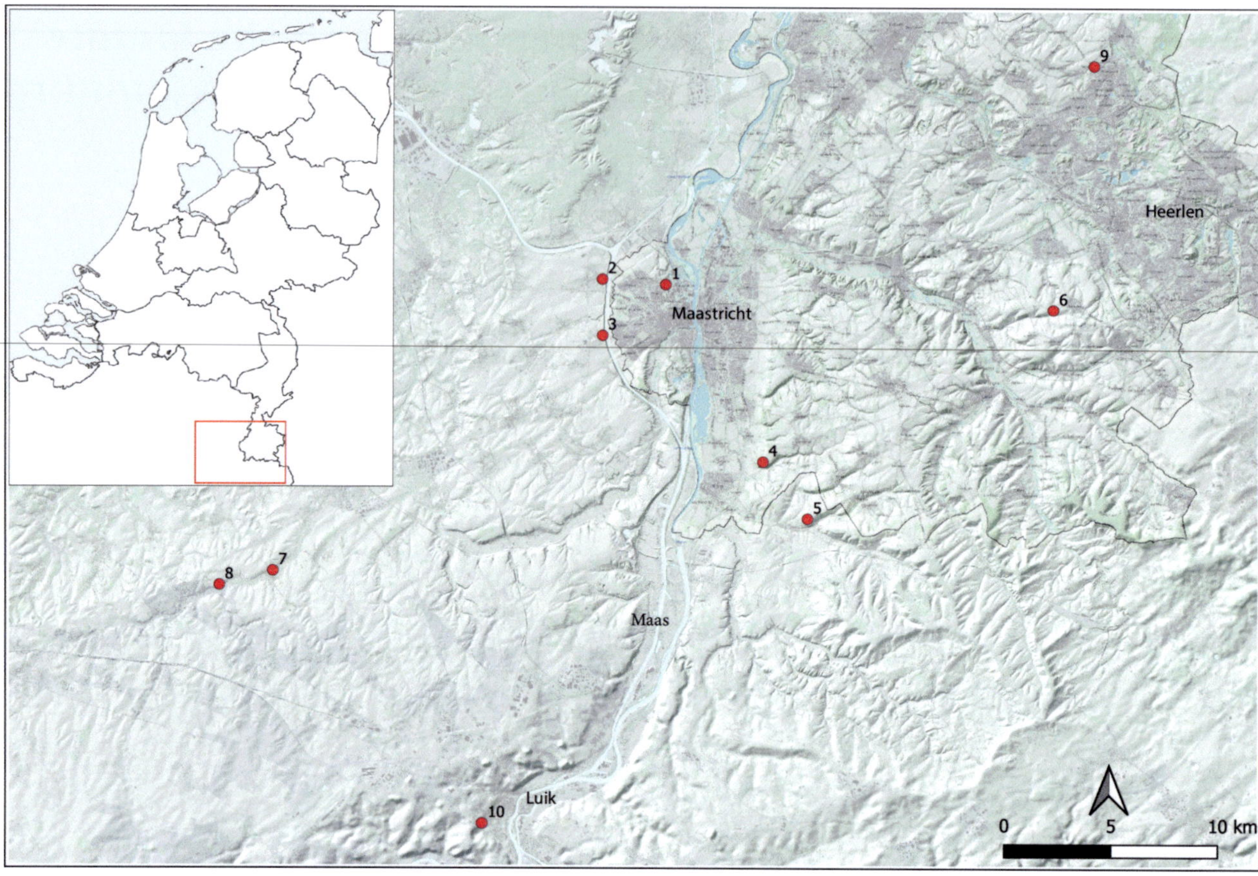

Figure 1. A map based on the digital elevation model (AHN; courtesy of Ivo van Wijk) of the relevant part of the Netherlands and Belgium, listing the sites mentioned in the text: 1: Maastricht-Belvédère, 2: Veldwezelt-Hezerwater, 3: Kesselt-Op de Schans, 4: Sint Geertruid-De Kaap, 5: Voeren-Snauwenberg, 6: Colmont-Ponderosa, 7: Lauw, 8: Otrange, 9: Amstenrade-Allée, 10: Comtes de Mean and Court Saint-Hubert.

time averaged surface sites South Limburg is known for (see the appendix in Roebroeks 1988; Henk 2006; Paulussen and Raczynski-Henk 2014; Raczynski-Henk *et al.* 2018; Glauberman 2006; 2014; 2016), and which have been the main interest of the avocational community.

The upland surface sites of South Limburg have also been research interests for some his graduate students. In fact, all the systematic Palaeolithic fieldwork which has been done in South Limburg, has involved one or more of them.

The graduate and later professional work of two of the authors of this Chapter, Phil Glauberman and Yannick Raczynski-Henk, was inspired by the predictions and recommendations W. Roebroeks (1980; 1988), and D. De Loecker (2006) made in their PhD dissertations, as well as by Kolen *et al.* (1999), who put forward that if we want to learn more about Palaeolithic hominin land use in the region, it is paramount to include upland surface sites in our research. The avocational archaeological community had also forecasted this as a necessity; that upland surface sites should be considered in any hypothesis-driven investigation into how Palaeolithic hominins utilized the Limburgish landscape (*e.g.* Groenendijk and De Warrimont 1995; Blezer *et al.* 1996).

Here, we summarize the results of research at some of these upland sites and synthesize results of comparison among upland and lowland fluvial sites. This new synthesis builds on our previous research and re-assesses lithic artefact assemblage formation in terms of hominin technological organization and land use behaviours. These behaviours are indicated by the lithic artefact assemblages that accumulated at different localities throughout the landscape, in a variety of geomorphic contexts. Subsequently, the current state of avocational contributions to prehistory and stewardship of the Palaeolithic record in Limburg are briefly reviewed from the perspective of Public Archaeology.

Upland sites

Sint Geertruid-De Kaap

Wil Roebroeks grew up in Sint Geertruid, The Netherlands. In the village fields, avocational archaeologists had been collecting Palaeolithic artefacts for nearly a century by the time Wil made his first public observations of Palaeolithic surface finds with the collaboration of a group of local artefact enthusiasts (Roebroeks 1980).

The rich site of De Heij or De Kaap is a south-facing, eroding loess hillside overlooking the Maas River valley from the high, third terrace. It is mantled by 2-20 meters thick loess-silt loam deposits, that are eroding on the south facing slope, where Palaeolithic artefacts appear due to alluvial and plough erosion. Palaeolithic artefacts were first documented at De Kaap in 1881 (De Puydt 1887; Ubaghs 1887; De Puydt *et al.* 1912). Thousands of flint artefacts have since been collected there (Roebroeks 1980; 1988; Groenendijk and De Warrimont 1995; De Warrimont 1997; 2002; Glauberman 2002; 2006; 2016; De Loecker 2006) and active artefact collection continues to this day.

Between 2007 and 2013, Glauberman analysed 666 artefacts from De Kaap from museum and private collections as part of his PhD research. While hundreds of artefacts from Dutch and Belgian Limburg have been illustrated, described, and published over the years (Wouters 1980; Roebroeks 1980; Groenendijk and De Warrimont 1995; Kolen *et al.* 1999), that study was the first to collect detailed techno-typological data on the surface finds from De Kaap (Glauberman 2014; 2016).

In 2011, thanks to a collaborative initiative with the Netherlands Rijksdienst voor Cultureel Erfgoed (RCE; Dutch State Service for Cultural Heritage), we conducted an extensive augering survey to study the stratigraphy at De Kaap, and excavated a 10 m x 5 m, x 5 m deep trench in search of Palaeolithic artefacts in stratigraphic context (Van Baelen *et al.* 2012, 2017). Sint Geertruid De Kaap is the first surface locality in Limburg to have undergone excavation and chronostratigraphic analysis.

Figure 2. Glauberman and Raczynski-Henk in the trench at Sint Geertruid in 2011. Photo L. Amkreutz.

We uncovered a 4.5 m thick loess/silt-loam-paleosol sequence, starting with Saalian, and ending with last glacial maximum sediments and a Holocene luvisol (Figure 2). Three artefacts were recovered from the base of the Weichselian sediments, in a complicated depositional and erosional context (Figure 3). The artefacts were retrieved from fine-grained, clayey loam, with some evidence of pedogenesis, overlying a gravelly layer. This in turn overlaid a compound paleosol.

A series of OSL samples were analyzed from throughout the sequence. The artefacts were found in/on units forming the lower parts of a truncated Bt horizon, with OSL dates ranging from 149,000 to 200,000 BP. A correlation with the Rocourt Pedocomplex is most likely. The overlying sediments are dated to 30,000 BP, which makes them younger than the presence of archaic humans in the area. This indicates a significant disconformity in the loess sequence and makes assigning an age to the artefacts problematic.

Figure 3. The Levallois core from the 2011 trench at Sint Geertruid-De Kaap (Van Baelen *et al.* 2017: 67).

An early Weicheslian age is most probable, since the area was likely not very densely populated during the warmest phase of MIS 5e, the Eemian (Roebroeks and Speleers 2002; Roebroeks *et al.* 1992, but see Pop *et al.* 2016 and Sier *et al.* 2015). The global climate was relatively cool in the early Weichselian, with temperatures up to 10 ^{0}C below the current average off the coast of Norway (Lowe and Walker 2014).

The artefacts are likely not in primary context. They retain variable patinas, with a retouched denticulate tool having a white porcelain patina, while a discoid core recovered not far away had only little grey-white powdery patina, suggesting weathering in different micro-environmental settings (Glauberman and Thorson 2012; Van Baelen *et al.* 2017).

An early Weichselian age is consistent with the composition of the surface and excavated artefact assemblages. Both contain Levallois cores and blanks, and also a series of bifaces. These artefact types have been observed in sites dated to the same time range in surrounding areas (Cahen and Peuskens 1977-79; Antoine *et al.* 2003; Depaepe 2010; Di Modica 2010; Locht *et al.* 2010, 2017). It has long been noted that bifaces resembling those associated with the so-called Mousterian of Acheulean Tradition (MTA) identified in France during MIS 5-3, and also those that resemble the *Keilmesser* observed in Micoquian assemblages to the east, co-occur in the De Kaap and other Limburgish surface artefact collections (Roebroeks 1980, 1988; Wouters 1980; Groenendijk and De Warrimont 1995; Blezer *et al.* 1996). Some have viewed this a 'border zone' between eastern and western Middle Palaeolithic groups, those who made Keilmesser, and those that did not (*e.g.* Ruebens 2013). However, this surface material very likely derives from chronostratigraphically distinct contexts and may represent many thousands of years of artefact deposition by several hominin groups.

That is not to say that one cannot view the assemblages in terms of artefact discard patterns related to technological organization. Landscape-sale core reduction sequence segregation characterizes Palaeolithic site and assemblage formation (*e.g.* Geneste 1988; Turq *et al.* 2013). Taking a wide spatio-temporal view of 'site' (*e.g.* Wagstaff 1991), one can observe what elements of reduction sequences were most commonly discarded over time at sites, test for variability among them, and hypothesize implications for land use behaviours. This approach sees palimpsests as inherent to all archaeological site formation (*cf.* Binford 1981) and utilizes their diachronic accumulation to view long-term depositional trends (*e.g.* Barton and Riel-Salvatore 2014). We do not assume that the Limburg archaeological landscape was created contemporaneously by a single group of hominins. We do recognize that every archaeological site is a former surface site (*cf.* Binford 1981), and by viewing them as situated in a unique depositional context, we can investigate long-term patterns. The following summarizes results of excavations at upland surface and lowland fluvial localities. While not reviewed here, there are several upland surface sites that require test excavation and dating, including the site of Snauwenberg in the Voerstreek region of Belgian Limburg. Like De Kaap, it has yielded thousands of MP artefacts but has never seen any attempts at excavation and chronostratigraphy.

Colmont-Ponderosa

Colmont-Ponderosa is situated at ~ 175 masl on the south-facing slope of the Vrakelberg dry valley, on the 'Island' of Ubachsberg, a plateau in central South Limburg in the Netherlands, that remained outside the influence of the laterally migrating Maas River throughout the Pleistocene (Kuyl 1980). Surface artefacts were found in high density there over a 1,600 m^2 area, in a clay-rich gravelly plow soil (Blezer *et al.* 1996; Kolen *et al.* 1999). Around 1,670 artefacts with Middle Palaeolithic techno-typological characteristics were collected from the late 1980s to the early 1990s. Results of the first study of 611 artefacts from this collection indicated a relatively uniform assemblage, including high frequencies of notch and denticulate tools and unsystematically reduced cores, and a few with disc and discoidal forms (Blezer *et al.* 1996). Test excavation was conducted in 2001, in three trenches over a 30 m^2 area just upslope from where artefacts appeared on the surface. 181 artefacts were excavated at the diffuse contact of the basal weathered limestone residue (clay-with-flints) and overlying loess-derived colluvial and slope deposits (Langbroek *et al.* 2001). Based on the stratigraphy and basal topography the artefacts are likely associated with a karstic doline formation, a sediment trap, that if water-holding, may have been attractive to Middle Palaeolithic hominins and other fauna (Glauberman 2002), a site context frequently observed in Eurasia (Van Andel and Runnels 2005; Glauberman 2006; Diez-Martín 2010; Friesem *et al.* 2014). No chronometric dates are available but estimates for the site converge around the early Weichselian or last glacial cycle (Blezer *et al.* 1996; Langbroek *et al.* 2001; Glauberman 2002; Glauberman 2016). Re-analysis of a sample of the surface collection and excavated artefacts corroborated previous assessments. However, currently ongoing lithics analysis is re-evaluating the artefact assemblages and has found techno-typological characteristics consistent with now better-studied Quina assemblages from France (Abruzzese 2022). Colmont-Ponderosa clearly needs chronometric age assessment and is suitable for luminescence dating.

Lauw

In Belgian Limburg, the site of Lauw is situated on the southwest-facing slope of a plateau at 115 masl, overlooking the Jeker/Geer River, a tributary of the Maas, roughly 20 m above the current floodplain. Discovered in 1979, it underwent systematic surface collection of a 14,062 m^2 area, and test excavations (Gijselings and Doperé 1983). This fieldwork yielded 1,462 artefacts from the surface, and 121 from a gravelly stratum overlying a Bt horizon, expected to be of Eemian (MIS 5e) age. That is, the gravel truncating the interglacial paleosol must post-date 130 ka, while chronometric age estimates are currently unavailable for the sediments. A sample of 320 artefacts from the surface collection and excavated assemblage was re-analysed by Glauberman (Glauberman 2014, 2016). Statistical comparison of excavated and surface artefact dimensions, artefact class frequencies, flaking technology, and patina types showed no significant differences, suggesting that the sample of surface artefacts is representative of the excavated assemblage (Glauberman and Thorson 2012). The artefact assemblages include one *Keilmesser*, typical Mousterian scraper forms, and retouched flakes including notches, denticulates, and bifacially worked pieces, alongside complete and broken flakes. Scar pattern analysis suggests that discoidal and Levallois core reduction techniques were the most common. Again, this site requires further chronostratigraphic assessment.

Otrange

Discovered in 1947, Otrange, Belgium, is located on a south-facing slope above the Jeker/Geer River at ~ 130 masl, and about 20 meters above the current floodplain (Thisse-Derouette and Destexhe-Jamotte 1948; de Heinzelin 1950). Around 2,000 artefacts were collected from one

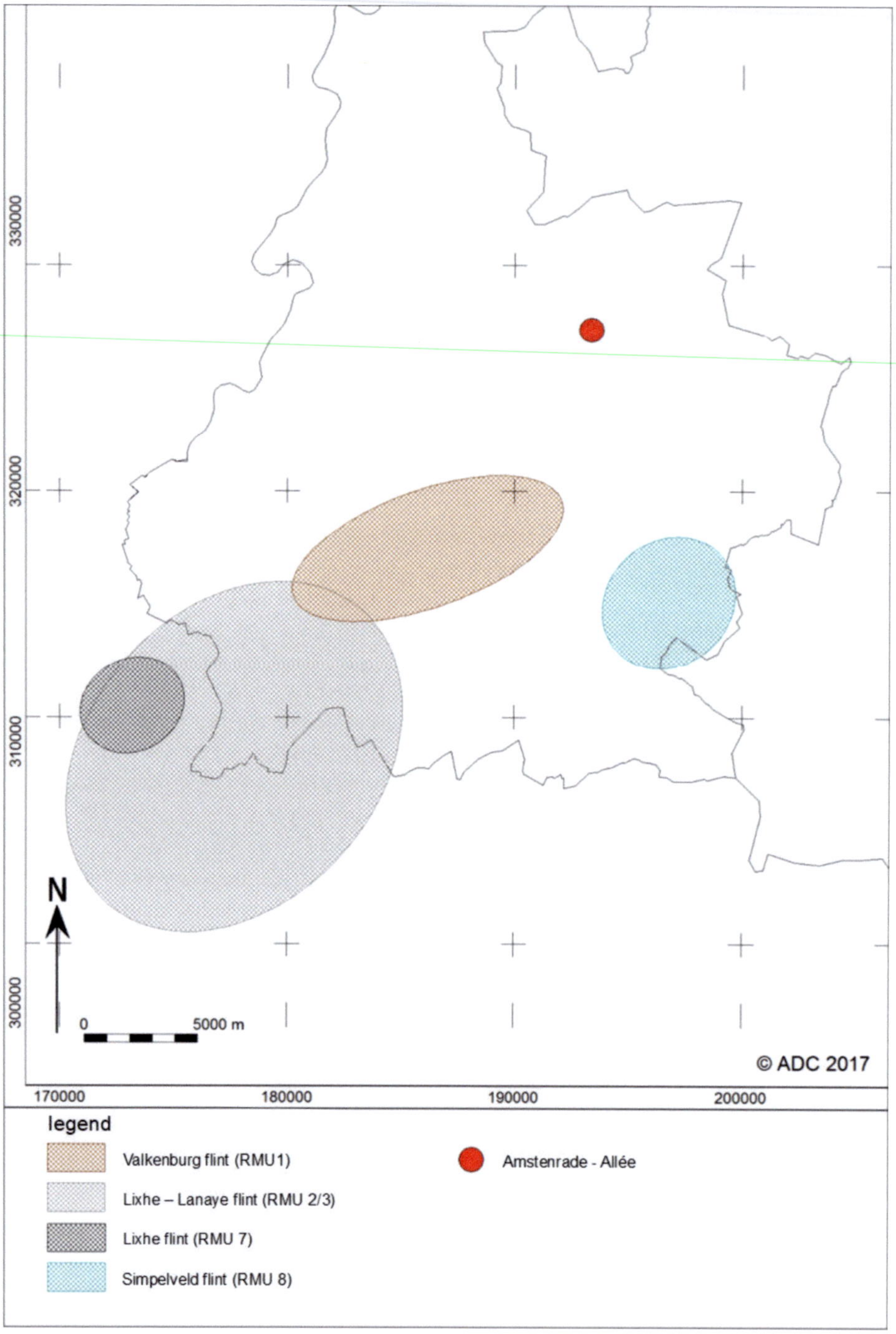

Figure 4. Location of the flint source areas relative to the site of Amstenrade-Allée, red dot. After Raczynski-Henk et al. 2018: 87.

field parcel, in an area of 330,000 m^2 (Ulrix-Closset 1975; Di Modica and Jungels 2009). In 1948, three trenches were excavated upslope of the main surface artefact concentration, exposing loess-derived deposits overlying Neogene sands. Five stratified archaeological horizons yielded a total of 4,650 artefacts (de Heinzelin 1950). Each of these horizons yielded artefacts at variable densities, in different alluvial and colluvial deposits, some with evidence of soil formation, intercalated with gravel-lag deposits. The lowest excavated artefact assemblage indicates a focus on raw material procurement and core preparation and reduction, while the uppermost assemblage appears to relate more to tool use and discard, indicating changing site function over time (Jungels 2005; Di Modica and Jungels 2009; Di Modica 2010). A sample of 317 surface finds amassed during the 1980s was analysed by Glauberman (Glauberman 2014). Comparative analysis of the surface and excavated assemblages indicates that the surface material is a geologically cumulative palimpsest that naturally sampled of all of the excavated archaeological horizons and represents dynamic use of place during the Saalian – early Weichselian (de Heinzelin,1950; Ulrix-Closset, 1975; Di Modica and Jungels, 2009). However, no chronometric age estimation has been undertaken.

Amstenrade

Amstenrade-Allée, The Netherlands, (89 masl) was discovered during cultural resources management (CRM) work preceding road construction. An augering campaign revealed a sediment trap with traces of a preserved, interglacial paleosol about 6 m below the surface (Paulussen and Raczynski-Henk 2014). In subsequently machine-dug test trenches, flint artefacts were found on or in a brown luvisol and at the erosional disconformity between these deposits and overlying loess. The remaining area in the road cut was excavated (Raczynski-Henk *et al.* 2018).

This resulted in the exposure of an archaeological site in deposits attributed to the Saalian and Early Weichselian, in which the Rocourt and Warneton Pedocomplex formed between 115,000 and 71,000 BP. The artefact-bearing layers have been dated by OSL to around 85,000 BP. The site possibly dates to the slightly older Amersfoort/Brørup Interstadial (MIS 5c), or the somewhat younger Odderade Interstadial, periods when the loess area was partly forested. The location of the site was probably chosen because of its proximity to wood and surface or spring water; it lies near the edge of a plateau, close to a valley which likely once held the adjacent Oirsbeek stream.

The site consists of a large, low-density scatter of artefacts (n=94) over about 3,300 m^2, in which two clusters were distinguished. In one, a small number of artefacts of Valkenburg flint could be refitted and attributed to a single knapping event indicative of core preparation. The almost complete absence of cortex shows that initial core preparation took place elsewhere. A small number of discarded tools (n=4) and one hand axe resharpening flake were found, but most of the material consists of debitage (n=71) and three small core fragments. Almost a third of the artefacts are broken, but only two broken pieces could be refitted. Eight distinct raw material units (RMUs; *cf.* Roebroeks 1988) were observed, but only four could be sourced: Valkenburg flint (the largest group with 23 artefacts), then one single artefact of Lixhe flint, and two Rijckholt artefacts which fall in the Lixhe-Lanaye group (Figure 4). The last RMU consists of one artefact of Simpelveld flint. Although the unidentified flints will probably skew this image, it appears that the occupants of the location were fairly local people. The discovery of a small fragment of haematite expands their expected mobility range somewhat. The closest sources are in the Weser/Vesdre river valley in the Belgian Ardennes, roughly 34 km distant (Goemaere *et al.* 2016).

A scatter of charcoal with roughly the same outline as the artefact scatter was found, with a few patches of higher density along the valley edge, coinciding with the areas of highest artefact density. No (other) traces of fire were found. Charcoal is an unreliable indicator of fire use on its own, but some of it is vitrified, which can be indicative of the production of birch tar (Kubiak-Martens 2008; Hayek *et al.* 1990; McParland *et al.* 2010; Kubiak-Martens *et al.* 2013). The use of birch tar by Pleistocene hominins in northwest Europe has been attested by the discovery of a flake partially covered in birch tar from the North Sea (Niekus *et al.* 2019). On the whole, we hypothesize that the charcoal can be linked to one or more phases of site occupation. One poorly preserved bone fragment was discovered just outside one of the artefact clusters. It was impossible to categorize it any more precisely than a 'long bone of a large mammal'. A link with the lithic assemblage and/or the charcoal scatter cannot be established.

The deeply buried site of Amstenrade-Allée is unique in the context of Limburgish upland sites: it is the only site without a (current) surface artefact counterpart, although there are similarities with the sites of Comtes de Mean and Court Saint-Hubert in the city of Liège; low density sites on the edge of the Maas that also lacked a surface component (Van der Sloot *et al.* 2011). It is similar to the sites mentioned earlier in the sense that, even though it was buried, it was not *stratified*. Most of the artefacts come from an erosional disconformity. It is however unique: it is the only Middle Palaeolithic site in the country that was not discovered by avocational archaeologists, illustrating the importance of their efforts for the palaeolithic record of the Netherlands.

Summary

While only four upland Middle Palaeolithic surface localities in Limburg have been excavated, we can summarize a few commonalities in buried artefact assemblage contexts:

- artefacts are associated with buried gravel lenses and deposits that vary in thickness, continuity, grain size, and slope angle (*e.g.* Lauw).
- artefacts are found in unconformable contexts with and without gravels, associated with truncated paleosol catenas (*e.g.* De Kaap and Amstenrade).
- artefacts are associated with gravels and clayey deposits that overlie or incorporate material from basal substrata including river gravels (*e.g.* Amstenrade), weathered limestone deposits and Neogene sands (*e.g.* Colmont and Otrange) (de Heinzelin 1950; Di Modica and Jungels 2009; see also De Puydt *et al.* 1912; Roebroeks 1981).

Excavated upland artefact assemblages have been consistently found in erosional lithostratigraphic contexts, indicating that their appearance on the modern surface constitutes at least a tertiary setting, the artefacts having undergone cycles of erosion and redeposition. The refits at the site of Amstenrade-Allée however, demonstrate that even under these conditions, (elements of) these sites can be found in primary context.

Lowland Sites

As described earlier, in the wider Maas River valley floor lithic artefacts have been recovered from exposed litho-pedostratigraphic sequences, often at brickmaking quarries, documenting climatic oscillations of the Middle-Late Pleistocene (van Kolfschoten and Roebroeks, 1985; Roebroeks, 1988; Vandenberghe *et al.* 1993; Bringmans 2006; Van Baelen *et al.* 2007, 2008; Meijs 2002, 2011, Meijs *et al.* 2012). Artefact assemblages at three key localities established the occupation of the region during the late Saalian (MIS 7) and Weichselian (MIS 5d-a) glacial cycles, or around 250-50 ka.

Maastricht-Belvédère

In the 1980s, Palaeolithic artefacts were discovered at the Belvédère quarry by avocational archaeologists in freshly cut pits on the Caberg or Middle Maas Terrace. A series of artefact concentrations exposed in fluvial Maas River deposits and buried plateau contexts were excavated under the direction of W. Roebroeks and colleagues. The majority was dated by thermoluminescence and lithostratigraphy to around 250 ka, while two localities, Sites J and E, were excavated in early Weichselian deposits (Roebroeks 1988; Roebroeks *et al.* 1992b; De Loecker 2006). Based on refitting and techno-typological analysis, several high-density artefact concentrations were found to be areas of core reduction and tool manufacture, superimposed on a background 'veil of stones' or low-density scatter of artefacts related to tool use and discard (Roebroeks *et al.* 1992). This pattern was predicted by surface surveys and excavations of lower Palaeolithic sites in east Africa that showed 'patches' and 'scatters' of artefacts that comprised archaeological landscapes (Isaac 1981).

The sites include the Saalian Site K, where nearly 40% of reduced cores were refitted, and demonstrated short-term occupation involving transport of flint nodules to the Maas riverbank for reduction, manufacture of tools, and their use and discard (De Loecker 2006). Sites G and N yielded assemblages of a different character, indicating only tool use and discard. To varying degrees, these assemblages indicate short-term, or episodic occupations. Sites J and E date to the early Weichselian. Site J is situated on a low plateau overlooking the Maas Valley. The assemblage is interpreted as resulting from a short-term butchering episode, with evidence of on-site core reduction and tool manufacture, use, and discard (Roebroeks *et al.* 1997). Site E is situated in gully infill sediments, stratigraphically between the Eemian interglacial paleosol and the cryoturbated marker horizon for the last glacial maximum (~ 24 ka). The lithic assemblage associated with faunal remains indicated evidence for tool use and discard (Roebroeks 1988).

The Maastricht-Belvédère deposits established the age of Palaeolithic occupations of Limburg, and are still regarded, perhaps alongside the site of Abric Romani in Spain, as some of the highest resolution 'snapshots' of MP behaviour to have ever been excavated. The excavations at Belvédère uncovered samples of an archaeological landscape buried under ~ 20 m of loess-derived deposits and demonstrated the flexibility and range of behavioural adaptations in the region. The site also set a foundational precedent, not only as the Palaeolithic 'type sequence' in the region, but also for collaborative prospection and excavation methods that would continue in the loess landscape, largely on the part of the avocational archaeological community.

Veldwezelt-Hezerwater

Prospection in the open loess quarries around the Maas River and its tributaries during the 1990s yielded another Middle Palaeolithic site in stratified context at the Nellisen quarry along the Albert Canal, at the Dutch and Belgian border near Maastricht (Bringmans 2006; Meijs 2011). It exposed a relatively complete litho-pedostratigraphic sequence spanning the Saalian, Eemian, and Weichselian climate cycles, including one of the most complete Eemian paleosols documented in the region, that can be correlated with sequences exposed at the site of Rocourt and several others in northern France that have also yielded Middle Palaeolithic artefacts (Antoine *et al.* 2003; Meijs 2011).

271 artefacts were retrieved at the VBLB site, in a stream valley slope setting, stratigraphically above two paleosol horizons likely dating to MIS 5b or 5a, or the early Weichselian. Based on refitting and lithics analysis, the assemblage is interpreted as resulting from short-term occupation where flakes were struck from a few cores, retouched tools were discarded, and a few cores were transported out of the excavated area (Bringmans 2006).

The WFL site was documented in pedogenically altered slope deposits overlying interglacial paleosol deposits. The artefact assemblage (n = 133) is interpreted as representing the remains of a few short core reduction episodes, with evidence of tool edge maintenance. The assemblage likely represents a tool use and discard locality, with some flake production (Bringmans 2006).

113 artifacts were recovered from the TLB layer (TL-R, TL-GF, and TL-W Loci). This stratum consists of loam valley slope deposits altered by weak pedogenesis during the middle Weichselian (Meijs 2011). The combined assemblage is interpreted as resulting predominantly from tool use, maintenance, and discard during short-term visits to the stream-side locality in the course of game hunting and butchery activities (Bringmans 2006).

Assemblage Formation and Hominin Land Use Behaviour

Glauberman collected and synthesized detailed techno-typological data on artefact collections and attempted to numerically investigate Middle Palaeolithic artefact assemblage formation and its relationship to land use (Glauberman 2014, 2016). That study assumed that the spatial accumulation of lithic artefacts at specific places resulted from thousands of years of artefact discard in the context of hominin land use and mobility, in line with prominent landscape archaeological theory and ethnoarchaeological and ethnographic observations of modern hunter-gatherers (Yellen 1977; Binford 1980; Foley 1981; Isaac 1981; Rossignol and Wandsnider 1992; Stern 1993, 1994; Kelly 1995; Holdaway and Wandsnider, 2008). This perspective expands to the landscape-scale, the 'scatters and patches' pattern of artefact assemblage formation, originally based part on surface material, and observed in excavated contexts at Maastricht-Belvèdére (Foley 1981; Isaac 1981; Roebroeks 1988; Roebroeks *et al.* 1992).

Essentially, the observable differences in artefact assemblages that suggest the degree of fragmentation of core reduction sequences and intensity of site occupation, can be boiled down to variability in assemblage diversity. That is, they differ in richness, or the number of artefact types, and evenness, or the frequency of specimens in each category (*e.g.* Jones *et al.* 1983; Leonard and Jones 1989; Rhode 1988). In ecology, diversity is a combined measure of evenness and richness in sampled species assemblages, and several formulas and diversity indices exist (*e.g.* Simpson 1949). Diversity measures can be applied to artefact assemblages to provide a singular numerical measure of the diversity of cores, flakes, and tools (*cf.* Leonard and Jones 1989). The ratios of these artefact classes reflect accumulative artefact discard patterns related to fragmentation of core reduction sequences that occurred through lithic technological organization (Nelson 1991; Kuhn 1994, 1995; Turq e*t al.* 2013). Modelling of palimpsest assemblage formation predicts that we should expect a negative correlation of artefact density and tool or retouch frequency. That is, at low density sites we should expect high frequencies of tools discarded in the course of use and maintenance at sites occupied at low intensity, whereas at high density sites we should expect low tool frequencies and higher proportions of cores and flakes derived from tool manufacture and relatively higher site occupational intensity (Barton and Riel-Salvatore 2014; Clark and Barton 2017; Glauberman 2016).

A problem in quantifying assemblage formation is that as sample size grows, so does richness (Peet 1974; Pielou 1975; Jones *et al.* 1983; Leonard and Jones 1989). The more a site is sampled, the number of artefact

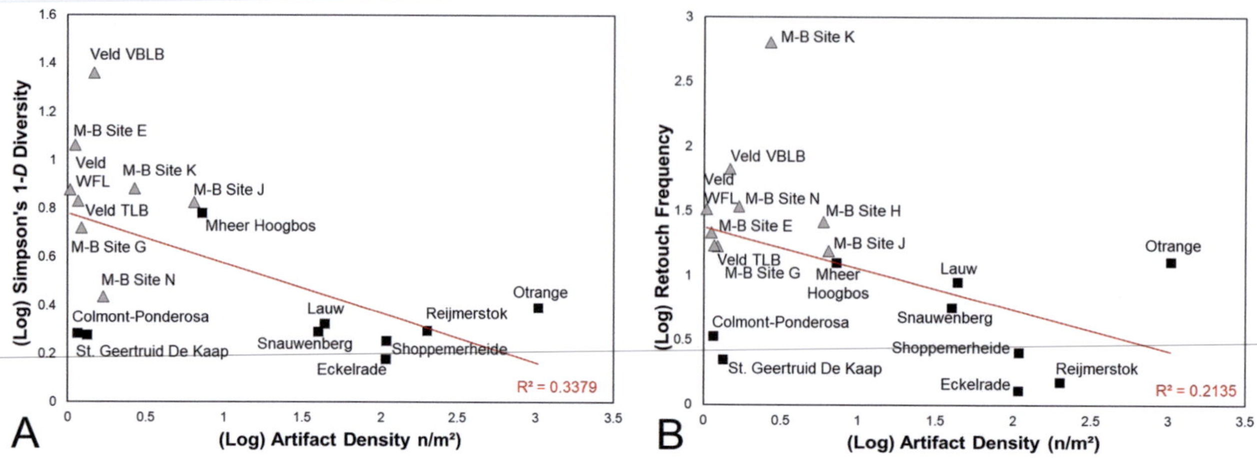

Figure 5. Bivariate plots showing A) artefact density vs. diversity (Simpson's 1-D index), and B) artefact density vs. tool or retouch frequency for lowland (triangles) and upland (squares) sites. The plots show the expected negative correlation between artefact density and retouch frequency and diversity predicted by Barton and Riel-Salvatore (2014). The Dutch and Belgian Limburg MP data shows that sites were occupied at a patterned range of intensity and with differential artefact discard behaviours. Lowland sites tend to preserve higher tool frequencies at lower densities, suggesting short term, infrequent and perhaps task-specific activities, while upland sites tend to have lower tool frequencies at higher densities, suggesting long-term or frequent re-occupations, and a range of core reduction and tool discard behaviours. (M-B = Maastricht-Belvedere; Veld = Veldwezelt-Hezerwater).

types grows until it reaches a representative amount, and no new categories are observed. Such sample size effects can be investigated if one analyses diversity in terms of sample size (*e.g.* Rhode 1988). For this reason, the earlier study chose to use sample size to measure and compare assemblage diversity among sites. Sample size showed no positive correlation with artefact assemblage richness or diversity, indicating that sample size effects were minimal among sample artefact assemblages from upland surface sites and lowland excavated sites (Glauberman 2016).

The analysed dataset included assemblages excavated in the 'lowland' fluvial parts of the landscape at Maastricht-Belvédère and Veldwezelt-Hezerwater, and surface material from 9 upland terrace and plateau localities. Results demonstrated patterned differences in lowland and upland assemblage formation. Upland sites preserved evidence of repeated, frequent occupations with relatively high assemblage diversity, i.e., a mixture of cores, flakes, and tools. Lowland sites on the other hand, had lower assemblage diversity, and higher tool frequencies (Glauberman 2016). This suggests that lowland sites saw more infrequent occupations, and perhaps task-specific behaviours. This pattern may result from site formation processes, where sites in the lowlands in floodplain or riverbank settings would naturally be buried quickly, perhaps seasonally. This depositional setting precludes continuous occupation of sites in fluvial contexts, and archaeological materials should then be stratified. In the erosional uplands however, artefact deposition could recur at single localities, and later denudation and erosion of sediments can create the appearance of high-density sites on the surface, where reworked occupation horizons can be mixed, as indicated at Otrange.

Palimpsest formation differences among sites notwithstanding, a clear separation was observed between upland and lowland sites in terms of artefact discard patterns. This makes sense, as much of the available flint crops out in upland settings, limestone beds, or as eroding nodules in colluvial slope deposits. Many of the upland surface sites are situated near such flint sources, and the origins of artefacts made from primary chalk flint and colluvial or fluvial contexts can be determined based on cortex characteristics (Roebroeks 1980; Groenendijk and De Warrimont 1995; Glauberman 2002; Fernandes *et al.* 2007). Such characteristics among surface finds and excavated materials analysed in the study suggest that Middle Palaeolithic hominins utilized flint from all of these procurement contexts throughout their occupation of Dutch and Belgian

Limburg (Roebroeks 1980, 1988; Groenendijk and De Warrimont 1995; Blezer *et al.* 1996; De Loecker 2004; Glauberman 2014, 2016).

Here, we re-evaluate the upland and lowland site artefact diversity study with reference to the artefact assemblage formation models. A problem with this is assessing the volume of archaeological deposits to calculate artefact density. To accommodate the often unknown thickness of strata from which surface artefacts originated, we chose to calculate artefact density in two dimensions using site and excavation surface area. In a way, this equalizes the sites, and treats them as synthetic cumulative palimpsests.

When we plot artefact diversity and retouch frequency against artefact density, we find the same separation among upland and lowland sites as we did in the first sample size-based study (Figure 5). This firstly validates the previous assessment based on sample size that demonstrated important differences between upland and lowland sites. Secondly, it further shows numerically that we should expect low intensity, perhaps infrequent or short-term occupations involving mainly tool use and discard in the lowlands, while at upland sites we should find sites occupied at higher, and probably variable, frequencies and intensities, for a range of activities, many involving tool manufacture and core reduction (*cf.* Glauberman 2006, 2014, 2016). Both analyses provide strong evidence of segregation of core reduction sequences among lowland and upland localities, indicating long-term hominin behavioural adaptations to the Dutch and Belgian Limburg or Maas River valley landscape (Roebroeks 1988; Groenendijk and De Warrimont 1995; De Loecker 2006; Glauberman 2016).

Searching for those who search for finds

The above clearly demonstrates the important groundwork avocational archaeologists laid for placing Palaeolithic archaeology on the map, something that in the early 20th century was deemed unlikely (Holwerda 1918). Apart from several pioneers (Van Ginkel and Theunissen 2009), most avocational archaeologists were active in the 20th century contributing to hallmark discoveries of Dutch key sites such as Maastricht-Belvédère, Eyserheide (Rensink 2010), and many important isolated finds (Figure 6). As indicated, they also actively shaped the construction of theoretical frameworks, at least for the southern loess area. Undeniably the forged discoveries of Vermaning and the dubious role of the former cleric Wouters (De Vries *et al.* 2022) in some instances drove a wedge between amateurs and professionals. But it also fuelled enthusiasm and debate, as demonstrated by the association APAN (Actieve Praktijk Archeologie Nederland / Active Practical Archaeology Netherlands) focusing on binding together mostly stone age amateur archaeologists and collectors. However, as much of the turmoil surrounding this affair settled into dust, so, it seems, have most avocational archaeologists. They appear to be a dying breed.

This is strange. Interest in history and our human past is booming. The popularity of publications such as 'Sapiens' (Harari 2015) and television productions such as 'Het verhaal van Nederland (The story of the Netherlands)' point to a shared human interest in what connects us and 'how we got here'. Publications on stone age archaeology have been received with much enthusiasm and good reviews, including Wil Roebroeks' well-known 'Oermensen in Nederland (Prehistoric man in the Netherlands)' (1990), but also 'Onze vroegste voorouders (Our earliest ancestors)' (Louwe Kooijmans 2017), 'Vuursteen Verzameld (Flint collected)' (Amkreutz *et al.* 2016) and the successor to Roebroeks' book: 'De eerste mensen in de Lage Landen (The first humans in the Low Countries)' (Dusseldorp and Raczynski-Henk 2022). There is also a reviving interest in the exploration of nature and the outdoors. Activities such as orienteering and geocaching see young people involved in pastimes that include elements of discovery, context and 'treasure' hunting (Skinner *et al.* 2018). These ingredients also apply to stone age archaeology. However, muddy boots and scouring rainy fields only appear to appeal to enthusiasts of metal detecting, which since the COVID-pandemic has become a popular hobby. The Danish 'Danefae' and Dutch Portable Antiquities scheme are overstretched (Dobat and Jensen 2016; Heeren *et al.* 2022), and in the UK there is even a hugely popular sitcom 'the Detectorists'. All the while avocational stone age archaeologists are becoming rare.

We could accept it is a sign of the times. However, the above demonstrates we actually need avocational archaeologists to discover new sites. According to an estimation by Raczynski-Henk (2014), up to 27,000 Middle Palaeolithic sites could be hidden in the Limburg Heuvelland area, yet only one, Amstenrade, has been discovered and investigated within current CRM-archaeology (Raczynski-Henk *et al.* 2018, also see Verhart 2007). This makes it necessary to investigate the reasons why our 'eyes and ears' are no longer there. Some can be mentioned. Stone age and lithic archaeology are less of a treasure hunt: they require skill and practice and benefit from organisation in groups, sharing experience and knowledge. This has become less common in recent times in general. Many organisations struggle to draw in younger people. They in turn appear mainly drawn to more individualistic experiences and their expression in social media.

How should we then move forward? A strong motivation may be found in the recent development of 'citizen science', which aims to include wider audiences in scientific projects. Archaeologically, this entails a step beyond 'Malta-Archaeology', which propagates public outreach of scientific results (Article 9), towards the Faro Treaty, where scientific research is explicitly opened up and people are requested to cooperate, or even co-create (Van Londen *et al.* 2021). Often this taps into feelings of community and localism (ibid.). A number of projects have already successfully adopted such an approach. For instance, many people participated in LiDAR-based research, allowing them to detect burial mounds from behind their desktop (Lambers *et al.* 2019). The project won the 2022 European Europa Nostra heritage award. The project 'Expeditie Vuistbijl' (Verhoeven *et al.* 2020) also actively sought out amateur archaeologists to report finds and sites. In the north of the Netherlands, archaeologist Marcel Niekus, in close collaboration with the avocational archaeology community, organised a systematic survey in Drenthe with the aim of finding Palaeolithic sites. In 2007, they hit the proverbial mother lode, discovering a rich site near Assen that preserves in situ archaeological deposits (Niekus *et al* 2016). On a larger scale, recent interest in North Sea Doggerland led to an increasing number of finds from artificially built-up beaches. Public involvement in this topic has been massive with many participating, sharing and discussing discoveries, in particular on social media (Peeters *et al.* 2019). Here also, most important discoveries were made by avocational archaeologists, including the first Dutch Neanderthal fossil (Hublin *et al.* 2009), 33 handaxes from a submerged Norfolk site (*e.g.* Davis *et al.* 2023), a decorated Late Palaeolithic aurochs' bone (Amkreutz *et al.* 2018), and a rare Neanderthal tool with birch bark tar (Niekus *et al.* 2021).

An exhibition in the 'Rijksmuseum van Oudheden' (National Museum of Antiquities) in Leiden showcased the efforts of these citizen scientists and their most special finds (Amkreutz and Van der Vaart-Verschoof 2022). This community also has a burgeoning digital infrastructure including Facebook pages and the Naturalis website www.oervondstchecker.nl, designed to identify fossils and finds. A current NWO (Dutch Science Foundation) project, 'LegaSea', will train AI to identify fossils and artefacts based on their large collections. All of the above indicates much potential, but also that the parameters have changed. In order to rekindle avocational stone age archaeology, and stewardship of the Pleistocene archaeological record, we need to find different ways of approaching potentially interested groups and individuals.

So how do we find those that find the finds and sites, or perhaps even create or enthuse those that might? Here we suggest a number of ideas:

- We need to move away from the antiquarian perspective of the enthusiastic collector and *his* collection of lithics. This is an outdated phenomenon that moreover has been both very white (in skin *and* hair colour), and very male. We should find ways of enthusing younger people, male and female (many beachwalkers are female) and from diverse backgrounds.
- We must acknowledge the strong motivation in local and regional history. Localism (Van Londen *et al.* 2021) is something we should embrace in terms of sparking the interest in research and investigation of the oldest history of a place or region.
- We must adopt social media as a powerful tool for connecting and communicating, seeking out partnerships with frameworks that are already in place, such as Portable Antiquities of the Netherlands, oervondstchecker.nl, and Archeohotspots. We must develop educational tools for burgeoning interest, such as introductory materials, identification and drawing courses, finds-recording etc.
- We need to invest in creating Faro-projects that have a clear scientific goal and ample room for meaningful contributions by (potential) amateur archaeologists.
- We must underline the importance of deep history and the rapport with us as humans today, to lift interest above amassing collections.
- We are convinced that rather than hoping for a return of the 20th century lithics afficionados, it is more realistic to invest in project-based research that clearly is participatory and allows people and communities to contribute. This creates a fertile environment from which eventually new collectors, avocational archaeologists and our eyes and ears in the field may arise.

Conclusion

Wil Roebroeks profoundly influenced Middle Palaeolithic research in Limburg and abroad. Whether they were involved or not in his many projects, so did the many avocational archaeologists, whose experience and knowledge far surpassed that of several generations of fledgling Palaeolithic archaeology students. The ideas, theoretical framework, and methodology of the fieldwork at Amstenrade, the first (and only!) Palaeolithic site investigated in Dutch CRM-archaeology, can trace its lineage back to the classrooms at the Faculty of Archaeology at the Reuvensplaats in Leiden, to fieldwork at Colmont-Ponderosa and Sint-Geertruid, to endless discussions on a usually cold and windy field in Belgian or Dutch Limburg, and to that short, wild excursion to the Somme River valley sometime in the early 2000s. Together with the inspiration provided by those that study, collect and document these sites this forms an important foundation of the Palaeolithic archaeology in the Low Countries.

The fact that there must be hundreds, if not thousands of (Middle) Palaeolithic sites in the loess area of South Limburg alone (Raczynski-Henk 2014), is grounds for concern. If we want to investigate Palaeolithic sites, we need Palaeolithic archaeologists. Or rather, if we want to find and investigate Palaeolithic sites, we need Palaeolithic enthusiasts. We hope that the beneficial avocational-professional relationships fostered over 40 years ago, rooted in a tradition going back another 100 years or so, will continue in the Dutch and Belgian loess region, by new generations collecting artefacts, discovering sites and collaborating intensively based on the precedent set by Wil and his colleagues.

References

Amkreutz, L.W.S.W. and A. Verpoorte. 2009. A Middle Palaeolithic scraper of Wommersom quartzite from Stein (prov. Limburg, the Netherlands). A short find report *Archäologisches Korrespondenzblatt* 39/4: 449-453.

Amkreutz, L. and S. Van der Vaart-Verschoof. 2022. *Doggerland. Lost world under the North Sea.* Leiden: Sidestone Press.

Amkreutz, L., A. Verpoorte, A. Waters-Rist, M. Niekus, V. Van Heekeren, A. van der Merwe, H. van der Plicht, J. Glimmerveen, D. Stapert and L. Johansen. 2018. What lies beneath… Late Glacial human occupation of the submerged North Sea landscape. *Antiquity* 92: 22-37.

Amkreutz, L., F. Brounen, J. Deeben, R. Machiels, M.-F. Van Oorsouw and B. Smit. 2016. *Vuursteen verzameld.* Amersfoort: Rijksdienst voor het Cultureel Erfgoed. (Nederlandse Archeologische Rapporten 50).

Antoine, P., P. Auguste, J.-J. Bahain, P. Coudret, P. Depaepe, J.-P. Fagnart, C. Falguéres, M. Fontugne, M. Frechen, C. Hatté, A. Lamotte, M. Laurent, N Limondin-Lozouet, J.-L. Locht, N. Mercier, A.-M. Moigne, A.-V. Munaut, P. Ponel and D.-D. Rousseau. 2003. Paléoenvironnements Pleistocenes et peuplements Paléolithiques dans le bassin de la Somme (Nord de la France). *Bulletin de la Sociètè Préhistorique Française* 100: 5-28.

Bailey, G.N..1983. Concepts of time in Quaternary prehistory. *Annual Review of Anthropology* 12: 165-192.

Bailey, G.N.. 2007. Time perspectives, palimpsests, and the archaeology of time. *Journal of Anthropological Archaeology* 26: 198-223.

Barton, M. and J. Riel-Salvatore. 2014. The formation of lithic assemblages. *Journal of Archaeological Science* 46: 334-352.

Binford, L.R. 1980. Willow smoke and dog's tails: hunter-gatherer settlement systems and archaeological site formation. *American Antiquity* 45: 4-20.

Binford, L.R.. 1981. Behavioral archaeology and the 'Pompeii Premise'. *Journal of Anthropological Research* 37: 195-208.

Binford, L.R.. The archaeology of place. 1982. *Journal of Anthropological Archaeology* 1: 5-31.

Binford, L.R.. 2016. Long-term land use patterns: some implications for archaeology. University of Michigan Anthropological Papers No. 72. In: R.C. Dunnell and D.K. Grayson (eds.), *Lulu Linear Punctuated: Essays in Honor of George Irving Quimby*: 27-54. Ann Arbor: Museum of Anthropology.

Blezer, L., D. De Loecker and J. Kolen. 1996. A Middle Paleolithic surface scatter at Colmont (Southern Limburg, The Netherlands). *Notae Praehistoricae* 16: 25-36.

Bringmans, P.M.M.A.. 2006. *Multiple Middle Palaeolithic Occupations in a Loess-soil Sequence at Veldwezelt-Hezerwater Limburg.* PhD diss., Katholieke Universiteit Leuven.

Cahen, D. and N. Peuskens. 1977-1979. Sites Paléolithiques entre Meuse et Geer au Nord de Liège. *Bulletin des Chercheurs de la Wallonie* 24: 17-49.

Conard, N.J. and D.S. Adler. 1997. Lithic reduction and hominid behavior in the Middle Paleolithic of the Rhineland. *Journal of Anthropological Research* 53: 147-175.

Davis, R., N. Ashton, R. Bynoe, J. Craven, R. Ferguson, I. Gardner, T. Grimmer, C. Harris, M. Hatch and C. Johnson. 2023. Middle Palaeolithic occupation of the southern North Sea Basin: evidence from the Sandscaping sediments emplaced on the beach between Bacton and Walcott, Norfolk, UK. *Journal of Quaternary science* 38: 866-890.

Heinzelin, J. de. 1950. Stratigraphie du gisement Paléolithique d'Otrange sur la base des résultats de la campagne de fouilles de 1948. *Bulletin de l'Institut Royal des Sciences Naturelles de Belgique* 26: 1-32.

De Loecker, D..2006. *Beyond the Site: The Saalian Archaeological Record at Maastricht-Belvèdére (The Netherlands)*. PhD diss., Leiden University.

Di Modica, K.. 2010. *Les Productions Lithiques du Paléolithique Moyen de Belgique: Variabilité des Systémes d'Acquisition et des Technologies en Réponse á une Mosaïque d'Environnements Contrastés*. PhD diss., Université de Liège.

Depaepe, P.. 2010. l'Apport des fouilles de grande superficie sur la connaissance du Paléolithique Moyen. In N. Conard and A. Delanges (eds.), *Settlement Dynamics of the Middle Paleolithic and Middle Stone Age, vol. III*: 357-372. Tübingen: Kerns Verlag.

De Puydt, M. 1887. Quelques constatations relatives à la Station Neolithique de Sainte Gertrude. *Publications de la Sociètè Historique et Archéologique dans la Duché du Limbourg* 24: 39-56.

De Puydt, M., J. Hamal-Nandrin and J. Servais. 1912. Liège Palèolithique. Le gisement de Sainte-Walburge dans le limon Hesbayen, *Bulletin de l'Institut Archéologique Liègeois* XLII: 139-215.

Diez-Martín, F.. 2010. Evaluating the effect of plowing on the archaeological record: the early Middle Palaeolithic in the river Duero Basin plateaus (North-Central Spain). *Quaternary International* 214: 30-43.

Di Modica, K. and C. Jungels. 2009. Le Gisement Paléolithique d'Otrange. In K. Di Modica and C. Jungels (eds.), *Paléolithique Moyen en Wallonie La Collection Louis Éloy*: 134-143. Bruxelles: Service du Patrimoine Culturel de la Communauté Française de Belgique (Collections du Patrimoine culturel de la Communauté Française 2).

Dobat, S.A. and A.T. Jensen. 2016. Professional amateurs. metal detecting and metal detectorists in Denmark, *Open Archaeology* 2: 70-84.

Dusseldorp, G. and Y. Raczynsky-Henk. 2022. De eerste mensen in de Lage Landen, Assen: Uitgeverij Koninklijke Van Gorcum.

Fernandes, P., F.X. Le Bourdonnec, J.P. Raynal, G. Poupeau, M. Piboule and M.H. Moncel. 2007. Origins of prehistoric flints: The neocortex memory revealed by scanning electron microscopy. *Comptes Rendus Palevol* 6: 557-568.

Foley, R.. 1981. A model of regional archaeological structure. *Proceedings of the Prehistoric Society* 47: 1-17.

Friesem, D.E., Y. Zaidner and R. Shahack-Gross. 2014. Formation processes and combustion features at the lower layers of the Middle Palaeolithic open-air site of Nesher Ramla, Israel. *Quaternary International* 331: 128-138.

Geneste, J.M..1985. *Analyse lithique d'industries Moustériennes du Périgord: Une approche technologique du comportement des groupes humaines au Paléolithique Moyen*. PhD diss., Université de Bordeaux.

Gijselings, G. and F. Doperé.1983. Een Midden-Paleolithisch site te Lauw. *Notae Praehistoricae* 3: 4-24.

Glauberman, P.J.. 2002. Middle Paleolithic surface sites in Southern Limburg, The Netherlands: The case study of Colmont-Ponderosa, and developing questions for future research. MA diss. Leiden University.

Glauberman, P.J.. 2006. Excavating surface sites, tapping a source of potential: the middle Palaeolithic surface scatters of Southern Limburg (NL) and the case study of Colmont-Ponderosa. In E. Rensink and H. Peeters (eds.), *Preserving the Early Past: Investigation, Selection, and Preservation of Palaeolithic and Mesolithic Sites and Landscapes*: 87-106. Amersfoort: Rijksdienst voor het Oudheidkundig Bodemonderzoek. (Nederlandse Archeologische Rapporten 32).

Glauberman, P.J..2014. *Middle Palaeolithic regional land-use and behavior in Dutch and Belgian Limburg: Integrating data from upland open-air Sites*. PhD diss., University of Connecticut.

Glauberman, P.J.. 2016. Middle Palaeolithic Land Use in Dutch and Belgian Limburg: Integrating Data from Upland Surface Sites. *Quaternary International* 411: 198-215.

Glauberman, P.J. and R.M. Thorson. 2012. Flint patina as an aspect of "Flaked stone Taphonomy": a case study from the loess terrain of the Netherlands and Belgium, *Journal of Taphonomy* 10: 21-43.

Goemaere, É., A. Katsch, I. Eschghi and R. Dreessen. 2016. Geological record and depositional setting of Palaeozoic oolitic ironstones in Western Europa. In: C. Billard, D. Bosquet, R. Dreesen, É. Goemaere, C. Hamon, I., Jadin, IH. Salomon and X. Savary (eds.), *Autour de l'hématite*: 23-43. Liège. (ERAUL 143 / Studia Praehistorica Belgica 8 / Anthropologica et Præhistorica 125 and 126).

Groenendijk, A.J. and J.P. de Warrimont. 1995. Middenpaleolithische Oppervlaktevindplaatsen in de Voerstreek, *Archeologie in Limburg* 65: 33-45.

Harari, Y..2015. *Sapiens*. New York: Harper.

Hayek, E., P. Krenmayr, H. Lohninger, U. Jordis, W. Moche, W. and F. Sauter. 1990. Identification of archaeological and recent wood tar pitches using gas chromatography/mass spectrometry and pattern recognition, *Analytical Chemistry* 62: 2038-2043.

Heeren, S., M. Kars, M. and V. van der Veen. 2022. *Van vondst tot verhaal: Archeologische vondsten in particulier bezit*. Utrecht: Matrijs.

Henk Y. .2006. *Scratching the surface: The potential of Middle Palaeolithic surface scatters and how (not) to find them. A case study of the Hej and the Henkeput near Sint-Geertruid (Limburg).* MA diss. University of Leiden.

Holdaway, S., J. Shiner and P. Fanning. 2004. Hunter-gatherers and the archaeology of discard behavior: an analysis of surface stone artefacts from Sturt National Park, Western New South Wales, Australia. *Asian Perspectives* 43: 34-72.

Holdaway, S., and L. Wandsnider. 2008. L. *Time in Archaeology: Time Perspectivism Revisited.* Ann-Arbor: University of Utah Press/Sheridan Books.

Holdaway, S., P. Fanning, P. and E. Rhodes. 2008. Assemblage accumulation as a time dependent process in the arid zone of Western New South Wales, Australia. In: S. Holdaway and L. Wandsnider (eds.), *Time in Archaeology: Time Perspectivism Revisited*: 110-133. Ann-Arbor: University of Utah Press/Sheridan Books.

Holwerda, J.H.. 1918. *Nederland's Vroegste Geschiedenis*. Amsterdam: Van Looy.

Hublin J.J., D. Weston, P. Gunz, M. Richards, W. Roebroeks, J. Glimmerveen J. and L. Anthonis L.. 2009. Out of the North Sea: the Zeeland ridges Neandertal. *Journal of Human Evolution* 57: 777-785.

Isaac, G.. 1981. Stone Age visiting cards: approaches to the study of early land use patterns. In: , I. Hodder, G. Isaac and N. and Hammond (eds.), *Pattern of the Past. Studies in Honour of David Clarke*: 131-155. Cambridge: Cambridge University Press.

Jones, G.T., D.K. Grayson and C.. 1983. Artefact class richness and sample size in archaeological surface assemblages. In: R.C. Dunnell and D.K. Grayson (eds.), *LuLu Linear Puntuated: Essays in Honor of George Irving Quimby,*: 55-73. Ann-Arbor : Museum of Anthropology. (Anthropological Papers 72).

Jungels, C.. 2004. *Étude du Comportement Moustérian á Otrange: Comparaison de Deux Entités Archéologiques*. PhD diss., Université de Liège.

Jungels, C.. 2005. Étude du Comportement Moustérian á Otrange: comparaison de deux ensembles archéologiques, *Anthropologica et Praehistorica* 116: 59-98.

Kelly, R.L.. 1983. Hunter-gatherer mobility strategies. *Journal of Anthropological Research* 39: 277-306.

Kelly, R.L.. 1995. *The Foraging Spectrum: Diversity in Hunter-Gatherer Lifeways*. Washington: Smithsonian Institution Press.

Kintigh, K.W.. 1984. Measuring archaeological diversity by comparison with simulated assemblages. *American Antiquity* 49: 44-54.

Kolen, J., D. De Loecker, A.J. Groenendijk and J.P. de Warrimont. 1999. Middle Palaeolithic surface scatters: how informative? a case study from Southern Limburg (The Netherlands). In: W. Roebroeks and C. Gamble (eds.), *The Middle Palaeolithic Occupation of Europe*: 77-191. Leiden: University of Leiden Press.

Kolfschoten, T. van, and W. Roebroeks (eds.). 1985. *Maastricht-Belvèdére: Stratigraphy, Paleoenvironment and Archaeology of the Middle and Late Pleistocene Deposits*. Haarlem: Rijks Geologische Dienst. (Mededelingen Rijks Geologische Dienst 39).

Kubiak-Martens, L.. 2008. *Voedseleconomie: parenchym en andere plantaardige macroresten (Hanzelijn Oude Land-Knooppunt Hattemerbroek)*. Zaandam. (BIAXiaal 386).

Kubiak-Martens, L., J. Langer and L. Kooistra. 2013. Plantenresten en teer in haardkuilen. In: T. Hamburg, A. Müller and B. Quadvlieg (eds.), *Mesolithisch Swifterbant. Mesolithisch gebruik van en duin ten zuiden van Swifterbant (8300-5000 v.Chr.) Een archeologisch opgraving in het tracé van de N23/N307, provincie Flevoland*: 341-360. Leiden: Archol.

Kuhn, S.L.. 1992. On planning and curated technologies in the Middle Paleolithic. *Journal of Anthropological Research* 48: 185-214.

Kuhn, S.L.. 1994. A formal approach to the design and assembly of mobile toolkits. *American Antiquity* 59: 426-442.

Kuhn, S.L.. 1995. *Mousterian Lithic Technology: an Ecological Perspective.* Princeton: Princeton University Press.

Lambers K, W.B. Verschoof-van der Vaart and Q.P.J. Bourgeois. 2019. Integrating remote sensing, machine learning, and citizen science in Dutch archaeological prospection. *Remote Sensing* 11: 794.

Langbroek, M., A. Verpoorte and B. Voormolen. 2001. *Colmont-Ponderosa: het Paleolithicum van de Vrakelberg en Omstreken, Een Verkennend Onderzoek. University of Leiden.* Leiden: University of Leiden.

Leonard, R.D., and G.T. Jones (eds.). 1989. *Quantifying Diversity in Archaeology. New Directions in Archaeology.* Cambridge: Cambridge University Press.

Locht, J.-L., É. Goval and P. Antoine. 2010. Reconstructing Middle Palaeolithic hominid behaviour during OIS 5 in Northern France. In: N.J. Conard and A. Delagnes (eds.), *Settlement Dynamics of the Middle Paleolithic and Middle Stone Age, vol. III*: 329-355. Tübingen: Kerns Verlag.

Locht, J.L., D. Hérisson and É. Goval. 2017. From the Middle to the Upper Pleistocene: origins and diversification of the Middle Palaeolithic in Northwest France. In: M. Pope, J. McNabb and C. Gamble (eds.), *Crossing the Human Threshold: Dynamic Transformation and Persistent Places during the Middle Pleistocene*: 225-230. New York: Routledge.

Londen, H. van, M. Van den Dries, and S. Landskroon, S. 2018. *Over lokalisme, liefdewerk en lonkend perspectief. Verkenning naar participatie en burgerinitiatief in de Nederlandse archeologie.* Amsterdam: Universiteit van Amsterdam.

Louwe Kooijmans, L.P.. 2017. *Onze vroegste voorouders. De geschiedenis van Nederland tot 5000 v.Chr.* Amsterdam: Bert Bakker.

Lowe, J., and M. Walker. 2014. *Reconstructing Quaternary Environments. 3rd Edition.* Harlow.

McParland, L., M. Collinson, A. Scott, G. Campbell and R. Veal, 2010. Is vitrification in charcoal a result of high temperature burning wood? *Journal of Archaeological Science* 37: 2679-2687.

Meijs, E.P.M..2002. Loess stratigraphy in Dutch and Belgian Limburg. *Eiszeitalter und Gegenwart* 51: 114-130.

Meijs, E.P.M.. 2011. The Veldwezelt site (Province of Limburg, Belgium): environmental and stratigraphical interpretations. *Netherlands Journal of Geosciences/Geologie en Mijnbouw* 90: 73-94.

Meijs, E.P.M., P. Van Peer and J.P. de Warrimont. 2012. Geomorphologic context and proposed chronostratigraphic position of Lower Palaeolithic artefacts from the Op de Schans Pit near Kesselt (Belgium) to the West of Maastricht. *Netherlands Journal of Geosciences/Geologie en Mijnbouw* 91: 137-157.

Nelson, M.C.. 1991. The study of technological organization. *Archaeological Method and Theory* 3: 57-100.

Niekus, M.J.L.Th., R.T. van Balen, J.M.G. Bongers, J.H.A. Bosch, D.C. Brinkhuizen, H. Huisman, J. Jelsma, L. Johansen, C. Kasse, S.J. Kluiving, H. Paas, E. Schrijer, and D. Stapert. 2016. News from the north: A late Middle Palaeolithic site rich in handaxes on the Drenthe – Frisian till plateau near Assen, the Netherlands; first results of a trial excavation. *Quaternary International* 411 A: 284-304.

Niekus, M.J.L.Th, P.R.B. Kozowyk, G.J.H., Langejans, D. Ngan-Tillard, H. van Keulen, J. van der Plicht, K.M. Cohen, W. van Wingerden, B. van Os, B.I. Smit, L.W.S.W. Amkreutz, L. Johansen, A. Verbaas and G.L. Dusseldorp. 2019. Middle Paleolithic complex technology and a Neandertal tar-backed tool from the Dutch North Sea, *PNAS* 116: 22081-22087.

Paulussen, R., and Y. Raczynski-Henk. 2014. *Buitenring Parkstad Limburg, Parkstad Gemeenten en Nuth. Inventariserend Veldonderzoek (IVO-O); Geo-archeologisch booronderzoek naar de verdiepte delen.* Eijsden: Archeopro.

Peet, R.K.. 1974, The measurement of species diversity. *Annual Review of Ecology and Systematics* 5: 285-307.

Peet, R.K.. 1975. Relative diversity indices. *Ecology* 56: 496-498.

Peeters, J.H.M., L.W.S.W. Amkreutz, K.M. Cohen and M.P. Hijma. 2019. *North Sea Prehistory Research and Management Framework (NSPRMF) 2019. Retuning the research and management agenda for prehistoric landscapes and archaeology in the Dutch sector of the continental shelf.* Amersfoort: Rijksdienst voor het Cultureel Erfgoed. (Nederlandse Archeologische Rapporten 63).

Pielou, E.C.. 1975. *Ecological Diversity.* New York: Wiley.

Pop, E., W. Kuijper, E. van Hees, G. Smith, A. García-Moreno, L. Kindler, S. Gaudzinski-Windheuser and W. Roebroeks. 2016. Fires at Neumark-Nord 2, Germany: An analysis of fire proxies from a Last Interglacial Middle Palaeolithic basin site. *Journal of Field Archaeology* 41-5: 603-617.

Raczynski-Henk, Y.. 2014. Neanderthaler-vindplaatsen in het Limburgse lössgebied: onmisbaar! *Grondboor en hamer* 4/5: 145-149.

Raczynski-Henk, Y., R. Paulussen, B. Weekers-Hendrikx, and R. Machiels. 2018. *Wie sjoeën oos Limburg waor. Een openluchtvindplaats uit het Midden-Paleolithicum in het tracé van de Buitenring Parkstad Limburg. Proefsleuvenonderzoek en een archeologische opgraving.* Amersfoort: ADC.

Rensink, E.. 2010. Eyserheide: a Magdalenian open-air site in the loess area of the Netherlands and its archaeological context. *Analecta Praehistorica Leidensia* 42.

Rhode, D.. 1988. Measurement of archaeological diversity and the sample-size effect. *American Antiquity* 53: 708-716.

Roebroeks, W..1980. De "Middenpaleolithische" Vindplaats sint geertruid (Limburg): Hypothesen voor nader Onderzoek. *Archeologische Berichten* 8: 7-37.

Roebroeks, W.. 1988. *From Find Scatters to Early Hominid Behaviour. A study of Middle Palaeolithic Riverside Settlements at Maastricht-Belvèdére (The Netherlands).* PhD diss., University of Leiden.

Roebroeks, W.. 1990. *Oermensen in Nederland.* Amsterdam: Meulenhoff.

Roebroeks, W., J. Kolen and E. Rensink. 1988. Planning depth, anticipation and the organization of Middle Palaeolithic technology: the Archaic natives meet Eve's descendants. *Helinium* 28: 17-34.

Roebroeks, W., D. De Loecker, P. Hennekens and M. van Ieperen. 1992. A veil of stones: On the interpretation of an early Middle Paleolithic low density scatter at Maastricht-Belvèdére (The Netherlands). *Analecta Praehistorica Leidensia* 25: 1-25.

Roebroeks, W., J. Kolen, M. van Poeke and A. van Gijn. 1997. 'Site J': an early Weichselian (Middle Paleolithic) Flint scatter at Maastricht-Belvèdére, The Netherlands. *Paleo* 9: 143-172.

Roebroeks, W. and B. Speleers. 2002. Last Interglacial (Eemian) occupation of the North European plain and adjacent areas. In: A. Tuffreau and W. Roebroeks (eds.), *Le dernier interglaciare et les occupations humaines du Paléolithique moyen*: 31-39. Lille.

Rossignol, J., and L. Wandsnider (eds.). 1991. *Space, Time, and Archaeological Landscapes.* New York: Plenum Press.

Ruebens K.. 2013. Regional behaviour among late Neanderthal groups in Western Europe: a comparative assessment of late Middle Palaeolithic bifacial tool variability. *Journal of Human Evolution* 65: 341-362.

Shott, M.J.. 1989. Diversity, organization, and behavior in the material record, *Current Anthropology* 30: 283-315.

Sier, M., J. Parés, P. Antoine, J.-L. Locht, M. Dekkers, N., Limondin-Lozouet and W. Roebroeks. 2015. Evidence for the Blake Event recorded at the Eemian archaeological site of Caours, France. *Quaternary International* 357: 149-157.

Simpson, E.H.. 1949. Measurement of diversity, *Nature* 166: 688-688.

Stern, N.. 1993. The structure of the lower Pleistocene archaeological record: a case study from the Koobi Fora formation. *Current Anthropology* 34: 201-225.

Stern, N.. 1994. The implication of time-averaging for reconstructing the land-use patterns of early tool-using hominids. *Journal of Human Evolution* 27: 89-105.

Skinner, H. D. Sarpong, D. and G.R.T. Wite. 2018. Meeting the needs of the Millennials and generation Z: gamification in tourism through geocaching. *Journal of Tourism Futures* 4/1: 93-104.

Thisse-Derouette, J. and J. Destexhe-Jamotte. 1948. Quelques précisions sur la découverte du gisement Paléolithique d'Otrange (Province de Limbourg). *Bulletin des Chercheurs de la Wallonie* 14: 3-17.

Turq, A., W. Roebroeks, L. Bourguignon and J.P. Faivre. 2013. The fragmented character of Middle Palaeolithic stone tool technology, *Journal of Human Evolution* 65: 641-655.

Ubaghs, C..1887. *Les Ateliers ou Station dits Prehistoriques de Ste-Gertrude et Ryckholt, près de Maestricht.* Liège: Vaillant-Carmanne.

Van Andel, T.H. and C.N. Runnels. 2005. Karstic wetland dwellers of Middle Paleolithic Epirus, Greece. *Journal of Field Archaeology* 30: 367-384.

Van Baelen, A., E.P.M. Meijs, P. Van Peer, J.P. de Warrimont and M. De Bie. 2007. An early Middle Palaeolithic site at Kesselt-Op de Schans (Belgian Limburg). Preliminary results, *Notae Praehistoricae* 27: 19-26.

Van Baelen, A., E.P.M. Meijs, P. Van Peer, J.P. de Warrimont and M. De Bie.2008. The early Middle Palaeolithic site of Kesselt Op de Schans (Belgian Limburg). excavation campaign 2008. *Notae Praehistoricae* 28: 5-9.

Van Baelen, A., Y. Raczynski-Henk, J.W. de Kort, J.W., H. Huisman, B. van Os, A.J. Versendaal, J. Wallinga, E.P.M. Meijs, and J. Deeben. 2017. *Onderzoek naar de stratigrafie van de lösssequentie en de daarin aangetroffen artefacten op De Kaap bij St. Geertruid.* Amersfoort: Rijksdienst voor het Oudheidkundig Bodemonderzoek. (Rapportage Archeologische Monumentenzorg 236).

Van Baelen, A., Y. Raczynski-Henk, P. Glauberman, G. Noens, J.W. de Kort and J. Deeben. 2012. Recent investigations at the Middle Palaeolithic site of St. Geertruid De Kaap (Limburg, The Netherlands). *Notae Praehistoricae* 32: 29-36.

Vandenberghe, J., W. Roebroeks, and T. van Kolfschoten (eds.). 1993. *Maastricht-Belvédère: Stratigraphy, Palaeoenvironment, and Archaeology of the Middle and Late Pleistocene Deposits, Part II.* Haarlem: Rijks Geologische Dienst. (Medelingen Rijks Geologische Dienst 47).

Van Der Sloot, P., P. Haesaerts and S. Pirson. 2011. Les sites du Mont Saint-Martin (Liège). In: M. Toussaint, K. Di Modica and S. Pirson (eds.), *Le Paléolithique moyen en Belgique. Mélanges Marguerite Ulrix-Closset*: 385-394. Liège. (Les Chercheurs de la Wallonie hors-série n° 4; Etudes et Recherches archéologiques de l'Université de Liège 128).

Van Ginkel, E. and L. Theunissen. 2009. *Onder heide en akkers: de archeologie van Noord-Brabant tot 1200.* Utrecht: Matrijs.

Verhart, L.B.M.. 2007. Als een roepende in de woestijn? enkele gedachten over steentijdonderzoek in Limburg, *Archeologie in Limburg* 105: 16-23.

Verhoeven, M., Y. Raczynski-Henk and R. Paulussen. 2020. *Expeditie Vuistbijl,* Weesp: RAAP Archeologisch Adviesbureau.

Vries, F. de, L. Postma, M. Niekus, H. de Kruijk, J. Timmner, and H. Kars. 2022. *Valsheid in gesteente – waarom zijn de vuistbijlen van Tjerk Vermaning vals en wie zit er achter?* Assen: Uitgeverij Koninklijke Van Gorcum.

Wagstaff, M.. 1991. The archaeological "site" from a geographical perspective. In: A.J. Schofield (ed). *Interpreting Artefact Scatters: Contributions to Ploughzone Archaeology*: 9-10. Oxford. (Oxford Monograph 4).

Warrimont, J.P. de. 1997. De "Biface Lancéole" van De Kaap in Sint Geertruid, *Archeologie in Limburg* 72: 28-30.

Warrimont, J.P. de. 2002. Middenpaleolithische Artefacten van Rijckholt-Sint Geertruid. *Archeologie in Limburg* 90: 2-7.

Wouters, A.. 1980. De Middenpalaeolithische Vindplaats Sint Geertuid (Limburg): afbeelding en Beschrijvingen van Middenpalaeolithische artefacten uit vondscomplexen uit de omgeving van Sint Geertruid. *Archaeologische Berichten* 8: 38-106.

Yellen, J.E..1977. *Archaeological Approaches to the Present: Models for Reconstructing the Past.* New York: Academic Press.

In Dusseldorp, G.L., Chu, W., Bakels, C.C. and M.A. Soressi (eds) 2025. Intent on the Paleolithic. Papers in honour of Prof. Dr. Wil Roebroeks. *Analecta Praehistorica Leidensia* 55. Leiden: Sidestone Press, pp. 57-70.

A handaxe made from a 'Meuse egg', found at Zandvoort in shell grit from the North Sea

D. Stapert, L. Johansen, M.J.L.Th. Niekus

Abstract

An amygdaloid handaxe made from a rolled flint pebble ('Meuse egg') is described and illustrated. It was found on a shell path at Zandvoort in 2019. Both the raw material and the (secondary) findspot are exceptional for a handaxe of more than 10 cm. The tool had been dredged up from the bottom of the southern North Sea in the context of shell extraction. It probably derives from fluvial deposits of the Kreftenheye Formation, as do many Middle Palaeolithic finds collected on Maasvlakte 2 and Zandmotor, and other replenished beaches along the Dutch coast. The best guess for its date is the Middle Weichselian (MIS 3). It is suggested that the tool had been transported by its owner over a long distance from the site of its manufacture (perhaps southern Limburg), because Meuse eggs of 12-15 cm – the probable size of the raw material – must have been very rare during Middle Palaeolithic times in what is now the southern North Sea.

Keywords: Middle Palaeolithic; handaxe; bifacial Quina-scrapers; 'Meuse eggs' (flint pebbles); North Sea; seashell dredging; shell paths

Introduction

In recent years, several Middle Palaeolithic artefacts have been found on paths in the western Netherlands among shells dredged up from the bottom of the southern part of the North Sea. One of these finds is described in this paper: an *amygdaloid* handaxe made from a flint pebble ('Meuse egg') found at Zandvoort in 2019. Another interesting find is a bifacially worked Quina-scraper, almost a leaf point, from Castricum (to be published elsewhere). These finds are of special interest because the only Neanderthal fossil known from the Netherlands, a small but characteristic skull fragment, was also dredged up in the context of shell extraction from the southern North Sea floor – together with flint artefacts including at least ten handaxes. The study of this Neanderthal fossil (nicknamed 'Krijn') from the Zeeland Ridges (*Middeldiep*), and its geological context, has been one of the many interesting cases that Wil Roebroeks has been involved in (e.g. Hublin *et al.* 2009). It is with pleasure that we dedicate this paper to him.

Shoreline shell gathering, for supplying lime kilns or for surfacing roads and paths, has been practised for several centuries. Nowadays seashells are suction-dredged for grit used mainly in poultry farming and underfloor insulation. A lot of this goes on in the Wadden Sea especially. Between 1950 and 1975 between 100,000 and 200,000 m^3 of seashells were dredged

D. Stapert
Groningen University, retired, Groningen, The Netherlands
D.Stapert@planet.nl

L. Johansen
Archaeological Drawings and Analyses
lykke.johansen.hn@gmail.com

M.J.L.Th. Niekus
Chairman of *Stichting STONE*, and external PhD-researcher Leiden University, Leiden, The Netherlands
marcelniekus@gmail.com / m.j.l.t.niekus@arch.leidenuniv.nl

up from the Dutch part of the Wadden Sea annually, while on average just ca 90,000 m^3 grow back (Eisma and Jansen 1976: 274). This unsustainable practice has – at least theoretically – been brought to a halt. The digital information desk of the national government on matters regarding the North Sea (*Noordzeeloket*, under the heading of *Oppervlaktedelfstofwinning*) says the following: "Shell dredging is permitted in areas up to 50 km offshore, from the -5 m depth contour and 3 nM (ca 5.5 km) seaward of the coastline (Lowest Astronomical Tide). Shell dredging may take place in the North Sea and the *Voordelta*, as well as in the tidal channels between the Frisian Isles and the shallow seaward ends of these channels. (…) The harvested amounts may not exceed the natural increment." This last condition of course delights the conservationists, but it is common knowledge that several companies illegally exceed this limit (see *e.g.* Pols 2017). The *Voordelta* is a 'Natura 2000' reserve, a sea area extending over 900 sq km in the North Sea off the southwestern isles belonging to the provinces of Zeeland and Zuid-Holland. To the south of it, another 'Natura 2000' reserve is present, called *Vlakte van Raan*, in the same type of location. A comparable nature reserve area is present to the north of the Wadden Islands: *Noordzeekustzone*. The protection of these areas is more limited than one might expect: Fishing is allowed, as is the exploitation of gas and oil in the subsoil.

Different regulations govern the truly colossal scale of sand dredging in the North Sea. For one thing, the sand-dredging areas lie further offshore. The *Noordzeeloket* says: "For sand dredging, an area between the 12-nM boundary (ca 22.2 km) and the continuous isobath of 20 m below NAP. In this area, covering 5,134 sq km, sand dredging has priority, while other uses are not excluded. Large-scale suction dredging for sand, dredging over 2 m into the sea floor, is permitted from 2 km seaward of the continuous -20 m isobath". The construction of *Maasvlakte* 2 (a coastal harbour extension near Rotterdam) beats every record in terms of artificially deposited sand: in 2009-2013 (see also Ministerie van Verkeer en Waterstaat 2008) amounting to ca 213,000,000 m^3! One may have doubts about this procedure (its impact on sea-floor life, for instance), but it has also produced numerous benefits for palaeontologists and archaeologists, including the recovery of a handaxe that is, astonishingly, made of Wommersom quartzite, ca 160 km from its source (Niekus *et al.* 2017b; 2021). Both on *Maasvlakte 2* and the *Zandmotor* (an artificial coastal sandbar northwest of Rotterdam), hundreds of artefacts from various phases of the Palaeolithic and Mesolithic, as well as many skeletal remains of Pleistocene and later animals, have been picked up by dozens of dedicated amateur archaeologists and palaeontologists (see e.g. Glimmerveen *et al.* 2004; Carmiggelt and Schiltmans 2021; Niekus *et al.* 2023). From the *Zandmotor,* one of them even recovered a flake hafted in a clump of birch tar, with a radiocarbon dating of 47,100 ± 500 BP (Niekus *et al.* 2019). We shall need many years to describe and interpret all these finds, which continue to come in regularly.

Important dredged-up assemblages have emerged from other parts of the Dutch part of the North Sea as well, such as Middeldiep (*Zeeland Ridges*), which produced the Neanderthal skull fragment mentioned above (e.g. Hublin *et al.* 2009; Amkreutz *et al.* 2010) and a series of artefacts (ca 70) including at least ten handaxes (Jan Glimmerveen pers. comm. 2009; see e.g. Verhart 2001; Mol *et al.* 2008; Hijma *et al.* 2011; 2012). This assemblage includes (sub)triangular, Micoquian and *bout coupé* forms, mostly made by experienced makers (see e.g. Hijma *et al.* 2011: fig. 1, for a photo of a beautiful subtriangular handaxe). However, we noted that at least three handaxes from this collection were probably produced by learners (children), two of which were described and illustrated in an earlier paper (Johansen and Stapert 2012: figs 5 and 6). The skull fragment and the handaxes from Middeldiep probably date from the Middle Weichselian (somewhere between 60-45,000 years ago), and we believe this also to be the most probable age of the Zandvoort handaxe described below.

Among the shell grit that surfaces the footpaths and cycleways in the Netherlands, one can almost always find flints as well. They mostly are small, black-patinated, glossy flints. This means that the shell dredgers not only sucked up seashells, but also washed-out flints that lay on the sea floor beneath the shell banks. Black-patinated flints are also regularly washed up on the beaches of the Frisian Isles, and elsewhere too, for instance at Cadzand (Stapert 1981a). Several dozen Neanderthal tools have been identified among them. For instance, from the beach of Vlieland we have a Levallois core (Stapert *et al.* 2011); and from the beach of Ameland, a bifacial tool that may be regarded as a failed preform of a leaf point (Stapert *et al.* 2013a and c). Middle Palaeolithic finds are known from Texel too (Stapert *et al.* 2013a). However, we believe the bulk of the flint finds from the beaches of the Frisian Isles to date from younger Stone Age periods.

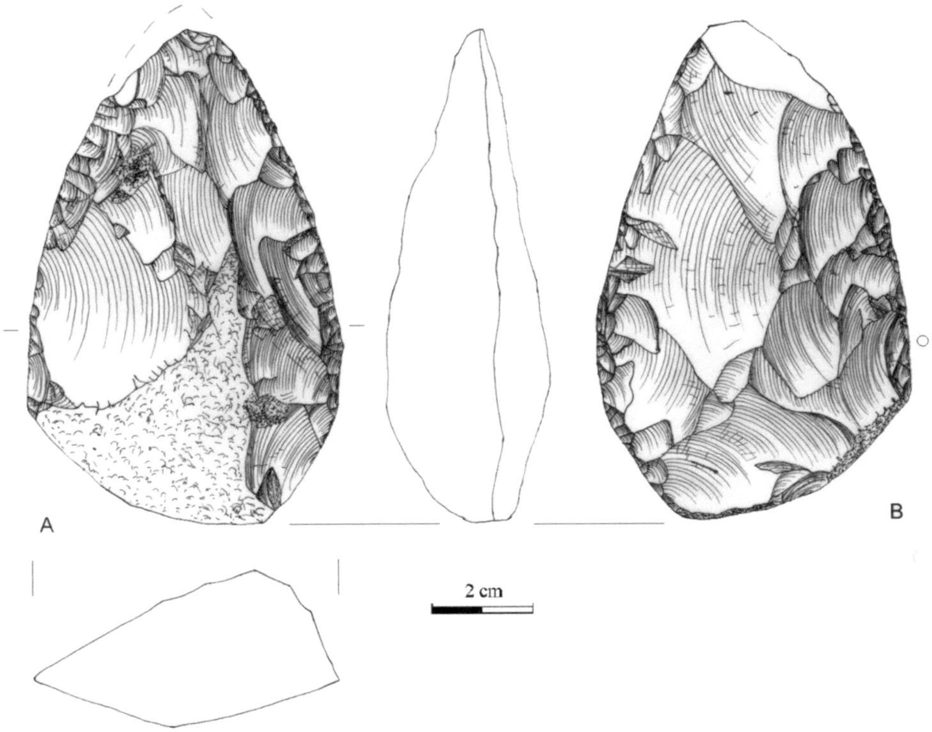

Figure. 1. The handaxe from the shell path at Zandvoort. Legend of the drawings: blank – recent damage; stippling – cortex; stippling with circles – heavily rolled ancient surface ('pseudo-cortex'); irregular cross-hatching – part of an ancient frost-split surface. The probable location of the point of percussion of the flake of which the handaxe is made is indicated by a circle with a question mark. Drawing Lykke Johansen, Haren.

The Middle Palaeolithic flints are distinguishable from the younger finds by their (sometimes) characteristic typology, but especially because they are covered by wind gloss: a somewhat greasy-looking gloss accompanied by minute pitting, produced by long-term exposure to drifting sand during glacial stages (Stapert 1976a). During the harsh Upper Pleniglacial of the last glacial, Middle Palaeolithic artefacts in the northern Netherlands could escape the formation of wind gloss only under exceptional conditions (such as lake shores in glacial basins). Most of the artefacts left behind by Neanderthals on the Friesland/Drenthe till plateau lay at the surface for long periods, often along the brows of valleys, where over the course of millennia they became incorporated into the so-called 'boulder sand': the weathering residue on top of the till; this has also been demonstrated *in situ*, at the site of Peest (see e.g. Niekus *et al.* 2015; 2017a). Flints in the boulder sand almost all have wind gloss, and furthermore display patinas and various traces of cryoturbation such as scratches and pressure cones (Stapert 1976a and b). In this context it should be remembered that Neanderthals became extinct ca 40,000 calendar years (ca 35,000 radiocarbon years) ago: thousands of years before the start of the Upper Pleniglacial (ca 27,000-14,500 calendar years ago).

The chances of finding prehistoric artefacts in shell paths probably are somewhat greater when the shells come from the North Sea than from the Wadden Sea, as off the west coast of the provinces of North and South-Holland the fluvial deposits from the last glacial lie relatively closely beneath the sea floor (see e.g. the E-W profile in De Mulder *et al.* 2003: 348; see also Hijma *et al.* 2012; Moree and Sier 2015; Laban and Van der Meer 2022; Niekus *et al.* 2023). Moreover, the shells lie more dispersed in the North Sea than in the Wadden Sea (Pols 2017), which means that more sand and gravel will be dredged up with them. Here we are in the main dealing with coarse fluvial deposits belonging to the Kreftenheye Formation, which in many places are known to contain Middle Palaeolithic artefacts (see Johansen *et al.* 2009). The bulk of the numerous Middle Palaeolithic finds from *Maasvlakte 2* and *Zandmotor* without doubt stem from layers of this Formation. Most dredged-up Middle Palaeolithic artefacts from the eastern and central Netherlands (e.g. from Zwolle and Empel), and from the North Sea, deriving from the Kreftenheye Formation, display no wind gloss – in contrast to

Figure. 2. The handaxe from Zandvoort. Photo Rijksmuseum van Oudheden, Leiden.

most Middle Palaeolithic items from the northern Netherlands. This is because they were relatively quickly incorporated into sediment as a result of their having been abandoned in the bed of a braided river system. However, occasional items which happened to have lain at the surface for longer before being washed about and buried in sediment ended up with a slight to medium degree of wind gloss. This goes for several artefacts from *Maasvlakte 2* and *Zandmotor*, and also for the bifacial Quina-scraper from Castricum, but not for the handaxe from Zandvoort described below.

The handaxe from Zandvoort

The handaxe (Figs 1 and 2) was found in July 2019 by Ms Sabine Schutte of Haarlem. She discovered it on a shell path on the premises of the rehabilitation centre 'Nieuw Unicum' (Zandvoortselaan, no. 165) at Zandvoort (a seaside town west of Amsterdam). The handaxe must have been suction-dredged from the sea floor together with the shells. Unfortunately, it is no longer possible to trace the area of its provenance (Ms Schutte unsuccessfully inquired at Nieuw Unicum; pers. comm. by telephone to the first author, 10 October 2019). The tool is kept in the *Rijksmuseum voor Oudheden* in Leiden (inventory no. g 2019/12.1).

It is remarkable for a handaxe to turn up in a shell path. Although flints (almost invariably with a black patina) frequently appear among the seashells, these tend to be quite small, rarely more than a few centimetres across. In terms of size this handaxe is an outlier, (originally) measuring ca 10.5 cm in length. Ms Schutte found several other (natural) flints on the path as well, all smaller than the handaxe. Still, in terms of size this artefact from Nieuw Unicum is not unique. From a different location in the same province, we have another sizeable bifacial tool which was recovered from dredged-up shell surfacing, near the 'Huis van Hilde' at Castricum (though this piece is less than 10 cm in length).

The Zandvoort handaxe was made from fairly fine-grained, grey flint with small and larger pale patches, the larger ones of which are of a slightly coarser texture. Examination with a magnifying glass reveals no convincing fossils of bryozoans (moss animals). These features suggest that the flint is of a southerly provenance. The handaxe has an unworked base, which has a convex surface of severely rolled and weathered 'cortex'. This is not the original flint cortex, however, for which reason it may be more appropriately termed 'pseudo-cortex'. It is characterised by very numerous, large, more or less circular impact marks, as we know them from shingle in beach ridges. The basic piece for the handaxe must have been a severely rolled flint pebble. We do not know of any exposed fossil shingle beaches in this region from the period in which the handaxe was left behind by Neanderthals: the first half or middle part of the Weichselian. Hence, we may assume that the pebble from which the tool was fashioned was selected from a gravelly fluvial sediment which contained rolled flints. An alternative explanation is that the handaxe was transported over a long distance from the site where it was manufactured, just as in the case of the handaxe of Wommersom-quartzite from *Maasvlakte 2*.

It seems obvious that we are dealing with an outsize *Maas-ei*, a 'Meuse egg'. Meuse eggs are flints which in the course of the Tertiary epoch, in multiple episodes (but mostly in the Miocene, 23-5 million years ago: see e.g. Umbgrove 1926) were rolled into rounded shapes on beaches, especially in the province of Limburg in the Netherlands and in Belgium, and that subsequently were carried along by rivers over large parts of the southern and middle Netherlands. These egg-shaped pebbles, also called 'blue flints', can be found in coarse Meuse deposits of the Pleistocene, but also in sediments of the Rhine dating from periods in which this river cut into and reworked older Meuse deposits, resulting in mixed gravel assemblages. Meuse eggs regularly occur in – among others – the gravelly layers of the fluvial Formations of Urk (ca 500,000 – 150,000 years old) and Kreftenheye (ca 125,000-15,000 years old; see e.g. De Mulder *et al.* 2003: 299).

It is of interest to note that there are also rolled flints with a northern provenance, mostly called *Wallsteine*. These often beautiful stones were shaped during the Palaeocene (66 to 56 million years ago) on a tropical beach in present-day Scandinavia, and much later carried to our part of the world in the Pleistocene ice cap (see e.g. Hucke 1967; Reinecke 2009; for some images, see www.budstones.de/feuerstein). They can be found in the Saalian till in the northern Netherlands, be it in very small numbers (near-negligible, see e.g. Van Capelle 1892; 1898; 1910; Maarleveld 1952; 1956).

In various episodes of prehistory, rolled flint pebbles were made into tools, or used as cores. We know such artefacts from as early as the 'Rhenen Industry', which dates from 250-200,000 years ago (on this Early Middle Palaeolithic industry, see e.g. Stapert 1981b; 1987; 1991; Niekus and Stapert 2005). These flints were gathered from coarse sediments of the Urk Formation, and eventually – in the form of discarded tools – ended up in these same fluvial sediments (see e.g. Ruegg 1981). The Rhenen Industry was created by people belonging to a late form of *Homo heidelbergensis* or an early form of *Homo neanderthalensis*, and these people frequented the then river valleys of the central and southern Netherlands – where among other things they collected and worked flint. Several thousands of artefacts of this tradition are known from more than ten sites. Palaeolithic handaxes usually were made from relatively high-quality flint, which Meuse eggs generally were not. But this is no hard and fast rule, and some early flint knappers (including learners) were not too choosy in this respect. Of course, there were areas where simply no high-quality flint was available, so people had to make do with second-rate material.

Remarkably, a few remnants of unweathered 'true' cortex are also present on the Zandvoort handaxe, off-white in colour, on face A (just over halfway up on the left; and to the right below this; see Fig. 1).

These remnants remained more-or-less intact because they lay in deep recesses of the surface and therefore were hardly exposed to weathering; they came close to the surface only in the course of the flint's transformation into a handaxe. Flint with deep, natural pitting in the cortex is also known as 'cavity flint' (*holtevuursteen*) and on the whole is more closely associated with southern gravel assemblages (especially those deposited by the Meuse) than with northern ones (glacial till). Once again: this is not an absolute distinction, as 'northern flint' is an umbrella concept covering many varieties; indeed the same goes for southern flint. On face A, there also is a small remnant of an old frost-split surface (centre, right) which displays wind gloss.

Our conclusion is that the raw material for this handaxe most probably was picked from a coarse fluvial deposit with a Meuse component among the gravel. The most likely sources are: 1. gravelly fluvial deposits of the Urk Formation, which surface at many places in the ice-pushed ridges of the provinces of Utrecht and Gelderland, and 2. gravelly fluvial deposits of the Kreftenheye Formation, occurring at some depth in large parts of the central Netherlands and in the North Sea. Although we cannot be sure of the provenance of the Zandvoort shell grit, the latter to us seems far more likely than the former – since the layers of the Kreftenheye Formation occur closer to the sea floor than do those of the Urk Formation. However, the Meuse egg from which the handaxe was made could also have been picked up from layers of the Urk Formation exposed in the ice-pushed hills of the central Netherlands.

The handaxe is likely to have been made from a flake, but this is not entirely certain. If indeed so, then face B is the original ventral side of the flake:

face A can be ruled out because it shows multiple cortex remnants. The cross section of the handaxe is asymmetrical: face B is fairly flat; face A, by contrast, quite convex – this too suggests that the tool was made from a flake. Because of the all-over working of face B, nothing or next to nothing is left of any ventral surface; only in the very lowermost part (right of centre, 2-3 cm above the base) has a possible remnant survived. The percussion bulb was retouched away by rather deep scars from the right edge of face B, some of which end in steps or hinges. A somewhat thicker part however could not be removed; it occurs ca 2 cm above the base. The most likely position of the striking point, if this indeed was a flake (as we believe to be the case), is indicated in the drawing by a small circle with a question mark. Owing to subsequent bifacial working, there is no remnant of any striking platform.

At the top, a part is missing as a result of severe recent impact. The damage will have arisen from a blow to the tool in the course of suction dredging or its transportation in the shell grit. As a result, its current length is not the original one; an estimated half a centimetre is missing. Elsewhere some smaller recent damage is also present.

Dimensions: length 9.9 cm; width 6.3 cm; thickness 3.0 cm. The original length would have been about 10.5 cm. The tool (currently) weighs ca 169 gr. The original nodule from which the handaxe was made must have been quite large, in our opinion at least 12 cm, but possibly even larger. Most Meuse eggs are only fist-sized, with maximum diameters of 3 to 6 cm being normal. The pebble from which the handaxe was made must have been much larger. Still, it is not absolutely necessary to assume long-distance transport of the Zandvoort handaxe, because larger Meuse eggs occasionally do occur in Pleistocene fluvial deposits. For example, Van Straaten (1946: 60-61) mentions the find of a Meuse egg in terrace gravels at Groot-Haasdal in southern Limburg with the following dimensions: 8 × 10 × 19 cm. Van Manen (1993: 50) found a Meuse egg with a maximum diameter of 13 cm at Vragender in the Achterhoek region (eastern Netherlands). Therefore, we assume that a Meuse egg of 12-15 cm theoretically could have been picked up from gravelly layers of the Kreftenheye Formation in what is now the southern North Sea. However, it should be noted that such specimens would have been extremely rare. In this connection it is of interest that Oostingh (1921: 108-109) never saw a Meuse egg larger than 8 cm during his extensive research. Therefore, we should definitely consider the possibility that the Zandvoort handaxe was transported over a long distance from the southeast (perhaps southern Limburg in the Netherlands or northeastern Belgium?) like the handaxe made of Wommersom quartzite found on *Maasvlakte 2* (Niekus *et al.* 2017b; 2021). In those areas, chances of finding Meuse eggs larger than 12 cm must have been much greater than in what is now the southern North Sea. An alternative, but less probable source might be one of the ice-pushed ridges in the central Netherlands.

The tool's unworked base is slightly oblique, rendering the piece asymmetrical. As a result, the handaxe is hard to classify within the system of Bordes (1961). On the basis of the estimated original length of 10.5 cm, the dimensions and indices as used by Bordes are as follows: L: 10.5; m: 6.25; a: 3.6; n: 5.65; e: 3.0; L/a: 2.92; n/m x 100 90.4; m/e: 2.08; L/m: 1.68. Although *cordiform* in outline, this handaxe (because m/e is less than 2.35) falls into the category of almond-shaped, amygdaloid handaxes. The presence of an oblique and non-cutting base is quite common with that type (see Bordes 1961: 52-53 and *Planches* 66-68). Interestingly, we know of several specimens from *Maasvlakte 2* which are similar in this respect; these are however mostly of smaller size (and partly probably workpieces by apprentice knappers; Niekus *et al.* 2021).

Amygdaloid handaxes are known from both the Acheulian and the Mousterian; in this case the latter seems the more probable, given the likelihood that the tool was made from a flake. Relatively small handaxes made from flakes are a fairly common occurrence in the Mousterian, but are seen more rarely in the European Acheulian. (In this context it is curious to note that Acheulian handaxes in Africa often *do* seem to be made from flakes, including truly large ones, over 25 cm long!).

The working of the handaxe is competent, but certainly not top-notch. The base has remained quite thick; attempts to thin it failed. On both faces, steps or hinges – some quite deep – have resulted from such attempts (*e.g.* halfway up the right side of face A). Yet from the side, both cutting edges look fairly straight over most of their length, with entirely serviceable cutting angles of ca 50 degrees. Most of the scars suggest that they were made by soft percussion, with a percussion tool of, for instance, bone. The right edge of face B was worked in greater detail than its counterpart on the left, with the aim of obtaining a more symmetrical lenticular cross-section in the relatively thin top part.

Figure. 3. Twelve of (at least) fourteen Meuse eggs from the shell path at Zandvoort, collected by Ms Sabine Schutte; the largest one has a maximum diameter of 7.9 cm. Photo Dick Stapert, Haren.

Since the flint itself is of a dark shade, it is hard to tell – but visible nonetheless – that the tool has a slight black patina on both faces (which is apparent in particular through the recent damage on the left edge of face A). Both faces show a slight gloss patina, but probably not windgloss, given the absence of the characteristic fine pitting (Stapert 1976a). In other respects the tool has come to us fairly unweathered; for instance, the ridges and edges are barely rounded. This indicates that the piece became embedded in sediment quite soon after it was discarded.

At the first author's request, Ms Schutte went on to collect a pocketful of other flints from the same path, as these might provide further evidence about the original geological context of the find. The assemblage also contained three shells: an oyster shell (max. diameter 8.0 cm), a large cockle shell (max. diameter 6.8 cm), and the shell of a moon snail, 2.4 cm high (identification on the basis of Entrop n.d.). The oyster shell is the only one with a slight black patina, and also is somewhat rounded. The moon-snail shell is of a beige colour, without any black patina. It is the shell of the common moon snail (*Polinices catena*).

This predatory snail has a shell up to 4.3 cm high. The moon-snail shell from 'Nieuw Unicum' is just 2.4 cm high, so probably it was a juvenile specimen. The (common) moon snail mostly lives at some distance from the coast, at depths to over 100 m, but usually at 5 to 20 m, and prowls around the sea floor in search of prey. It drills round holes in the shells of nuns, cockles and other molluscs, and sucks out the fleshy creature. Live moon snails are washed onto the shore only by severe storms, since they live well offshore. Shell dredging too is performed at a distance from the coast (as noted above), so the presence of a moon-snail shell on the shell path where the handaxe was found is not surprising.

The most conspicuous in Ms Schutte's flint collection is the high proportion of Meuse eggs, usually fragmented: at least 14 (fig. 3). Characteristic of these is the presence of innumerable small collision cones in the surface: (semi-)circular cracks with diameters of often more than 0.5 cm.

The largest Meuse egg from the shell path has a maximum diameter of 7.9 cm. Most of the Meuse eggs are patinated to a grey-black hue, but two are patchy off-white and yellow-brown. These include the largest piece, which was split down the middle in subrecent times. The broken surface shows black staining, which in this case seems to have penetrated the entire flint. The edges of the broken surface display a kind of 'stratigraphy' of successive later surface modifications: first to evolve was a thick white patina (about 2-3 mm deep); then a thin brown patina. The same can be seen on the broken surfaces of several other Meuse eggs. These flints were first patinated black through-and-through, after which some developed a white, and then a brown patina. This sequence seems to suggest that these flints saw a period of lying at or near the surface (Upper Pleniglacial?), after having been submerged for quite some time.

Black patina occurs not only on many flint artefacts that have recently been washed up on our beaches (e.g. on the Frisian Isles: Stapert *et al.* 2011; Stapert *et al.* 2013a, b), but also on Late Middle Palaeolithic artefacts (ca 125,000-35,000 years old) from inland deposits of the Kreftenheye Formation. For example, this is a familiar feature of flints from sand suction pits in the eastern and central Netherlands, such as those at Haerst and Sekdoorn near Zwolle (see e.g. Johansen *et al.* 2009).

As noted above, Meuse eggs also occur abundantly in gravelly fluvial deposits of the much older Urk Formation, from which we know the numerous Early Middle Palaeolithic artefacts of the 'Rhenen Industry' (Niekus and Stapert 2005; Stapert 1981a; 1987; 1991). The artefacts recovered from deposits of that formation, however, tend to have a brown patina rather than black. This is not an absolute but a statistical difference; relatively few (but not zero) flints from the Kreftenheye Formation are brown-patinated; relatively few (but not zero) from the Urk Formation are black-patinated. This difference in patination can be attributed to a difference in the prevailing type of river regime during the deposition of these layers: predominantly meandering in the case of the Urk Formation, predominantly braided in that of the Kreftenheye Formation (e.g. Stapert 1981b). In braided systems, flints have a much greater chance of periodically ending up in stagnant water without oxygen.

Quite a lot has been written by geologists about egg-shaped flint pebbles in the Netherlands and Germany. L. Meyn (1874) found such pebbles on the island of Sylt and named them *Wallsteine* (he erroneously interpreted them as fossil sponges; see Oostingh 1921). Meyn later also used this term for flint pebbles from Meuse deposits in the Netherlands which he found near Arnhem and elsewhere. We think it is terminologically useful to distinguish between southern flint pebbles, 'Meuse eggs', and northern specimens, *Wallsteine*, because these two types, though similar, originate from different Tertiary periods, consist of slightly different kinds of flint and also do not look exactly the same. Some important publications describing egg-shaped flint pebbles from Pleistocene deposits in the Netherlands are: Wichmann 1876; Martin 1878; Van Capelle 1892, 1898, 1910; Van Baren 1908-1915; Tesch 1915; Kruizinga 1918; Oostingh 1921; 1925; Umbgrove 1926; Van der Lijn 1935; Van der Lijn and Boekschoten 1973; Van Straaten 1946; and Maarleveld 1952; 1956. This is not the place to go into details about these studies; several interesting points have been mentioned above. Above all, it may be concluded that Meuse eggs larger than about 8 cm are extremely rare outside southern Limburg, and that northern equivalents are almost non-existent in Pleistocene deposits in the Netherlands. Because they are so rare, it is of interest that two *Wallsteine* (about 5 cm in diameter) were present at the Epi-Ahrensburgian site of Oudehaske in the province of Friesland; this site probably dates from the Preboreal (Johansen and Stapert 1997/1998: 23-24). The pebbles do not show any traces of use; perhaps they had been collected out of curiosity?

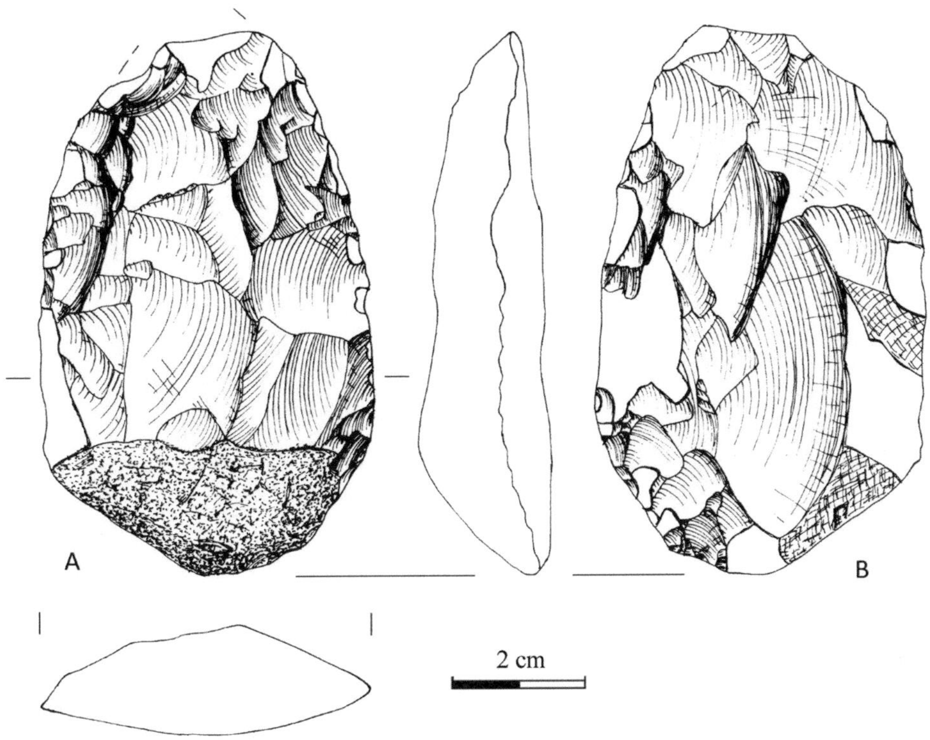

Figure. 4. The bifacial Quina-scraper from a shell path at *Huis van Hilde*, Castricum Drawing Lykke Johansen, Haren.

Apart from the Meuse eggs, Ms Schutte's collection from the shell path contains 22 further flints, almost all to a greater or lesser degree covered in black patina. Four small flints were heavily rolled without being classifiable as typical Meuse eggs. The 18 remaining ones were examined under a magnifying glass, with no trace of bryozoans. The grey flint with pale patches most probably is Meuse flint. The finds include several 'incertofacts', flake-like items with unweathered surfaces; these forms probably arose in the course of suction dredging and transportation of the material. Six pieces display not just black patina but also a relatively high gloss. On at least five of them, examination under magnification showed fine pitting, which indicates that the gloss may be interpreted as wind gloss.

All of this leads us to the conclusion that the flints from the shell path almost certainly have a southerly provenance, and that they come from gravelly fluvial layers of the Kreftenheye Formation. Hence the most likely date of the handaxe is the first half or middle part of the Weichselian. There are indications that the handaxe was made from a thick flake. Since in Europe handaxes made from flakes occur mainly within later phases of the Mousterian (see e.g. Soressi 2002, 2004), a date in the Middle Weichselian (MIS 3) is the most likely, also in view of the typology of the handaxes known from *Maasvlakte 2* (Niekus *et al.* 2021).

Some concluding remarks

This is not the only find of a sizeable Middle Palaeolithic artefact from shell paths in the Netherlands. An important find is the bifacial Quina-scraper from the 'Huis van Hilde' at Castricum (fig. 4), mentioned above; this piece was also acquired by the *Rijksmuseum voor Oudheden* in Leiden (inventory no. g 2016/2.1). As in the case of Zandvoort, the provenance of the shell grit among which it was found in 2015 cannot be established, though it certainly derives from the North Sea. The tool is slightly damaged, and had an original length of about 8.5 cm. The piece may be described as a *Racloir à retouche biface, type demi-Quina, passant à la pièce foliacée* in the typology of Bordes (1961; see e.g. *Planches* 28-30). It is clear that this tool dates from a late phase of the Middle Palaeolithic, somewhere between 45,000 and 50,000 years ago, given recent datings for sites such as La Quina (e.g. Jelinek 2013) and Roc de Marsal, French Quina (Guérin *et al.* 2017. In addition to other surface modifications (including brown patina), this tool displays windgloss.

Several other examples of bifacial Quina-scrapers are known in the Netherlands, for example from Mander in the east (Stapert *et al.* 2013b). It is of great interest that this site also produced a triangular handaxe made of a flake (Stapert 1982), on the basis of which the site was attributed to the *Moustérien de tradition acheuléenne*. Other Dutch sites with bifacial Quina-scrapers are Overloon-Biesplanken in the south, and Meppel and Tynaarlo in the north (about the implement from Meppel: Beuker *et al.* 2007). *Maasvlakte* 2 also produced several Quina-scrapers of which at least one is partly bifacial (Niekus *et al.* 2021: 60-61).

In addition to the finds at Zandvoort and Castricum, several other finds of Palaeolithic artefacts from shell paths are known. On a shell path along the river Amstel in Amsterdam, a flake of about 5 cm in diameter was found not long ago. A photo in the publication (Anon. 2022) shows that the flint is covered by both brown and black patina, and also by windgloss with small pits.

From a shell path in a garden at Rasquert (province of Groningen) a small flake (length 2.8 cm) was picked up by archaeologist Gijsbert R. Boekschoten (as recorded by the owner of the garden, geologist G.J. (Bert) Boekschoten, 2023); it has a platform remnant but does not exhibit any dorsal negatives and therefore some uncertainty about its artefactual character remains. This piece most probaby comes from the Wadden Sea in the north.

Last but not least we mention three Middle Palaeolithic artefacts (with black patina) found by amateur archaeologist Paul de Smet on a shell path at the town of Purmerend, north of Amsterdam (pers. comm. to the third author). His finds consist of a convex sidescraper (length 4.0 cm, width 2.5 cm), a distal fragment of a plunging flake (length 4.5 cm) and a core-edge flake (length 3.3 cm). Two of these pieces display clear wind gloss.

All artefacts from shell paths mentioned in this paper must have been transported together with shells from the bottom of the North Sea or the Wadden Sea. It is suggested that when treading shell paths in Dutch nature areas, one should remember to look down now and then. More finds may well await us on such trails!

Acknowledgements

We are indebted to Ms Sabine Schutte (Haarlem) for reporting her interesting find, and for her warm welcome when the first author visited her. We thank our friend Xandra Bardet (Groningen) for editing the English translation. Furthermore, we thank Luc Amkreutz (Rijksmuseum voor Oudheden, Leiden), and Paul de Smet.

References

Amkreutz, L.W.S.W., J.C. Glimmerveen, J.J. Hublin, W. Roebroeks and L. Anthonis. 2010. Een Neanderthaler uit de Noordzee. *Westerheem* 59: 50-59.

Amkreutz, L., M. Niekus, D. Schiltmans and B. Smit. 2017. Meer dan bijvangst! De prehistorische archeologie van de Noordzee. *Cranium* 34: 34-47.

Amkeutz, L., A. Verpoorte, A. Waters-Rist, M. Niekus, V. van Heekeren, A. van der Merwe, H. van der Plicht, J. Glimmerveen, D. Stapert and L. Johansen. 2018. What lies beneath... Late Glacial human occupation of the submerged North Sea landscape. *Antiquity* 92: 22-37.

Amkreutz, L. and S. van der Vaart-Verschoor (eds.), 2021. *Doggerland. Verdwenen Wereld in de Noordzee.* Leiden: Sidestone Press.

Amkreutz, L., M. Niekus, D. Schiltmans and B. Smit. 2022. Beyond catch. the prehistoric archaeology of Doggerland. In: D. Mol and B. Langeveld (eds.), The palaeontological and archaeological finds from the bottom of the North Sea. *Grondboor and Hamer* 76: 167-187.

Anon.. 2022. Het oudste van Amsterdam. *Tijdschrift van de Rijksdienst voor het Cultureel Erfgoed* 14: 25.

Baren, J. van. 1908-1915. *De bodem van Nederland.* Amsterdam: S.L. van Looy.

Beuker, J.R., L. Johansen, M.J.L.Th. Niekus and D. Stapert (met een bijdrage van S. Ziermans), 2007. Een werktuig van de Neanderthalers in de goot. *Nieuwe Drentse Volksalmanak* 124: 105-112.

Boekschoten, G.J.. 2023. review of: *Van Neanderthaler tot rendierjager* by Marcel Niekus en Yuri van Koeveringe. *Gea* 56: 28.

Bohmers, A.. 1951. Die Höhlen von Mauern. Teil I. Kulturgeschichte der altsteinzeitlichen Besiedlung. *Palaeohistoria* 1: 1-107.

Bordes, F.. 1961. *Typologie du Paléolithique ancien et moyen.* Bordeaux: Institut de Préhistoire de l'Université de Bordeaux.

Busschers, F.S., C. Kasse, R.T. van Balen, J. Vandenberghe, K.M. Cohen, H.J.T. Weerts, J. Wallinga, C. Johns, P. Cleveringa and F.P.M. Bunnik. 2007. Late Pleistocene evolution of the Rhine-Meuse system in the southern North Sea Basin; imprints of climate change, sea-level oscillations and glacio-isostacy. *Quaternary Science Reviews* 26: 3216-3248.

Cappelle, Herman van, 1892. *Het Diluvium van West-Drenthe*. Amsterdan: Johannes Müller. (Verhandelingen der Koninklijke Akademie van Wetenschappen te Amsterdam, Tweede Sectie, Deel I, No. 2.)

Cappelle, H. van. 1898. *Nieuwe waarnemingen op het Nederlandsche Diluviaalgebied*. Amsterdam. (Mededelingen omtrent de Geologie van Nederland, no. 23. Verhandelingen der Koninklijke Akademie van Wetenschappen, 2e sectie, deel VI, no. 2).

Cappelle, H. van. 1910. Bijdrage tot de kennis van de landijsvormingen in de provincie Friesland en van het oudere fluviatiele Diluvium in den ondergrond van Noord-Nederland. *Verhandelingen Koninklijke Nederlandse Akademie van Wetenschappen Afdeeling Natuurkunde 2e sectie* 16: 3-55.

Carmiggelt, A. and D.E.A. Schiltmans. 2021. Onderzoek naar de bewoning van Doggerland op het grondgebied van de gemeente Rotterdam. In: A. Carmiggelt and D.E.A. Schiltmans (eds.), *Doggerland en Rotterdam. Een inventarisatie van opgespoten paleolithische en mesolithische artefacten van (vuur) steen, bot en gewei van Maasvlakte (1 en 2) en Hoek van Holland (gemeente Rotterdam): een aanzet voor vervolgonderzoek.*:7-29. Rotterdam: Gemeente Rotterdam. (BOORnotitie 41)

Coles, B.. 1998. Doggerland: a speculative survey. *Proceedings of the Prehistoric Society* 64: 45-81.

Eisma, D. and F. Jansen. 1976. Delfstofwinning. In: J. Abrahamse, W. Joenje and N. van Leeuwen-Seelt (eds.), *Waddenzee. Natuurgebied van Nederland, Duitsland en Denemarken*: 273-281. 's-Graveland: Landelijke Vereniging tot behoud van de Waddenzee, Harlingen /Vereniging tot behoud van natuurmonumenten in Nederland..

Entrop, B.. n.d. *Schelpen vinden en herkennen*. Zutphen: W.J. Thieme and Cie.

Glimmerveen, J., D. Mol, K. Post, J.W.F. Reumer, H. van der Plicht, J. de Vos, B. van Geel, G. van Reenen and J.P. Pals. 2004. The North Sea Project: the first palaeontological, palynological, and archaeological results. In: N. Flemming (ed.), *Submarine prehistoric archaeology of the North Sea*: 43-52 York: Council for British Archaeology. (CBA Research Report 141).

Guérin, G., M. Frouin, J. Tuqoi, K. J. Thomson, P. Goldberg, V. Aldeias, Ch. Lahaye, N. Mercier, P. Guibert, M. Jain, D. Sandgathe, S. P. McPherron, A.Turq and H. L. Dibble. 2017. The complementarity of luminescence dating methods illustrated on the Mousterian sequence of the Roc de Marsal: A series of reindeer-dominated, Quina Mousterian layers dated to MIS 3. *Quaternary International* 433: 102-115.

Hijma, M., W. Westerhoff, W. Roebroeks, Th. van Kolfschoten and K. Cohen. 2011. Neanderthalers in de Noordzee? *ARCHEObrief* 15: 32-38.

Hijma, M.P., K.M. Cohen, W. Roebroeks, W.E. Westerhoff and F.S. Busschers. 2012. Pleistocene Rhine-Thames landscapes: geological background for hominin occupation of the southern North Sea region. *Journal of Quaternary Science* 27 (1): 17-39.

Hublin, J.J., D. Weston, P. Gunz, M. Richards, W. Roebroeks, J. Glimmerveen and L. Anthonis, 2009. Out of the North Sea: the Zeeland Ridges Neandertal. *Journal of Human Evolution* 57: 777-785.

Hucke, K.. 1967. *Einführung in die Geschiebeforschung (Sedimentärgeschiebe). (Nach dem Tod des Verfassers herausgegeben und erweitert von Ehrhard Voigt.)* Oldenzaal: Nederlandse Geologische Vereniging.

Jelinek, A.J., 2013. *Neandertal lithic industries at La Quina.* Tucson: The University of Arizona Press.

Johansen L. and D. Stapert. 1997/1998. Two 'Epi-Ahrensburgian' sites in the northern Netherlands: Oudehaske (Friesland) and Gramsbergen (Overijssel). *Paleohistoria* 39/40: 1-87.

Johansen, L., M.J.L.Th. Niekus and D. Stapert. 2009. Zwarte vuurstenen uit het Midden-Paleolithicum in Nederland. *Paleo-Aktueel* 20: 1-8.

Johansen, L. and D. Stapert. 2012. Some workpieces by Middle Palaeolithic apprentice flintknappers. In: M.J.L.Th. Niekus., R.N.E. Barton, M. Street and Th. Terberger (eds.) – *A mind set on flint. Studies in honour of Dick Stapert* 49-76. Groningen: Barkhuis and Groningen University Library. (Groningen Archaeological Studies, volume 16.)

Kruizinga, P., 1918. Bijdrage tot de kennis der sedimentaire zwerfsteenen in Nederland. (Zwerfsteenen van Baltischen oorsprong, uitgezonderd die, welke in en bij de stad Groningen en bij Maarn zijn gevonden.) *Verhandelingen van het Geologisch-Mijnbouwkundig Genootschap voor Nederland en Koloniën, Geologische Serie* deel IV: 1-271.

Laban, C. and J.J.M. van der Meer. 2022. Geological development of the Southern North Sea during the Quaternary. In: D. Mol and B. Langeveld (eds.), The palaeontological and archaeological finds from the bottom of the North Sea. *Grondboor and Hamer* 76: 188-199.

Lijn, P. van der and G.J. Boekschoten. 1973 (1923). *Het Keienboek (6th edition)*. Zutphen: Thieme.

Lijn, P. van der. 1935. *Nederlandse zwerfstenen.* Zutphen: Thieme.

Maarleveld, G.C., 1952. Over rolstenen. Opmerkingen bij een voorlopige kaart van de geografische verspreiding van rolsteenassociaties in de midden-pleistocene fluviatiele afzettingen in Nederland. *Tijdschrift van het Aardrijkskundig Genootschap* 69: 405-414.

Maarleveld, G.C., 1956. *Grindhoudende Midden-Pleistocene sedimenten. Het onderzoek van deze afzettingen in Nederland en aangrenzende gebieden.* Maastricht: Ernest van Aelst. (Mededelingen van de Geologische Stichting, Serie C, VI, No. 6).

Manen, A.R. van. 1993. Over het zoeken van zuidelijke zwerfstenen in de Oost-Achterhoek. *Grondboor en Hamer* 47: 49-50.

Martin, K.. 1878. *Niederländische und Nordwestdeutsche Sedimentaergeschiebe.* Leiden.

Meyn, L.. 1874. Silurische Schwämme und deren eigenthümliche Verbreitung, ein Beitrag zur Kunde der Geschiebe. *Zeitschrift der Deutschen geologischen Gesellschaft* 26: 41-58.

Ministerie van Verkeer en Waterstaat, 2008. *Beschikking. Ongrondingenwet/vergunning voor het winnen van zand in de Noordzee in verband met de aanleg van Maasvlakte 2.* (https://www.commissiemer.nl/docs/mer/p14/p1450/1450-371beschikking.pdf)

Mol, D., K. Post, R. Bakker, B. van Geel, J. Glimmerveen, H. van der Plicht and K. Post. 2008. *Mammoeten, neushoorns en andere dieren van de Noordzeebodem. Kleine encyclopedie van het leven in het Pleistoceen.* Diemen: Veen Magazines.

Moree, J.M. and M.M. Sier. 2015. Twenty metres deep! The Mesolithic period at the Yangtze Harbour site – Rotterdam Maasvlakte, the Netherlands. Early Holocene landscape development and habitation. In: J.M. Moree and M.M. Sier (eds.), *Interdisciplinary archaeological research programme Maasvlakte 2, Rotterdam*: 7-350. Gemeente Rotterdam/RCE/Port of Rotterdam, Rotterdam. (BOORrapporten 566).

Mulder, E.F.J. de, M.C. Geluk, I.L. Ritsema, W.E. Westerhoff and Th.E. Wong. 2003. *De ondergrond van Nederland.* Groningen/Houten: Wolters-Noordhoff.

Niekus, M.J.L.Th. and D. Stapert. 2005. Het Midden-Paleolithicum in Noord-Nederland. In: J. Deeben, E. Drenth, M.-F. van Oorsouw and L. Verharf (eds.), *De Steentijd van Nederland*: 91-118. Stichting Archeologie, Zutphen. (Archeologie 11/12).

Niekus, M.J.L.Th., R.T. van Balen, J.M.G. Bongers, J.H.A. Bosch, D.C. Brinkhuizen, H. Huisman, J. Jelsma, L. Johansen, C. Kasse, S.J. Kluiving, H. Paas, E. Schrijer and D. Stapert. 2015. News from the North: A late Middle Palaeolithic site rich in handaxes on the Drenthe-Frisian till plateau near Assen, the Netherlands; first results of a trial excavation. *Quaternary International* 411: 284-304.

Niekus, M.J.L.Th., G.R. Boekschoten, L. Hoven, L. Johansen, H. Paas, D. Stapert and P. Wiersma. 2017a. Vijf bijzondere vuistbijlen uit de omgeving van Assen: enkele Neanderthaler-vondsten van Dick C. Brinkhuizen (1946-2016). *Grondboor and Hamer* 71: 68-83.

Niekus, M.J.L.Th., L.W.S.W. Amkreutz, L. Johansen and D. Stapert. 2017b. Een bijzondere vuistbijl uit de Noordzee gevonden op Maasvlakte 2 bij Rotterdam. *Grondboor and Hamer* 71: 162-169.

Niekus, M.J.L.Th., P.R.B. Kozowyk, G.H.J. Langejans, D. Ngan-Tillard, H. van Keulen, J. van der Plicht, K.M. Cohen, W. van Wingerden, B. van Os, B.I. Smit, L.W.S.W. Amkreutz, L. Johansen, A. Verbaas and G.L. Dusseldorp. 2019. Middle Paleolithic complex technology and a Neandertal tar-backed tool from the Durch North Sea. *PNAS* 116: 22081-22087.

Niekus, M.J.L.Th., L. Johansen and D. Stapert. 2021. Een inventarisatie van opgespoten vuurstenen artefacten van Maasvlakte 2 en Hoek van Holland (gemeente Rotterdam). In: A. Carmiggelt and D.E.A. Schiltmans (eds.), *Doggerland en Rotterdam. Een inventarisatie van opgespoten paleolithische en mesolithische artefacten van (vuur)steen, bot en gewei van Maasvlakte (1 en 2) en Hoek van Holland (gemeente Rotterdam): een aanzet voor vervolgonderzoek*: 31-123. Rotterdam: Gemeente Rotterdam. (BOORnotitie 41).

Niekus, M.J.L.Th., K.M. Cohen, L. Johansen and D. Stapert. 2023. *Opgezogen, opgespoten en opgeraapt. Vuurstenen en hardstenen artefacten van de Zandmotor en hun sedimentaire context.* Amersfoort: Rijksdienst voor het Cultureel Erfgoed.

Oostingh, C.H.. 1921. *Bijdrage tot de kennis der zuidelijke zwerfsteenen in Nederland en omgeving.* Wageningen: Veenman. (Mededeelingen Landbouw Hoogeschool, Wageningen 19).

Oostingh, C.H.. 1925. De fluviatiele afzettingen langs de Maas in België en Frankrijk. In: C.L. van Balen (ed.), *Gedenkboek ter herinnering aan den 70sten verjaardag van R. Schuiling, 27 mei 1924*: 219-236. Groningen: P. Noordhoff.

Pols, G.. 2017. Schelpenvissers verdacht van illegale visserij op Waddenzee. *Volkskrant* 29 maart 2017.

Reinecke, R.. 2009. *Steine am Ostseestrand.* (4. Auflage.) Demmler Verlag, Ribnitz-Damgarten: Demmler Verlag.

Ruegg, G.H.J.. 1981. Ice-pushed Lower and Middle Pleistocene deposits near Rhenen (Kwintelooijen): sedimentary-structural and lithological/ granulometrical investigations. *Mededelingen Rijks Geologische Dienst* 35 165-177.

Soressi, M.. 2002. *Le Moustérien de tradition acheuléenne du sud-ouest de la France. Discussion sur la signification du faciès à partir de l'étude comparée de quatre sites: Pech-de l'Azé I, Le Moustier, La Rochette et la Grotte XVI.* Ph.D. thesis, Bordeaux: Université Bordeaux I.

Soressi, M.. 2004. From the Mousterian of Acheulian Tradition Type A to Type B: a change in technical tradition, raw material, task, or settlement dynamics? In: N.J. Conard (ed.), *Settlement dynamics of the Middle Paleolithic and Middle Stone Age. Volume II*: 343-366. Tübingen: Kerns Verlag.

Stapert, D.. 1976a. Some natural surface modifications on flint in the Netherlands. *Palaeohistoria* 18: 7-41.

Stapert, D.. 1976b. Middle Palaeolithic finds from the Northern Netherlands. *Palaeohistoria* 18: 43-72.

Stapert, D.. 1981a. Middle Palaeolithic finds from the beach at Cadzand (province of Zeeland). *Berichten van de Rijksdienst voor het Oudheidkundig Bodemonderzoek* 31: 293-302.

Stapert, D.. 1981b. Archaeological research in the Kwintelooijen pit, municipality of Rhenen, the Netherlands. *Mededelingen Rijks Geologische Dienst* 35-2/7: 202-222.

Stapert, D.. 1981c. Middle Palaeolithic finds from the Wijchense Meer (province of Gelderland) and several other finds from Pleistocene fluviatile deposits in the surrounding region. *Berichten van de Rijksdienst voor het Oudheidkundig Bodemonderzoek* 31: 273-291.

Stapert, D.. 1982. A Middle Palaeolithic artefact scatter, and a few younger finds, from near Mander, NW of Ootmarsum (province of Overijssel, the Netherlands). *Palaeohistoria* 24: 1-33.

Stapert, D.. 1987. A progress report on the Rhenen Industry (Central Netherlands) and its stratigraphical context. *Palaeohistoria* 29: 219-243.

Stapert, D.. 1991. Archaeological research in the Fransche Kamp pit near Wageningen (Central Netherlands). *Mededelingen Rijks Geologische Dienst* 46: 71-88.

Stapert, D., G .J. Boekschoten and L. Johansen. 2011. Een Levallois-kern en een bijzondere geslepen bijl van Vlieland (Fr.). *Paleo-Aktueel* 22: 1-8.

Stapert, D., L. Johansen and M.J.L.Th. Niekus. 2013a. A bifacial tool of the Neanderthals from Ameland, the Netherlands. *Netherlands Journal of Geosciences/ Geologie en Mijnbouw* 92: 181-192.

Stapert, D., M. Niekus, D. Schlüter and L. Johansen. 2013b. De Neanderthaler-site bij Mander (Ov.) krijgt contouren. *Paleo-Aktueel* 24: 13-22.

Stapert, D., L. Johansen, M. Niekus, M. Diertens and E. Knol. 2013c. Een bijzonder middenpaleolithisch werktuig van Ameland (Fr.). *Paleo-Aktueel* 24: 23-31.

Straaten, L.M.J.U. van. 1946. *Grindonderzoek in Zuid-Limburg.* Proefschrift Universiteit van Leiden (Mededeelingen van de Geologische Stichting, C-VI, No. 2).

Tesch, P.. 1915. Rolsteenen van de Doggersbank. *Handelingen van het 15e Nederlandsch Natuur- en Geneeskundig Congres.* Haarlem: 525-530.

Umbgrove, J.H.F.. 1926. Bijdrage tot de kennis der stratigraphie, tektoniek en petrographie van het Senoon in Zuid-Limburg. *Leidsche Geologische Mededeelingen* 1: 255-332.

Verhart, L.B.M.. 2001. De zee neemt, de zee geeft. Spectaculaire vondsten uit de Noordzee. *Westerheem* 50: 102-107.

Verhart, L.B.M., 2004. The implications of prehistoric finds on and off the Dutch coast. In: N. Flemming (ed.), *Submarine prehistoric archaeology of the North Sea* 57-61. York: Council for British Archaeology. (CBA Research Report 141.)

Wichmann, A., 1876. Über Puddingstein. *Neues Jahrbuch für Mineralogie, Geologie und Palaeontologie*: 907-918.

In Dusseldorp, G.L., Chu, W., Bakels, C.C. and M.A. Soressi (eds) 2025. Intent on the Paleolithic. Papers in honour of Prof. Dr. Wil Roebroeks. *Analecta Praehistorica Leidensia* 55. Leiden: Sidestone Press, pp. 71-84.

Faunal exploitation at Laugerie Basse, Dordogne, France: The Legacy of Alain Roussot

Richard Cosgrove

Abstract

This chapter outlines the results of a selected study of faunal material excavated by Dr Alain Roussot from the Laugerie Basse rock shelter in 1980 and 1984. Before his intervention, at least eight excavations had taken place at the site. The first by Lartet and Christy in 1863 and the most extensive by Maury in 1912 (1924). Large quantities of faunal and lithic remains were recovered during these early excavations, but no detailed analyses were undertaken. Most references to the fauna were descriptions of engraved bone with figures of horses, reindeer, lizards, and birds (Maury 1924). It follows a forlorn tradition of unrealised potential of some of the most famous French Palaeolithic rock shelter sites. O'Hara *et al.* (2015) have for example described the plundering of antiquities from archaeological sites such as Abri de la Souquette. Although much of the archaeological record was lost to the cavalier pursuit of saleable antiquities, their subsequent radiocarbon dating, and limited excavation retrieved some useful information on Aurignacian occupation (Roussot 1982). Laugerie Basse has, in many respects, suffered a similar fate, although an unknown quantity of this famous deposit remains intact under large collapsed limestone blocks. Alain Roussot undertook a more systematic approach than his predecessors in 1980. He aimed to stabilize the profile and to carry out a detailed stratigraphic reanalysis of the Abri des Marseilles section through limited excavation. His work was largely successful and meticulous: collecting fallen debris, cleaning the exposed profile and wet-sieving all the excavated material. His small faunal collection is the focus of this chapter, providing some insight into human subsistence activities and an understanding of the taphonomic processes that led to their deposition. This analysis provides some small recompense for the past poor treatment of this famous site. The overall interpretation of human behaviour at the site, that includes a study of the vast lithic industry, is to be determined.

Keywords: faunal exploitation; Laugerie Basse; Dordogne, France; Alain Roussot

Background

Laugerie Basse is a large limestone rock shelter located about 50m west of the Vézère River, near the town of Les Eyzies-de-Tayac in France (Fig. 1a). It was first recognised as an archaeological site around 1863 and excavated by Lartet and Christie, and subsequently by individuals such as Vibraye, Hardy, Riviere and Girod.

Richard Cosgrove
La Trobe University, Melbourne

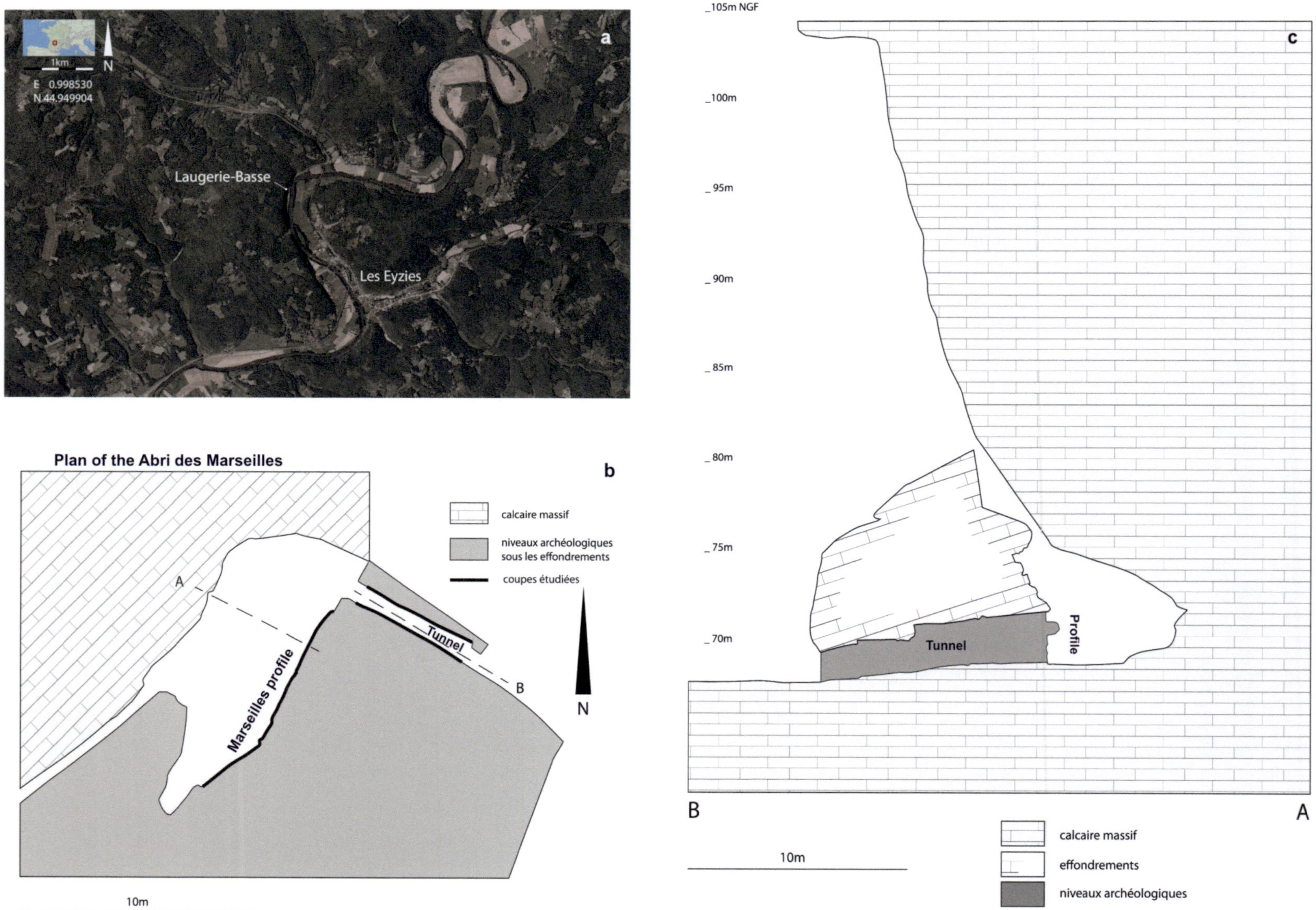

Figure 1. a) Location of Laugerie-Basse (credit B. Chadele). b) Map of the Abris des Marseilles shelter. The profiles of the Aris and the tunnel are indicated by the thick line (credit B. Chadelle). C) Sagittal section of the Abris des Marseilles area, showing the huge collapse blocks that cover the Magdalenian deposits (credit B. Chadelle).

Otto Hauser excavated from 1907 until 1912 and then Maury until 1920 (Laville 1964: 222). During this time no systematic scientific excavations were carried out and few detailed investigations were made of the stratigraphy or the cultural remains. Sedimentological analyses were carried out by Laville (1973, 1975), although Roussot (1986) states that it is difficult to resolve their descriptions as no diagrams were made to accompany the analysis. More recent work by Camus (2018) as part of our larger study into late Pleistocene occupation of Combe Saunière and Laugerie Haute, provided a highly detailed description of the stratigraphy and geomorphology. New radiocarbon and stable $\delta^{13}C$ and $\delta^{15}N$ isotope studies have improved our understanding of Laugerie Basse chronology and climate fluctuations in the Vézère River basin during human occupation (Cosgrove *et al.* 2015; Chadelle *et al* 2017; Verpoorte *et al.* 2018).

The history of the excavations and research by Roussot has been described before (Roussot 1986; Cosgrove *et al.* 2016; Chadelle *et al.* 2017; Camus 2018) and will not be reiterated here. In summary, he undertook an evaluation of the stratigraphy after gaining permission to erect a grid in front of the Abri des Marseilles (Fig. 1b). The work consisted of systematically re-numbering all the layers to replace the previous level names. This allowed a much clearer notation of layers above and below the major rock fall (Fig. 2, level 10; Camus 2018: Fig. 5) for future stratigraphic investigations. During his 1980 work he collected all the debris that had been dislodged by looters as well as material that had been knocked from the tunnel walls by tourists and natural erosion. He undertook cleaning (< 5cm) and straightening of the section along the exposed profile (Fig. 1c). He collected 14C samples for dating, composed of a mixture of charcoal and soil from two hearths, one in square J12 labelled 15a (GIF-5386) and one from J10 labelled 15b (GIF-5387). A total of 240g was obtained for the GIF-5386 sample while 85g was collected from the GIF-5387 hearth. They returned uncalibrated ages of 12,500±250 BP and 13,850±160 in squares J12 and J10 respectively (Fig. 2).

Among his collection was an assemblage of faunal remains that had not been analysed (Roussot 1985: 7). In his report "Recent Research at Laugerie Basse", Roussot (1985: 7) describes how all the osseous material was retrieved by 'water sieving', suggesting that most of the smallest animal bones were collected. He does not however, describe the size of the sieve mesh used. Presumably it was small because bone elements such as mice teeth and fish vertebrae were recorded.

The focus of this study is the material collected from the clean-up of the Abri des Marseilles profile (Fig. 1b). A tunnel dug under the deposit in 1912 reveals exposed bones, lithics and charcoal in abundance in the upper Neolithic layers (Camus 2018) with a similar chronology to that of the Abri des Marseilles (Cosgrove *et al.* 2016; Chadelle *et al.* 2017). According to Maury (1924: 16), Le Bel's excavations revealed a range of fauna such as horse, reindeer bone and antler, bison, red deer, ibex, Saiga antelope, wolf, fox and pig. A mammoth tooth was apparently found in the 'inferior' layers of the Marseilles section. It is not known to this author what happened to Le Bel's excavated material.

Chronology

Our recent detailed study of the site confirmed a late Pleistocene age of the deposits (Cosgrove *et al.* 2016; Chadelle *et al* 2017). The results are based on collagen extracted from reindeer bone in couches 16 to 7 and charcoal found in the Neolithic layers of the tunnel (Fig. 2).

They have a high reliability because of their good carbon to nitrogen ratio (R. Wood, pers comm; Fig. 3b). These deposits appear to have accumulated relatively rapidly, beginning about 18,070-17,750 years ago (Fig. 3b), with three major occupation periods based on Bayesian chronological modelling (Fig. 3a). These are accompanied by several episodes of rock fall that are evidenced in the tunnel and Marseilles profiles (Camus 2018). At least three major collapses of the overhanging cliff face (Fig. 1c) are recognised. Angular debris in couche 15 suggest a small episode of rock fall after c. 17,000 BP, while couche 12 represents a larger fall of blocks prior to c. 15,000 BP. Couche 10 and 6 represent a catastrophic event that finally blocked the cave from outside access after c. 13,000 and c. 11,000 BP (Camus 2018: 21-23). Camus also suggests the probability of episodic collapses under couche 16, the base of the excavation that overlies additional archaeological deposit.

From our new radiocarbon dates, analytical units were compiled so that archaeological materials could be compared within equitable time units. The rationale for the choice of these analytical units was based on eight additional radiocarbon dates obtained from the exposed section in the Abri des Marseilles and analysed by Dr Rachel Wood, (Australian National University Radiocarbon Dating Laboratory; Fig. 3a). A further five dates denoted in red text in Fig. 3b

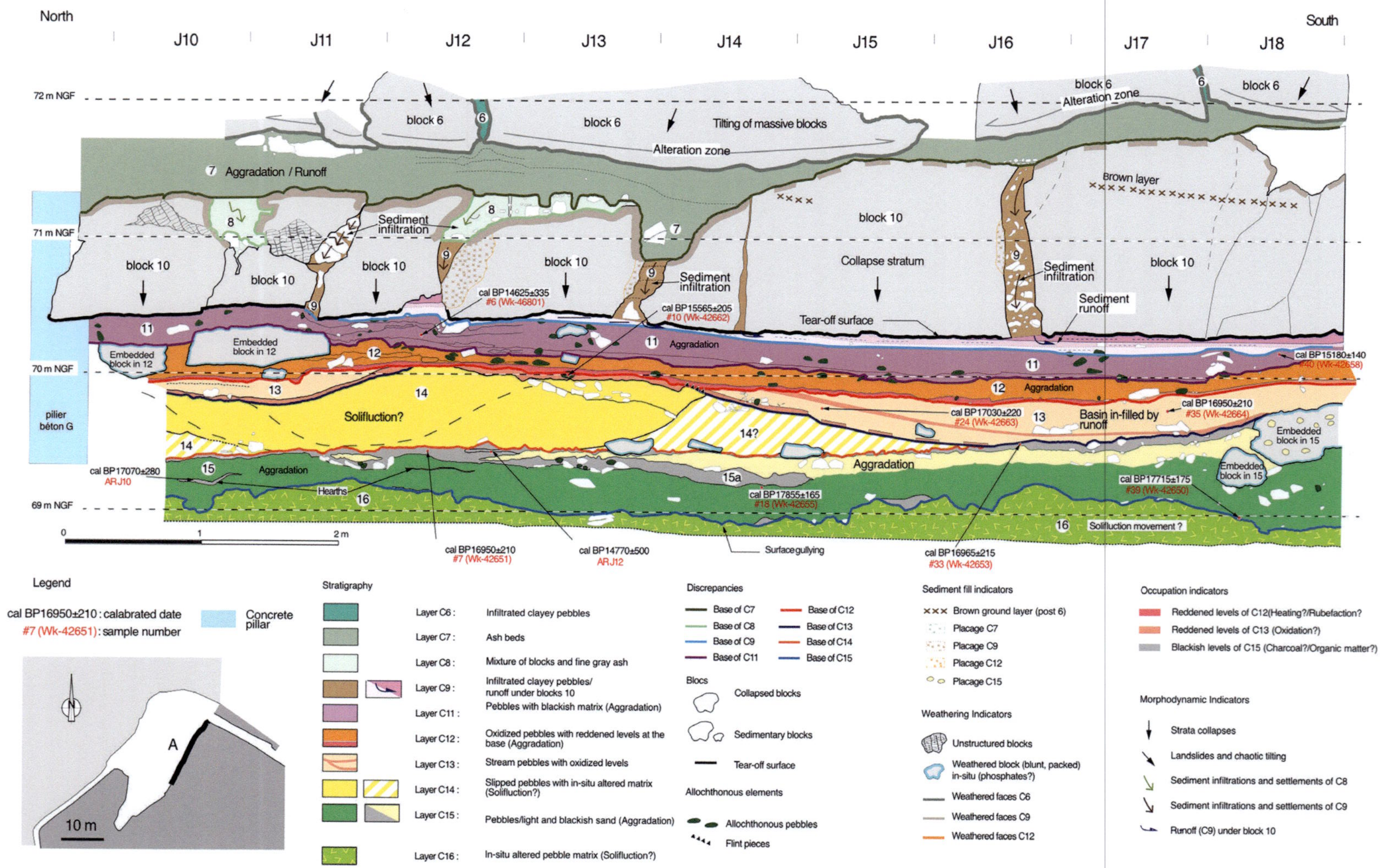

Credit: Geomorphology and interpretation by Hubert Camus (2018). Translation R. Cosgrove

Figure 2. Abris des Marseilles in Laugerie-Basse, showing the excavated section of archaeological deposits based on the original drawing of Alain Roussot. Geomorphological analysis and illustrated by H. Camus (2018), *Stratigraphic context and deposit dynamics of the Laugerie-Basse deposit, commune of Eyzies-de-Tayac, Dordogne,* page 30. Translated by R. Cosgrove.

a

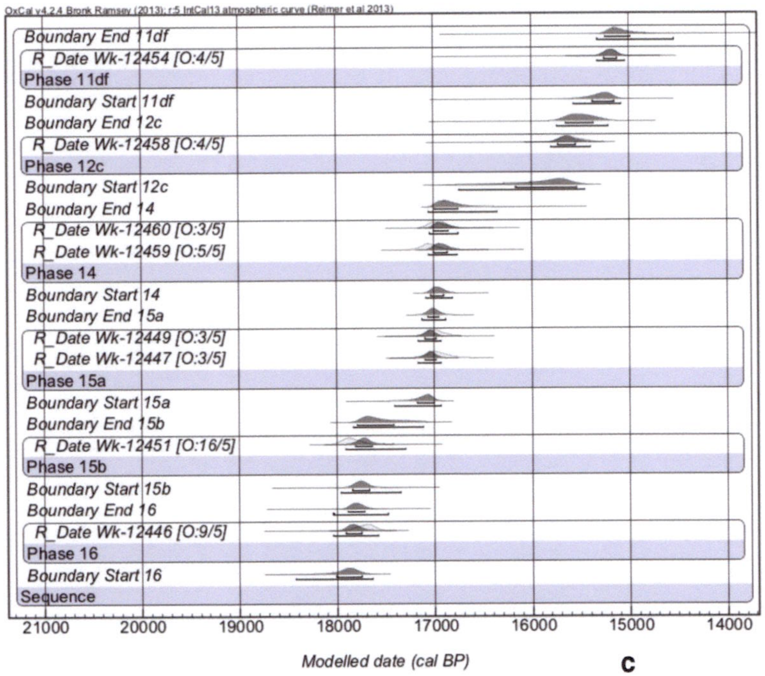

b

Sample ID	ID No.	Age 14C uncal	Age calBP range 95.4% Prob	Age calBP	C:N
42649	LB39b				3.30
42650	LB39a	14527±33	17890-17540	17715±175	3.35
42651	LB07	13977±32	17160-16740	16950±210	3.39
42653	LB33	13987±37	17180-16750	16965±215	3.30
42654	LB13				3.25
42655	LB18	14674±42	18020-17690	17855±165	3.19
42656	LB09				3.25
42657	LB03				3.25
42658	LB40	12738±35	15320-15040	15189±140	3.29
42659	LB20				3.31
42660	LB08				3.32
42661	LB12				3.74
42662	LB10	13020±29	15770-15360	15565±205	3.35
42663	LB24	14031±37	17250-16810	17030±220	3.24
42664	LB35	13977±32	17160-16740	16950±210	3.35
42665	LB31				3.34
42666	*LB54	4322 ± 20			
42667	LB55	9985±22	11510-11290		3.37
42668	*LB57	5221 ± 20			
42669	LB61				3.31
42670	LB63	14724±35	18070-17750		3.28
42671	LB70	13921±31	17060-16640		3.29

c

d

e

Figure 3. a) Bayesian model of the dates of the Magdalenian levels of the Marseilles shelter. There are three main periods of occupation centred around 17800 cal. BP, 17000 cal. BP and 15500 cal. BP. (image R. Wood) b) Results of radiocarbon dating with indication of calibrated ages for comparison with uncalibrated dates. NB: All C:N values were measured on bone gelatin. c) engraving of a fish from couche 15 (image J-P Chadelle), d) distal end of ungulate tooth root that has been cut in half (image R. Cosgrove), e) close-up of the cut-off showing the marking of a saw. Image R. Cosgrove.

were obtained from the tunnel section. Only the dates in the Marseilles section excavated by Roussot were used in this modelling.

The analytical units are divided into four chronological periods and ignore the smaller stratigraphic subdivisions applied by Roussot in his original report. This is because of the dynamic nature of the soil formation, as identified by Camus (2018) in his geomorphological report. Many small lenses of sediment were introduced over hundreds, if not thousands of years. Some in couche 13 and 14 have small amounts of archaeological material, while couche 10 and 12 are periods of natural roof fall containing little in the way of cultural material. Some bones have fallen into the cracks between the blocks and Magdalenian material from an occupation found on top of the couche 10 roof fall has also migrated below into the cracks. Some pottery from Neolithic layers above couche 10 was also identified. The analytical units (AU) are described below.

AU 1: the youngest period is represented by Neolithic and later archaeology that includes pottery, lithics and faunal remains. They are represented by couches 6, 7, 8, and 9 with dates of 11,400 calBP [Wk-42667], 6085±95 calBP [Wk-46407], 5967±23 calBP [Wk-42668] 5375±75 calBP [Wk-46408] and 4880±25 calBP [Wk-42666]. These dates came from radiocarbon samples in the two tunnel walls (Camus 2018: Fig. 6 and 7).

AU 2: is represented by a period of occupation with dates of 15,565±205 calBP (Wk-42662), 15,180±140 calBP [Wk-42658] and 14,625±335 calBP [Wk-46801]. They are represented by couche 11 and couche 12 in the Abri des Marseilles section (Camus 2018: Fig. 5).

AU3: is represented by couches 13, 14 and 15, although according to Camus (2018), couche 13 and, particularly 14, contain evidence of solifluction. Couche 14 is said to be sterile, without cultural material, and this is supported by this bone analysis. The small number of bones and other items recorded have probably been introduced from further above. This AU corresponds to dates from bone collagen of 17,030±220 calBP [Wk-42663] and 16,950±210 calBP [Wk-42664]. These bones may have moved into this AU, but their surfaces appear relatively unmodified, suggesting minimal transport.

AU4: this is the oldest unit, lying directly on top of unexcavated deposit. According to both Camus, and in an earlier report by Roussot, there may be older cultural material below in the unexcavated layers of the rock shelter. This AU is represented by couche 16, with dates of 17,715±175 calBP [Wk-42650] and 17,855±165 calBP [Wk-42655].

These analytical units differ from the couche divisions identified by Camus but are chronologically distinct through statistical modelling to represent periods of significant human occupation. The whole deposit appears to have been built up in less than 3000 years. An alternative division of AU by couche, based on the geomorphic formation process described by Camus (2018), could be used as well to construct different analytical units. However, this may not make the outcome of the faunal analysis much different. For the purposes of this report, the AU based on the Bayesian modelling are used here.

Faunal Analysis

Analyses were restricted to squares J11, J12, J13, J17 and J18 from the Abri des Marseilles archaeological layers (Fig. 2) due to the large number of bones collected, a detailed analysis and limited time. This nevertheless provided an initial indication of the range of animals deposited in the rock shelter. Roussot's collection had been stored in the Musée du Aquitaine, but is now housed at Pole Mixte de Recherches Archéologiques, Campagne. This is the material that forms part of this analysis.

The methodical approach was both qualitative and quantitative, with bone elements given a new consecutive identification number and systematically recorded. Any original labelling given to the bones by Roussot was preserved, along with any labelling he had written on the bones. Each bone was then re-bagged with information written on the label. Each bone recorded was identified with the square it came from, the couche, class of animal (i.e. mammal, bird, fish, amphibian mollusc), taxon, body part, element, specimen, number of bones, maximum length, weight, symmetry (left, right, medial), natural and anthropological modifications, age (where possible) and any comments such as whether any photos were taken, conjoined bones, and other details. Where the animal could not be identified a series of decreasing taxonomic descriptors was used to categorise the bone. Categories included an 'indeterminant', 'microfauna' (weighing between 0.01kg – 0.05kg), 'small fish/ bird' (weighing between 0.01 kg – 0.9kg), 'very small mammal' (>0.05 <5kg), 'small mammal' (>5kg <20kg) medium mammal' (weighing between >20kg <150kg), 'large mammal' (>150kg).

	Analytical Unit													
	1			2			3			4			Total	
	Frequency			Frequency			Frequency			Frequency			Frequency	
Taxon	**Sum**	**Row %**	**Column %**	**Sum**	**Row %**	**Column %**	**Sum**	**Row %**	**Column %**	**Sum**	**Row %**	**Column %**	**Sum**	**% of Total**
Rana sp.	104	10	4.94%	0			0			0			**104**	**2.64%**
Bovid	2	10	0.09%	0			0			0			**2**	**0.05%**
Cervus elaphus	6	10	0.28%	0			0			0			**6**	**0.15%**
Indeterminate	457	43.94%	21.69%	183	17.60%	22.99%	394	37.88%	38.29%	6	0.58%	10	**1040**	**26.41%**
Large Mammal	14	82.35%	0.66%	2	11.76%	0.25%	1	5.88%	0.10%	0			**17**	**0.43%**
Medium Bird	1	10	0.05%	0			0			0			**1**	**0.03%**
Medium Mammal	1130	50.79%	53.63%	521	23.42%	65.45%	574	25.80%	55.78%	0			**2225**	**56.50%**
Microfauna	15	93.75%	0.71%	1	6.25%	0.13%	0			0			**16**	**0.41%**
Microtus sp.	3	5	0.14%	0			3	5	0.29%	0			**6**	**0.15%**
Murid	19	79.17%	0.90%	2	8.33%	0.25%	3	12.50%	0.29%	0			**24**	**0.61%**
Oryctolagus cuniculus	3	75.00%	0.14%	0			1	25.00%	0.10%	0			**4**	**0.10%**
Rangifer tarandus	7	10.61%	0.33%	16	24.24%	2.01%	43	65.15%	4.18%	0			**66**	**1.68%**
Small Bird	16	88.89%	0.76%	2	11.11%	0.25%	0			0			**18**	**0.46%**
Small Fish	83	57.64%	3.94%	61	42.36%	7.66%	0			0			**144**	**3.66%**
Small Mammal	112	91.80%	5.32%	6	4.92%	0.75%	4	3.28%	0.39%	0			**122**	**3.10%**
Small Ungulate	2	5	0.09%	0			2	5	0.19%	0			**4**	**0.10%**
Snail	0			0			1	10	0.10%	0			**1**	**0.03%**
Sorex sp.	1	10	0.05%	0			0			0			**1**	**0.03%**
Very Small Mammal	132	96.35%	6.26%	2	1.46%	0.25%	3	2.19%	0.29%	0			**137**	**3.48%**
Total	**2107**		**99.98%**	**796**		**99.99%**	**1029**		**10**	**6**		**10**	**3938**	**100.02%**

Table 1. NISP and percentage of identified taxon and indeterminate species by Analytical Unit. The Grand Total column includes all taxon that could be identified to a size class (AU1-4), and those bones that were so badly damaged they could not be assigned to taxon or size class.

Analytical Unit	1			2			3			4			Total	
	Weight (g)			Weight (g)			Weight (g)			Weight (g)			Weight (g)	
Taxon	Sum	Row %	Column %	Sum	Row %	Column %	Sum	Row %	Column %	Sum	Row %	Column %	Sum	% of Total
Rana sp.	3.624	10	0.31%	0			0			0			3.624	0.17%
Bovid	2.55	10	0.22%	0			0			0			2.55	0.12%
Cervus elaphus	11.24	10	0.98%	0			0			0			11.24	0.54%
Indeterminate	137.79	54.43%	11.97%	40	15.80%	9.34%	74.33	29.36%	14.51%	1.03	0.41%	10	253.15	12.10%
Large Mammal	191.48	93.23%	16.63%	10.71	5.21%	2.50%	3.19	1.55%	0.62%	0			205.38	9.81%
Medium Bird	0.42	10	0.04%	0.			0.			0			0.42	0.02%
Medium Mammal	669.094	53.18%	58.12%	307.48	24.44%	71.80%	281.6	22.38%	54.99%	0			1258.174	60.12%
Microfauna	0.24	96.00%	0.02%	0.01	4.00%		0			0			0.25	0.01%
Microtus sp.	0.08	17.39%	0.01%	0.			0.38	82.61%	0.07%	0			0.46	0.02%
Murid	0.79	66.95%	0.07%	0.09	7.63%	0.02%	0.3	25.42%	0.06%	0			1.18	0.06%
Oryctolagus cuniculus	1.83	20.04%	0.16%	0.			7.3	79.96%	1.43%	0			9.13	0.44%
Rangifer tarandus	27.56	11.59%	2.39%	67.69	28.47%	15.81%	142.5	59.94%	27.83%	0			237.75	11.36%
Small Bird	1.41	78.77%	0.12%	0.38	21.23%	0.09%	0			0			1.79	0.09%
Small Fish	3.11	78.34%	0.27%	0.86	21.66%	0.20%	0			0			3.97	0.19%
Small Mammal	23.82	91.93%	2.07%	1.01	3.90%	0.24%	1.08	4.17%	0.21%	0			25.91	1.24%
Small Ungulate	5.06	84.05%	0.44%	0			0.96	15.95%	0.19%	0			6.02	0.29%
Snail	0			0			0.32	10	0.06%	0			0.32	0.02%
Sorex sp.	0.01	10		0			0			0			0.01	
Very Small Mammal	71.19	99.76%	6.18%	0.02	0.03%		0.15	0.21%	0.03%	0			71.36	3.41%
Total	1151.298		10	428.25		10	512.11		10			10	2092.688	100.01%

Table 2. Distribution of taxa by weight and percentage (row, column and total) by Analytical Unit.

Results

Number of Identified Specimens (NISP)

A grand total of 3,938 bones were analysed that could be divided into 19 categories of identified taxa within the four analytical units (Table 1). Of these 1,258 could be identified to either genus, species, or size class. The Grand Total column includes those bone elements that were so fragmented that they could not be assigned to a taxa or size class and were grouped together, counted, and weighted. Bone preservation was generally very good with small unidentified fish vertebrae and bird bones well-preserved. However, about 26.41% (n=1040) were unidentifiable because they were either too badly damaged by erosion, burnt, fragmented and/or, did not have any identifying features to indicate a genus or species. Although this was the case, many more bones could be attributed to a weight category. Animal size was estimated by measuring the length, thickness, and characteristics of the bone. In this category, medium mammals made up 56.44% (n=2225) of the total assemblage. The next most numerous bones were from small fish (9.78%; n=144), followed by small mammal (3.74%; n=122), very small animal (2.54%; n=115), small birds (1.43%; n=18), large mammal (1.35%; n=17), very small mammal (0.40%; n=22) and medium bird (0.03%; n=1).

Only seven genera/species could be identified reliably. Of these, four (*Microtus* sp., *Sorex* sp., *Rana* sp. and murids) fall within the very small animal category, *Rangifer* sp. and *Cervus* sp. are in the large mammal category, one (*Oryctolagus cuniculus)* in the small size category. It is likely that some of the bones classified as medium-sized probably belonged to the larger-sized group because of the absence of any other medium-sized mammal such as the deer, fox, or ibex as noted by Maury (1924). Because the fauna sample from the Roussot clean-up is limited, the range of identifiable species is correspondingly small. Unlike faunal assemblages from other French Magdalenian cave sites where horse, bison and Saiga antelope are present (e.g., Valensi and Boubles 2018), none of these were recorded in the Roussot faunal collection under discussion.

However, the fauna in the Tunnel Section does have bones of larger size that are very weathered, with smoothed surfaces perhaps due to the action of water. The Musée du Aquitaine collection stored at Pole Mixte de Recherches Archéologiques, Campagne has additional species and elements that were absent from the squares analysed. These include carnivore canines, large fish vertebrae, shell fossils, horse teeth, bovid long bones, engraved bone, eyed needles, and perforated shells. However, few axial elements such as vertebrae, ribs, pelvis or crania of large prey were recovered from his excavation.

From Roussot's inventory notes, he also recorded human teeth, fish, red ochre and microfauna from couches 7 and 8. A similar range of faunal remains were recorded from couches 10-14, but few details were given, often being described as 'bone fragments'. Further work on these collections could expand the range of animals found that were used by people at the site of Laugerie Basse.

Species distribution

The purpose of these analyses is to highlight the most commonly occurring taxa found in the Roussot collection. Surprisingly, fish, frogs, small mammals and reindeer are the most common species. Although the sample analysis is limited to five out of eight squares, the aim is to understand the human subsistence / taphonomic patterns of taxa accumulation. Table 1 lists a range of species within each analytical unit that include small and large mammals, the former probably introduced by carnivores, raptors, or natural deaths. This is reinforced by the fact that all identifiable elements are represented in the skeletal remains with crania (21%), forequarter (10%), hindquarter (28%) thoracic (19%) bones suggesting that they were introduced as complete specimens, a hallmark of raptor predation (Andrews 1990). They contribute 5.3% (n=191) of the total weight of the assemblage whereas the medium and *Rangifer tarandus* together total c.70% of the total weight of bone (N=2291).

The most common anthropological alteration to bone is burning, with the majority (n=381, 30%) of them occurring on medium-sized mammal bone. At least seven reindeer bones had significant cut marks, indicating butchering of long bones and cranial areas.

Fish

Fish are relatively well represented in couches 9, 11 and 12, dating to between 15,565±205 [Wk-42662] and 15,189±140 [Wk-42658]. Roussot recorded a concentrated layer of fish bones in the section in couche 11 that suggests a spatially discrete deposit (Fig. 2). A total of 123 fish bones were recorded, of which 85% were vertebrae (n=105), while eight cranial elements (2.84%) made up the rest. There has been some speculation about the origin of fish remains in sites, whether they represent human prey or were introduced through natural taphonomic pathways (Guillaud *et al.* 2017; Guillaud *et al.* 2018). At Laugerie Basse it is impossible to know because there was no systematic excavation that mapped the distribution of these fish remains. We therefore cannot know what association they may have had with human activities. At least 20% had been burnt (n=25), all found together in couche 9, but it is hard to say whether these were found in hearths or became burnt through incidental fire placement. The fact that they were not randomly distributed and found together in discrete loci may suggest human deposition.

At the Magdalenian site of Taillis des Coteaux cave (Vienne, France), Guillaund *et al.* (2018) analysed the spatial distribution and density of fish bones across the site. The bony elements showed signs of butchering, burning and modifications consistent with human consumption. They also used sclerochronological methods on fish scales to estimate whether the catch was seasonal. Results showed that most were caught during the spawning period, between March and June. Although these patterns were found on salmon and large grayling, it might be possible using these methods to understand the origin of fish remains at Laugerie Basse. DNA analysis of vertebrae also remains a possibility (Ferrari *et al.* 2021). There are at least two engravings of fish on mammal long bone from the site (Fig. 3c), suggesting they were an important animal in the lives of the people living at Laugerie Basse.

Rana sp. (Frog)

104 bones of frogs were identified in analytical unit 1. They date to the early to mid-Holocene period, between 11,400 calBP (Wk-42667) and 4880±25 calBP. 89 bones, or 85.5%, come from couche 6, while the remaining 15 (14.4%) were found in couches 7a and 7b, 8 and 8b. Thus, the majority of the Laugerie Basse frog bones are from the Neolithic period but unfortunately, we only have a general provenance for their location within the upper stratigraphy. Despite this limitation, we know that couche 6 is composed of infiltrated clay gravel from layers above (Camus 2018: Fig. 6), located between the large fallen limestone blocks. Thus, the infilling by sediments occurred after the blocks fell sometime after c. 11,500 years ago. It is probable therefore that some of the bones were introduced into couches 7 and 8 by the gravitational movement of sediment from above.

This distribution of different bone elements is interesting. Most of the body parts are from the hindquarter (n=77; 74%), where the femur (n=28), tibia (n=24) and ilium (n=22) dominate. The combined crania, forequarter, thoracic and unidentified bones make up just 16%. This is significant because frogs have been identified in several archaeological sites in Europe dating to the Neolithic (Kyselý 2008). They have been interpreted as being human food items rather than the result of taphonomic process such as introduction by carnivores or raptors. Kyselý (2008) found a high proportion of the bones were from the hindquarter, particularly the femur, tibia, and ilium at the site of Kutná Hora-Denemark in the Czech Republic. Although many of the bones were burnt, Kyselý suggested that the burning was a natural consequence of humans using fire to 'clean' the site before re-occupation.

Kyselý (2008) proposed five hypotheses that could explain the presence of frogs at Kutná Hora-Denemark; (1) hibernation, (2) predator kills, (3) natural traps, (4) human consumption, or (5) human association with frogs other than food (2008: 145).

We can rule out the first two hypotheses because it has been experimentally shown that a full range of bone elements can be expected if raptors like owls are regurgitating pellets (Dodson and Wexlar 1979; Andrews 1990), while carnivores would not only eat the entire frog, but we would also expect a range of surviving body parts in their scats. The fragile bones would also be affected by acid in the gut, show signs of teeth puncture marks, cracking and would be highly fragmented.

It is possible that Laugerie Basse acted as a natural trap, with the animals attracted to a permanent *sortie d'eau* that runs under the site. However, all frog bones were found in the upper layers, and it seems unlikely that the sediments would form ponds since the *blocs effondrés* and *eboulis calcaire* in the upper deposit are very permeable. Therefore, it is unlikely that water would stay on the surface for any length of time to form a pond.

The specific body part representation would suggest that hypothesis 4 is most likely, that they have been used as food for human consumption (Kyselý 2008). The relatively large number of bones found at Kutná Hora-Denemark suggested to Kyselý that they were gathered systematically during the mating season, around March or April. While the sex was not identified from the bones at Laugerie Basse, the dominance of adult males at Kutná Hora-

Denmark reflected this seasonal collecting activity. While it was difficult to determine the sex from the Laugerie Basse sample, a similar explanation could be argued for the frog bones found here in the Neolithic strata at Laugerie Basse.

Rangifer tarandus (reindeer)

A total of 66 reindeer bones were found in the collection. The majority were from the hindquarter (38%, n=25). They were found mostly in couche 15, dated to before the collapse of the limestone blocks, 16,965±220 [Wk-42653] (Camus 2018: Fig. 5). Antler is present in the lowest layers (n=8; 12.3%), while cranial elements, represented by eight teeth, and forequarter bones are in very low numbers (n=22, 24%). The other notable statistic is the very low number of reindeer axial skeletal elements. Only two ribs were positively identified, and no vertebrae were recorded. This is discussed below. The other notable feature is the over-representation of the hindquarter bones, suggesting a preferential selection of the lower limb bones, probably for marrow processing. Of these, the metatarsal was most common long bone element with 20 (n=30%) with a medium length of 36mm.

Reindeer is the most common animal species hunted during Magdalenian times (Castel *et al.* 2014) and has been shown to provide substantial resources for hunters during the height of the Upper Palaeolithic glacial stage (Castel *et al.* 2002; Castel 2003). The appendicular skeleton, in many instances, seems to be overrepresented in Upper Palaeolithic sites, where systematic breakage and the extraction of marrow from long bones appears to be commonplace.

Marrow forms an important nutritional part of the human diet, especially access to oleic acid, an important 'soft' fat found in the extremities of reindeer (Morin 2007). Oleic acid makes up at least 65% to 73% of the marrow found in the lower limb bones of several animals, including marsupial macropods (Morin 2007: Fig. 7; Garvey 2011). In the absence of vegetable carbohydrates, linoleic and oleic acids assist in metabolising lean red meat that avoids potential human protein poisoning when consumed in large quantities (Meng et al. 1969; Speth and Spielman 1983, Noli and Avery 1988; Cosgrove *et al.* 1990). The quality can also be affected by seasonal changes in ungulates, but less so or minimally in marsupials (Garvey 2011). Thus, consumption of marrow and selection of bone type is important in understanding the composition of archaeological bone assemblages and the behaviour behind it (Cosgrove and Allen 2002; Pike-Tay and Cosgrove 2002; Allen *et al.* 2016). It is significant that the metatarsal is the most numerous reindeer bone in the assemblage. From Morin's (2007) work cited above, it suggests that Laugerie Basse people were targeting the lower extremities of the reindeer to obtain a nutritious source of essential dietary fat.

The other interesting aspect is the lack of reindeer vertebrae and ribs in the Roussot assemblage. With regards to vertebrae, Paola Villa *et al.* (2002: 38) have argued that bone burning for fuel as a substitute for wood was a common practise during the Upper Palaeolithic. At the site of Cuzoul de Vers, most of the small calcined and burnt material was identified as originating from spongy bone such as epiphyses, vertebrae centra, tarsals and carpals. While no large quantities of burnt spongy bone have been recorded in the Roussot sample, the lack of reindeer vertebrae may be explained in several ways. Firstly, despite wet-sieving of the bone, tiny pieces of burnt spongy bone may have been discarded; that selective butchery was carried out away from the Abri des Marseille; or that the fires fed by cancellous bone were placed further out under the overhang and remain undiscovered. These spongy bones may also have been ground up to produce a source of pemmican, apparently a practice in Magdalenian sites in the Paris Basin and Portugal (Audouze and Enloe 1991; Manne 2014). Only two rib fragments were identified in the entire collection, suggesting that specific body part selection was being practiced away from the Marseilles deposit.

In a study of the fluvial deposits below Laugerie Haute and Laugerie Basse rock shelters, Roebroeks *et al.* (2011: 20) found large (n=452) and small mammal (n=538) remains in association with a Magdalenian flint industry. The small mammal fauna was dominated by *Microtus* species (n=403) and the larger category by deer and reindeer (n=153). It is notable that Roebroeks *et al.* (2011) found large quantities of mandibles, crania of reindeer and horse, as well as worked red deer antlers. Considerable numbers of metatarsals and metacarpals were found along with hindquarter and forequarter elements. Significantly, axial elements were rare. The faunal remains, despite being on a lower river deposit, were adjudged in situ with minimal working, with refits of bone and teeth evident (Roebroeks et al. 2011: 23). Their faunal assemblage, both in terms of species and element composition, compares favourably with Laugerie Basse, c. 400m to the southwest, although skulls and mandibles are only minimally represented in the Roussot collection. This comparison suggests that the axial skeleton was being treated differently to the other body parts. In addition, higher numbers (NISP) of other animals such as horse were being hunted and butchered away from Laugerie Basse. However, the processing of long bones resembles the breakage patterns on similar elements recorded at Laugerie Basse. It is also clear that Magdalenian people live outside rock shelters and were not permanent 'cave dwellers'.

Bone industry

Evidence of the manufacture of personal ornamentation was found in the form of a modified distal end of an ungulate incisor (Fig. 3d-e). Roussot may have identified this in square J13, couche 15b, but appears not to have described it further. The proximal end of the tooth has been removed by cross cutting the tooth root from the upper tooth. This craft work has parallels with the Upper Magdalenian site of Peyrazet, where considerable quantities of bead work were recovered (Rigaud *et al.* 2019). Similar cross-cutting marks are prominent on numerous reindeer and chamois teeth that removes the lower part of the root, allowing drilling into the remaining base of the upper part of the tooth to produce beads. This activity appears to have been practised at Laugerie Basse and the marks suggest the tooth was cut while in the jaw, at the gingiva.

There are additional teeth and long bones of this species in the other unexamined squares that may have additional evidence of this craft work. Thus, the remaining faunal material from squares J10, J14, J15, J16 may add a finer level of detail to an understanding of this ornament production.

Conclusion

This study of the small Roussot faunal collection and subsequent geomorphological studies (Camus 2018) has gone some way to improving our understanding of human behaviour, chronology, and site formation process at Laugerie Basse. The questions raised in this study should be addressed by a further interrogation of the remaining lithics and faunal material. Unfortunately, much of the rich archaeological record at Abris des Marseilles has been destroyed by the past indiscriminate removal of this vast archaeological deposit. It may be that the catastrophic cliff face collapse was a blessing in disguise, as it has protected the remaining archaeological deposit from further looting and destruction. The systematic study of relatively smaller archaeological faunal collections has shown its potential in furthering questions about site function and chronological change. This may be the real legacy of Alain Roussot.

Acknowledgements

This analysis evolved out of a larger project between Australian, Dutch and French archaeologists. Our thanks to the Australian Research Council for funding the work under the DP120100580 project "Forty-two degrees of latitude: Comparative Archaeologies of Southwest Tasmania and Southwest France during the last Ice Age". Our team is indebted to Ms. Nathalie Fourment, Regional Curator of Archaeology, Mr. Germinal Peiro, President of the Dordogne Departmental Council, owner of the deposit, Mr. André Barbé, Director of SEMITOUR, operator of the site, Ms. Mathilde Régeard, Head of the Departmental Department of Archaeology, Messrs. Nabil Hamzaoui, as well as Ms. Danielle Robin who gave our team access to Alain Roussot's documents and ground plans. Without them, I would not have been able to carry out this faunal analysis. I also thank Dr. Rachel Wood of the Radiocarbon Analysis Laboratory of the Research School of Earth Sciences at the Australian National University in Canberra, who conducted the Bayesian analysis of the Laugerie-Basse dates. I would also like to thank Dr Jean-Pierre Chadelle for his friendship, his invaluable advice and providing access to the faunal material, Ms Magen O'Farrell whose fieldwork and translation of Roussot's site report was of immeasurable help, Ms. Frances Cosgrove for the database input, as well as Ms. Françoise Demaison-Gorce and her colleagues in Laugerie-Basse for their knowledge of the site and their welcome. Collagen extraction was accomplished by Mr Zak Jones for which I thank him. Mr Wei Ming is acknowledged for redrawing and digitising Roussot's original section illustration. And finally, Mr. Bastien Chadelle for the photogrammetry of the tunnel section, site maps, and Mr Dylan Cosgrove who assisted me greatly with the faunal analysis and with his companionship.

References

Allen, J., R. Cosgrove and J. Garvey. 2016. Optimality models and the food quest in Pleistocene Tasmania. *Journal of Anthropological Archaeology*. Vol 44, Part B: 206-215.

Andrews, P.. 1990. *Owls, Caves and Fossils*. Chicago: University of Chicago Press.

Audouze, F. and J. Enloe. 1991. Subsistence strategies and economy in the Magdalenian of the Paris Basin,

France. In: R.N.E Barton, A.J. Roberts and D.A. Roe (eds.), *The Late Glacial in North-west Europe: Human adaptations and environmental change at the end of the Pleistocene*, Council for British Archaeological Research Report 77: 63-71. London.

Camus, H. 2018. *Context stratigraphique et dynamiques dépôts du gisement de l'abri des Marseilles, Laugerie Basse, Commune des Eyzies-de-Tayac, Dordogne, Campagne 2017*. Unpublished geomorphological report Pro-R-2017-06.

Castel, J-C., L. Despina, V. Laroulande, F-X Chauviere, J-P. Chadelle, A Pike-Tay and J-M. Geneste. 2002. Solutrean animal resource exploitation at Combe Saunière (Dordogne, France). In: M. Maltby (ed.), *ntegrating zooarchaeology : proceedings of the 9th Conference of the International Council of Archaeozoology, Durham, August 2002*: 138-152. Oxford: Oxbow Books.

Castel, J.-C., 2003. Economie de chasse et d'exploitation de l'animal au Cuzoul de Vers (Lot) au Solutréen et au Badegoulien. *Bulletin de la Société Préhistorique Française* 100: 41-65.

Castel, J.-C., M. Boudadi-Maligne, S. Ducasse, C. Renard, F.-X. Chauvière, D. Kuntz and J.-B. Mallye. 2014. Animal exploitation strategies in Eastern Aquitaine (France) during the Last Glacial Maximum. In: F.W.F. Foulds, H.C. Drinkall, A.R. Perri, D.T.G. Clinnick and J. W. P. Walker (eds.), *Wild Things: Recent advances in Palaeolithic and Mesolithic research*:160-174. Oxford: Oxbow Books.

Chadelle, J-P., R. Cosgrove and M. O'Farrell. 2017. Laugerie-Basse, Abri des Marseilles, Rapport de Sondage 2016. Arrêté préfectoral ALPC-AQ-2016-118 Numéro national d'opération: OA 02-6772. Unpublished report to Conseil Départemental de la Dordogne.

Cosgrove, R, J. Allen and B. Marshall. 1990. Palaeoecology and Pleistocene human occupation in South Central Tasmania. *Antiquity* 64: 59 – 78.

Cosgrove, R., J.-P. Chadelle and M. O'Farrell, 2016. Les Eyzies de Tayac, Dordogne Laugerie-Basse, Abri des Marseilles, Rapport de Sondage, Arrêté préfectoral SD-2015-112. Unpublished report to Conseil Départemental de la Dordogne.

Dodson P. and D. Wexlar,.1979. Taphonomic investigation of owl pellets. *Paleobiology* 5: 27 5-284.

Garvey, J.. 2011. Bennett's wallaby (*Macropus rufogriseus*) marrow quality vs quantity: evaluating human decision-making and seasonal occupation in late Pleistocene Tasmania. *Journal of Archaeological Science* 38: 763-783.

Guillaud, E., P. Béarez and J. Primault. 2018. Magdalenian Fishing: Difficulty to identify an highly suspected behaviour. [Présenté par J. Primault]

Guillaud, E., R. Elleboode, K. Mahé and P. Béarez. 2017. Estimation of age, growth and fishing season of a Palaeolithic population of grayling (*Thymallus thymallus*) using scale analysis. *International Journal of Osteoarchaeology* 27: 683-692.

Ferrari, G., A. Cuevas, A. T. Gondek-Wyrozemska, R. Ballantyne, O. Kersten, A. H. Pálsdóttir, I. van der Jagt, A. K. Hufthammer, I. Ystgaard, S. Wickler, G. F. Bigelow, J. Harland, R. Nicholson, D. Orton, B. Clavel, S. Boessenkool, J. H. Barrett and B. Star. 2021. The preservation of ancient DNA in archaeological fish bone. *Journal of Archaeological Science* 126: 105317.

Laville, H.. 1964. Recherches sédimentologiques sur la paléoclimatologie du Würmien récent en Périgord, *L'Anthropologie* 68: 1-48, 219-52.

Laville, H.. 1973. *Climatologie et chronologie du Paléolithique en Périgord: étude sédimentologique de dépôts en grotte et sous abris*. PhD Thesis n° 400, Université de Bordeaux I.

Laville H., 1975. Climatologie et chronologie du Paléolithic en Périgord. Etude sédimentologique de dépôts en grottes et sous abris. Marseille, Éditions du Laboratoire de Paléontologie humaine et de Préhistoire.

Kyselý, R.. 2008. Frogs as a part of the Eneolithic diet. Archaeozoological records from the Czech Republic (Kutná Hora-Denemark site, Řivnáč Culture). *Journal of Archaeological Science* 35: 143-157.

Maury J.. 1924. *Laugerie-Basse. The Excavations of M. J.-A. Le Bel*. Le Mans: Monnoyer. Edition translated by Arthur Edmonds.

Meng, M.S., G.C. West and L. Irving. 1969. Fatty acid composition of caribou bone marrow. *Comparative Biochemistry and Physiology* 30: 187-191.

Morin, E.. 2007. Fat composition and Nunamiut decision-making: a new look at the marrow and bone grease indices. *Journal of Archaeological Science* 34: 69-82.

Manne, T.. 2014. Early Upper Paleolithic bone processing and insights into small-scale storage of fats at Vale Boi, southern Iberia. *Journal of Archaeological Science* 43: 111-123.

Noli, D. and G. Avery. 1988. Protein poisoning and coastal subsistence. *Journal of Archaeological Science* 15: 395-401.

O'Hara J. F., R. White, Z.S. Garrett, T. Higham and A. Roussot.. 2015. The Aurignacian site of the Abri de la Souquette (commune de Sergeac, Dordogne): A history of archeology. In: R. White R. and R.

Bourrillon (eds.) with the collaboration of F. Bon, *Aurignacian Genius: Art, Technology and Society of the First Modern Humans in Europe. Proceedings of the International Symposium, April 08-10 2013, New York University*. *Palethnology*, 7: 98-11.

Pike-Tay, A. and R. Cosgrove. 2002. From reindeer to wallaby: recovering patterns of seasonality, mobility, and prey selection in the Paleolithic Old World. *Journal of Archaeological Method and Theory.* 9: 101-146.

Rigaud, S., S. Costamagno, J.-M. Pétillon, P. Chalard, V. Laroulandie and M. Langlais. 2019. Setlement Dynamic and Beadwork: New Insights on Late Upper Paleolithic Craft Activities. *PaleoAnthropology*: 137-155.

Roebroeks, J. W. M., H. Kamermans, J.A. Mol, A. Turq and M. van Kolfschoten. 2011. Watching the river flow: A small-scale survey of the floodplain deposits in the Vézère valley, between Le Moustier and Les Eyzies (Dordogne, France). *Analecta Praehistorica Leidensia* 41: 1-40.

Roussot A.. 1982. *Abri de la Souquette, commune de Sergeac Dordogne. Rapport de fouilles. Récapitulatif, 1980-1982*, Unpublished administrative report submitted to the Direction des Antiquités préhistoriques d'Aquitaine.

Roussot A.. 1986. Indices Artenaciens à Laugerie-Basse, (Les Eyzies, Dordogne). *Bulletin de la Société Préhistorique Française* 83: 119-124.

Speth, J.D. and K.A. Spielmann. 1983. Energy source, protein metabolism, and hunter-gatherer subsistence strategies. *Journal of Anthropological Archaeology* 2: 1-31.

Valensi, P. and N. Boulbes. 2018. The Magdalenian fauna from Roc-aux-Sorciers, a sculptured rock shelter (Vienne, France): Main archaeozoological results. *Quaternaire* 29(1): 75-80.

Verpoorte, A, R. Cosgrove, R. Wood, F. Petchey, A. Lenoble, J.-P. Chadelle, C. Smith, H. Kamermans, W. Roebroeks. 2018. Improving the chronological framework for Laugerie-Haute Ouest (Dordogne, France). *Journal of Archaeological Science: Reports* 23: 574-582.

Villa, P., F. Bon and J.-C. Castel. 2002. Fuel, fire and fireplaces in the Palaeolithic of western Europe. *The Review of Archaeology* 23: 33-42.

In Dusseldorp, G.L., Chu, W., Bakels, C.C. and M.A. Soressi (eds) 2025. Intent on the Paleolithic. Papers in honour of Prof. Dr. Wil Roebroeks. *Analecta Praehistorica Leidensia* 55. Leiden: Sidestone Press, pp. 85-100.

Zaozer'e – the site of the beginning of the Upper Palaeolithic in northeastern Europe (western foothills of the Middle Urals)

Pavel Pavlov

Abstract

The article presents various data on the Paleolithic site Zaozer'e and its setting. Based on the technological and typological characteristics of the stone artifact assemblage's radiocarbon age and the presence of bone/antler tools and personal ornaments, we propose to assign the Zaozer'e site to the beginning of the Upper Palaeolithic. The Zaozer'e materials evidently demonstrate that modern humans reached the northeast of the East European plain and the Urals (i.e., Subarctic zone of the Eastern Europe) at least 40 ka.

Keywords: Northeastern Europe; Middle Urals; human occupation; beginning of the Upper Palaeolithic

Introduction

The history of the origins and evolution of the Upper Palaeolithic in Eastern Europe and the settling of various climatic zones by modern humans constitutes an important research topic in the archaeology of northern Eurasia of the last decade (see Hublin 2015; Hoffecker 2011). A significant aspect of this problem is determining the chronology and the characteristics of human inhabitation of the regions with extreme climatic conditions, including the Arctic and Subarctic. The data obtained in Northeastern Europe (Pechora and Upper Kama River basins 58-67 °N; 50-60 °E) (Fig.1) – one of the few regions of Northern Eurasia where sites of the beginning of the Upper Palaeolithic are known – are relevant for this research (Kanivets 1976; Pavlov *et al.* 2004).

Two cultural-chronological groups of sites in this region, obviously relating to different episodes of occupation during the first half of the Upper Palaeolithic, are revealed: 43-35 ka and 33-30 ka (Pavlov 2021, 2022). The first group (with the sites of Mamontova Kurya and Zaozer'e) date to the beginning of the Upper Palaeolithic. The second group is represented by the Streletskian site Garchi I and the assemblage of the site Byzovaya (Pavlov 2022: 136-140; Slimak *et al.* 2011).

Among the sites of the first chronological group, the site of Zaozer'e is of significant interest to scholars, as determined by its geographical position, comparatively early age, and unique features of the its assemblage. The assemblage of Mamontova Kurya, the second site in this group, is too small for reliable analysis as only a few stone artefacts were found (Pavlov *et al.* 2004). The primary goal of this paper is a presentation of various data on the site and determining its place among the sites of the beginning of Upper Palaeolithic in Europe and Northern Asia.

Pavel Pavlov
Federal Research Centre, Komi Science Centre Urals Division RAS, Institute of Language, Literature and History, Syktyvkar, Russia
Ppavlov120@gmail.com

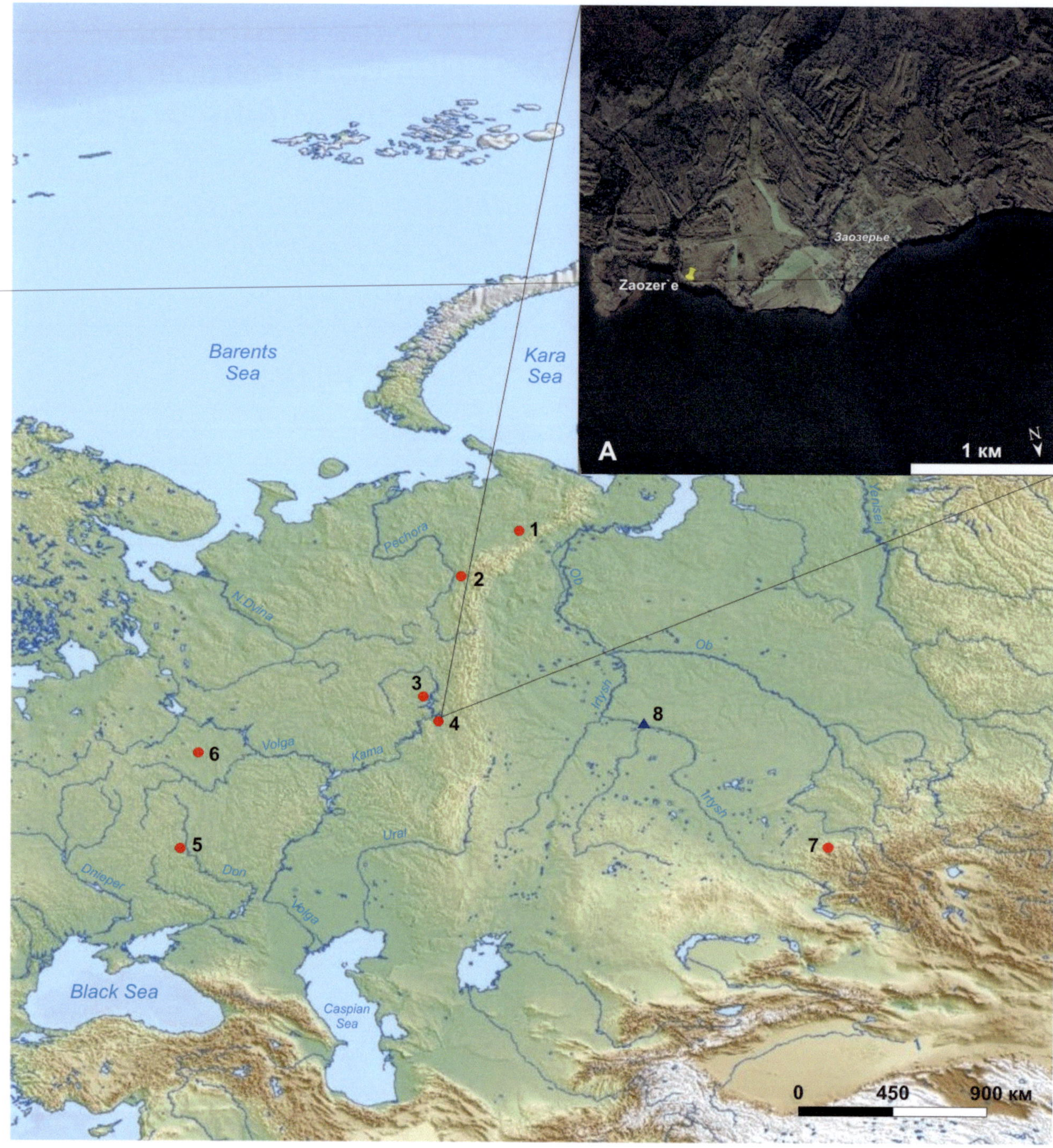

Figure 1. The map of the distribution of the main sites of the beginning of the Upper Palaeolithic and Early Upper Palaeolithic in the East European Plain and in the Altai. 1: Mamontova Kuria; 2: Byzovaya; 3: Garchi I; 4: Zaozer'e; 5: Kostenki; 6: Sungir; 7: Denisova and Kara-Bom; 8: Ust-Ishim Homo sapiens sapiens. A: topographical position of the Zaozer'e site (GoogleEarth image).

Material	Date in years BP	Laboratory number
Horse tooth (AMS)	30 700 ± 400	Poz-5124
Charcoal (AMS)	30 140 ± 190	GrA-22751
Bone (AMS)	33 150 ± 410	Poz-5075
Bone (AMS)	33 450 ± 420	Poz-5076
Bone (AMS)	35 140 + 390/ – 310	(GrA-28187)
Bone (AMS)	33 720 + 310/ – 280	(GrA-28191)
Bone	31 000 ± 700	GIN-11501
Bone	31 000 ± 400	GIN-11499
Bone	31 000 ± 500	GIN-11500
Bone	31 500 ± 500	GIN-11498

Table 1. Radiocarbon dates for bone and charcoal samples from Zaozer'e. The GIN-dates are conventional C14 dates from the laboratory of the Geological Institute of the Russian Academy of Sciences; the others are AMS dates from the laboratories at Groningen, The Netherlands (GrA) and Poznan, Poland (Poz).

The Zaozer'e site

The site situated in the western foothills of the Middle Urals on the Chusovaya River – the left inflow of the Kama River – at approximately 40 km east of the city of Perm in Russia. The site is situated on the left bank of the river (presently the Kama reservoir), at a 58°15' northern latitude, 56°59'eastern longitude, within the present-day southern taiga area. The site is 1,5 km to the west of the small village of Zaozer'e, on the concave side of a large meander of the Chusovaya River on a promontory of its third terrace (Fig.1 A). The height of the terrace is 15-18 meters above the present-day water level of the reservoir, and, more importantly, 27 – 30 meters above the level of the river before the reservoir was created: hence the site once dominated the 1,5 km wide floodplain of the river. The slope of the terrace is covered mostly with coniferous trees with some broad leaf trees (lime) interspersed, the general type of flora for the southern taiga.

Excavations were carried out by the author in 2001-2003, 2005 and 2008 (in 2003 and 2005 together with W. Roebroeks). During these five years the site was investigated over an area of ca. 200 square meters. Surveying work and the collecting of material from the bottom of the terrace allows us to estimate the total extension at ca. 2500 square meters. Of this total area only ca. 10% has been examined up till now.

The geology of the site was studied by a group of scientists headed by A. A. Velichko and Y. N. Gribchenko (Institute of Geography, Russian Academy of Sciences) in the period 2001 – 2003. In 2003 and 2005 additional geological studies were carried out by J. I. Svendsen (University of Bergen, Norway). C14-dating of bone and charcoal samples from the site was performed by L.D. Sulerzhitski (Geological Institute of Russian Academy of Sciences), H. van der Plicht (Groningen University, Netherlands) and T. Goslar (Poznan Radiocarbon Laboratory, Poland). The OSL samples were analyzed at the Nordic Laboratory for Luminescence Dating (Aarhus University) at the Risø National Laboratory (A. Murray), Denmark. Studies of a sample of the bone assemblages recovered were performed by P. A. Kosintsev (Institute of Plant and Animals Ecology Urals Division RAS), S. A. Kuzmina (Zoological Institute RAS) and E. A. Vangengeim (Geological Institute RAS). Palynological analysis of samples from the site's sections was performed by V.V. Pisareva (Geographical Institute RAS). Stable isotopes analysis of horse bones was performed by V.I. Silaev (Geological Institute, Komi Science Center, Urals Division RAS).

Stratigraphy and age

The Zaozer'e find layer deposited in the middle part of the humic bed, is covered by a loess-like sandy loam with a thickness of 2,5 – 3,8 meters (for details see Pavlov 2009: 7-8; Gribchenko 2017). The cultural finds were mostly recovered from a humified horizon with distinguishable fragments of multiple-aged soil formation (Gribchenko 2017: 128-133). Stone implements and animal bones are often confined to gleied lenses, indicative of excessive moistening periods at the time of human habitation. Such formations may be related to buried seasonal-thawing layer dated to the middle Valday (MIS 3) interstadial. After a period of human occupation an active eolian and eolian-diluvial accumulation of loess started that was interrupted at least twice by short periods of surface stability; those episodes were recorded by two levels of ephemeral soil formation and the development of large fault systems (Gribchenko 2017: 132-133). The occurrence of the soil complex in the site sections displays signs of weak downslope movements.

Bone and charcoal samples from the find layer and overlaying deposits (Poz-5124 and GrA-22751) were submitted for radiocarbon dating (AMS and conventional) methods used in three laboratories. The results are shown in Table 1.

The Poznan lab commented that Poz-5124 was on a quite weathered horse tooth, while the GrA-22751 charcoal sample was taken from the upper part of the find bearing unit. The dates indicate that the assemblage was formed in the 31 000 – 35 000 radiocarbon year interval. But AMS data obtained from the samples of horse bones which were taken from one living floor yielded the dates within the considerably narrower chronological interval of the existence of the object: 35.1-33.7 14C kyr BP (41-38 ka). The consistent series of OSL-dates obtained through the excavated sequence at Zaozer'e also supports the assumption that the deposition of the assemblage represents a relatively brief time span. These dates indicate that the find layer has an age in the range 40-35 ka which is in complete agreement with the calibrated radiocarbon ages (Svendsen *et al.* 2010: 3151).

Fauna and Flora

During the excavation approximately 1500 bone fragments were recovered, most of which are well preserved. In the assemblage horse (*Equus ferus*) remains dominant, (MNI: 8). Furthermore rhino (*Coelodonta antiquitatus*) (MNI: 1), reindeer (*Rangifer tarandus*) (MNI: 1), hare (*Lepus* sp.) (MNI: 2) were all present, while mammoth (*Mammuthus primegenius*) (MNI: 1), was only represented by human-modified tusk fragments. Bones from the find layer are heavily fragmented. Some of the horse bones show clear traces of human interference in the form of cut marks. Dense elements of low food value (vertebra, ribs) are under-represented in the inventory of skeletal parts, which apparently reflects selective retrieval of parts (limb and head) from the kill location. This assumption suggests that the place of hunting was in the nearest vicinity of the site. Along with the horse, the common animals which were hunted were small animals such as hares. Bones of this animal were found close to fireplaces within the living areas. Most likely, hares were hunted for the sake of fur, which was used for making warm clothes.

A palynological study of sediment samples performed by V. V. Pisareva indicate that boreal and northern boreal vegetation was typical for the time of the formation of the assemblage as well as for the layers above. Local vegetation included marsh or bog, water, and ruderal groups; a type of vegetation typical for floodplain environments which are turned into terraces by down cutting of the river. The vicinity of the site included partly bogged areas. The results obtained of stable isotope analysis of horse bones indicate that horses at the Zaozer'e site lived in a moderately cold boreal climate with a prolonged snowy winter and relatively short warm summers. The habitat of these animals was the area of transition from grassy woodlands to steppes, where meadow grasses, wild-growing cereals, spruce, pine, and birch grew (Silaev *et al.* 2019: 74-76).

Combining all palaeogeographical data available with the geomorphological setting of the site – i.e., its position deep within the river valley – we can infer that 1) the site was formed during moderately cold climatic conditions, 2) its immediate environment consisted of a forested river valley and 3) the site, in a wider perspective, was situated in an open forest – steppe-like type of landscape.

Distribution of the finds

The find layer is visible as a gray-brown loess-like sandy loam with many small (<3mm) particles of charcoal, spots of red ochre, stone artefacts, and bone fragments. In most of the excavated area, the vertical distribution of finds seemed to consist of two parts: an upper one, 0,25 – 0,35 m in thickness, containing small stone artifacts and tiny bone fragments, and a lower one, 0,03 – 0,05 m (in fact the thickness of the bones and stone implements), with larger bones and stone artifacts, usually in a horizontal position. The clear vertical distribution of the finds, with a separation of finds according to size class, was very probably the result of the freezing and thawing of the matrix of the finds. In some parts of the excavated area cracks of unknown origin (karst?) split up the sedimentary matrix of the finds into separate blocks, which in some cases moved vertically up to 0,50 m away from each other. Deformations of the find bearing strata are also represented by cryogenic deformations (ice-wedges cast) which tear up the

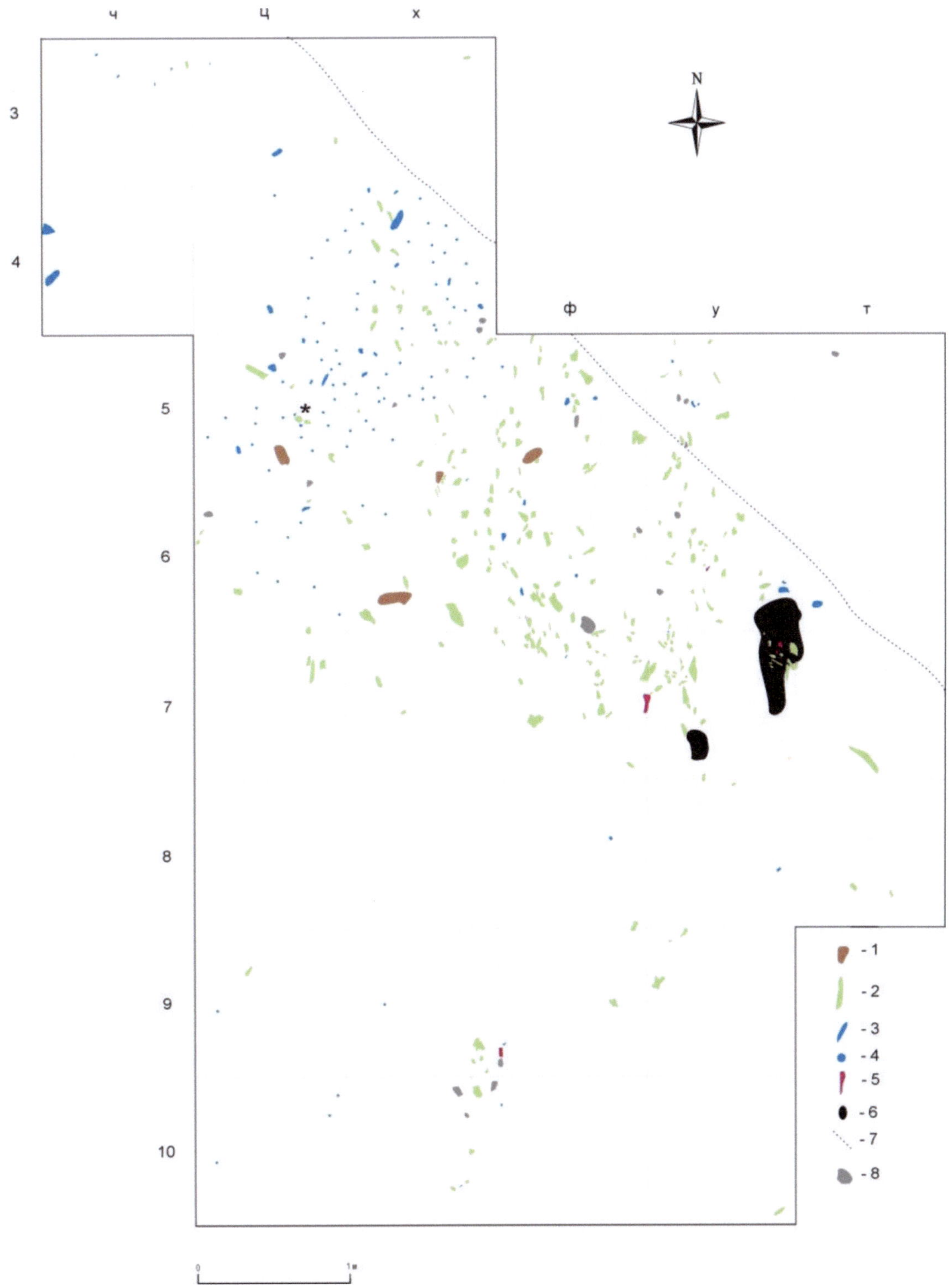

Figure 2. The plan of the cultural layer with living structure on the Zaozer'e site."1: "anvils," large pebbles of quartzite-sandstone; 2: bones; 3,4: lithic pieces more than 20 mm (3) and less than 20 mm (4) in size; 5: bone implement; 6: charcoal spots (several fireplaces or one large hearth); 7: fissure; 8: fragments of siliceous slate with thermal treating; 9: pendant (Svendsen *et al.* 2010).

cultural layer locally. This local cultural layer was also destroyed by run-off. The site is most likely in a primary archaeological context, and no significant reworking of the finds took place after primary deposition.

In spatial terms most of the finds were found in concentrations of different size and composition, some rich in stone artifacts, some in faunal remains, with few finds being present in the areas between the concentrations. Small stone flaking debris (<1 cm) is abundantly present at the site.

At least three find concentrations were recognized within the excavated area, and it appears as if each find category (e.g., bones, specific types of tools, and burnt materials) has a unique and restricted spatial distribution, suggesting some organized use of space during the period of human presence. This main concentration (Fig. 2) displayed three major characteristics: 1) a charcoal concentration at its southeastern boundary, 2) a zone with abundant faunal fragments to the northwest of the charcoal concentration, 3) an area with many stone tools and numerous pieces of thermally treated (?) siliceous slate (Pavlov 2009: 9-10).

The eastern boundary of the region, with abundant faunal remains, was characterized by gray charcoal-stained areas with charcoal and ash remains, tiny bone fragments, ochre particles, and a few very small flint chips. These charcoal concentrations are very probably the remains of former fireplaces.

The central part of the concentration yielded a dense cluster of bone fragments of horse (overwhelmingly prevalent), rhino, reindeer, and hare. This bone rich area yielded many charcoal and ochre fragments, about 40 siliceous slate blocks and fragments with thermal (?) treatment (up to 0,15 m in max. dimension) as well as some twenty stone tools. In this part of the concentration the bone fragments were clearly oriented (not the case elsewhere in the excavated area), with their long axis pointing towards the northwest, i.e., parallel to the former edge of the terrace. At its northwestern margin five quartzite boulders were found, the largest being 0,35 m in maximum dimension. All of them show abundant traces of impact, which suggests that they were used as a kind of anvil. Points and knives, (see below) and numerous (>100) blocks and fragments of siliceous slate were found here, as well as an ovoid pendant carved from a *Unio* shell (see Fig. 6).

One hearth which has been uncovered in one of the partly destroyed living floors deserves our particular interest. The initial size of the hearth was, apparently, not less than 2,00 × 2,00 m. The thickness of the hearth infilling, consisting of wood ash with small charcoal particles, was ca. 0,05 – 0,10 m. In the ash infilling stone artefacts, *Crinoidea* beads, small fragments of polished bone implement, bones that had been sliced and notched, numerous burned bone fragments as well as pieces of red ochre were found. This hearth is remarkable as an artificial construction – containing a ditch with a depth of ca. 0, 10 m, dug on its perimeter, located on the upper part of the ancient slope. In profile the ditch was wedge-like and was filled with an ash layer.

Despite the relatively small excavated area of the site, its overall layout can be set quite definitely. Zaozer'e in general, probably represents the remnants of several small temporary hunting camps with food production (as represented by the meat-bearing fore and hind legs of horses) and processing (the fireplace) activities, where stone and bone tools were produced as well as some personal ornaments. The site could have been visited for several seasons and was situated in a location convenient for hunting.

The stone artifacts

During the 1997, 2001 – 2003, 2005 and 2008 excavations 2.561 stone artifacts were recorded in total (Table 2).

The raw material of the assemblage is heavily composed of comparatively high-quality flint. The main variety of the raw material of the stone inventory (more than 50%) is the brownish flint which probably originated from the outcrops in local Upper Perm limestone cliffs. At the same time however, pebbles of siliceous slate from alluvial deposits were also used to produce artifacts (39%). Whereas quartzite (around 7%), gray-blue Devonian pebble flint (1%) and rock crystal (1%) were used to a considerably lesser degree. The origin of the remaining (11%) of raw material is not identified. Basically, this remainder is made up of a variety of siliceous rocks. Lithic technology is characterized by the prevalence of flat sub-parallel and volumetric debitage.

There are three performs of edge-faceted cores (quartzite and silicious slate) (Fig. 3 no. 1), blanks removed longitudinally on prismatic cores (Fig. 3 no. 2) and crested blades (Fig. 3 no. 4). Radial cores (Fig. 3 no. 5) are also present in the site's assemblage. Among blanks from systematic debitage, large (up to 13 cm) (Fig. 4 nos. 1-3,28) and medium sized blades (Fig. 4 nos 9-13, 22-23) dominate. These blades were detached following a sub-parallel reduction method. There are a few blades

Categories	Excavations 1997, 2001-2003, 2005, 2008 yrs.	Stray finds	Total
Flakes	755 (30,6%)	23	778
Blades and bladelets	154 (6,3%)	6	160
Chips	1151 (45,2%)	-	1151
Chunks	315(12,3%)	2	317
Splitted pebbles	36 (1,2%)	1	37
Cores and fragments	9 (0,3%)	1	10
Burin spalls	5 (0,2%)	-	5
Abrasive	3 (0,1%)	-	3
Retouched tools	97 (3,8%)	3	100
Total	**2525(100%)**	**36**	**2561**

Table 2. A composition of the stone artifacts from the Zaozer'e site.

Categories	Excavations 1997, 2001-2003, 2005, 2008 yrs. and stray finds	%
Retouched blades	26	26%
End-scrapers	24	24%
Points	14	14%
Side scrapers	6	6%
Bifaces and unifaces	6	6%
Burins	9	9%
Divers	15	15%
Total	**100**	**100%**

Table 3. Zaozer'e. Composition of the tool assemblage.

which show that they were struck from prismatic cores with two opposite striking platforms. There are several fragments of the straight narrow bladelets (Fig.4 nos. 5-8), and small curved lamellar chips without retouching. These are most likely by-products of shaping of a thick form of scrapers.

Whereas blades constitute only a small part of the total assemblage (c. 6%, see Table 2), they form, together with elongated flakes, about half of the blanks (47%) used for tool production.

The composition of the tool assemblage is shown in Table 3.

The toolset of the site's assemblage in general consists of two morphological clusters of implements – blades and bifaces. The blade cluster is represented by blades with marginal retouching, points, and burins. An important group is formed by massive blades with scalar and abrupt Aurignacian-like retouching (Fig. 4, nos 2, 13,15, 18-20). Blades with blunted edges occur too (Fig. 4 no. 14).

Points make up considerable part of the assemblage. Implements made on blades with steep retouching are the most common (Fig. 4 no. 30), including a double (?) Aurignacian-like point on a large and thick blade (Fig. 4 no. 28). One noteworthy group is the points on blades with retouching edges – the first convex and the second straight or slightly concave – with flat dorsal retouching on the tip (Fig. 4 no. 29).

Also it should be mentioned the presence of crescent-shaped and arch-backed tools in the Zaozer'e assemblage (Fig. 4 nos 24-26).

Group of burins consist of burins on breaks (Fig. 4 nos 4, 32) and burins on truncation with

Figure 3. The Zaozer'e site: Stone assemblage. 1: preform of edge-faceted core; 2: longitudal scar from prismatic core; 4: crested blade; 5: radial core; 3, 6-14: end-scrapers (Pavlov 2022).

retouched platforms (Fig. 4, nos 21, 27). Some implements with a burin spall in fact could be cores to produce straight narrow bladelets (Fig. 4 nos. 5-8). For this type of core, bladelets are detached from the edge and ventral face of the blade (Fig. 4 no. 27, 32).

The second major cluster of tools consists of bifacial implements, as well as flakes generated by their trimming, which include side scrapers (Fig. 5 nos 1-2) and knives (Fig. 5 nos 3-4). A distinguishing feature of most of the bifaces is their plano-convex retouch with trimming of the ventral face with wide flat scars, from which the dorsal side was retouched. This is the leading technological feature of the Middle Palaeolithic Kielmessergruppen industries (KMG) of

Figure 4. The Zaozer'e site: Stone assemblage. 1,3;9-13;22-23 blades; 5-8 -bladelets 2,14-20 retouched blades; 4; 21 burins; 24-25-lunates (?); 26 arch-backed tool; 27,32 burins-secondary cores (?) on blades; 28-31-points (Pavlov 2022).

Figure 5. The Zaozer'e site: Stone assemblage. 1-2, 5-side-scrapers; 3, 6-11 plano-convex bifaces and unifaces; 4 backed biface – knife.

Eastern Europe. Two smaller bifaces are biconvex (?) in cross-section (Fig. 5 no. 7), probably representing the final stages of reduction of plano-convex bifaces.

Together with the bifaces, a group of unifacial tools occurs (Fig. 5 nos. 8-11), of which only the dorsal face underwent total retouching, with the ventral side only displaying some proximal or distal retouch. With this group the group of end-scrapers with similar morphological elements could be considered. Rather large tools on flakes with ventral trimming are the most prevalent (Fig. 3 nos 6-7,9-10), including thick forms of end-scrapers with a straight edge (Fig. 3 nos 6-7), but the majority of the scrapers have a convex edge with planar retouch (Fig. 3 nos 11-13). Small scrapers made on rock crystal flakes with retouch on one of the longitudinal edges are also present (Fig. 3 no. 14). End scrapers on blades are practically absent: only single tools on the distal part of massive blades (?) are represented in the assemblage (Fig. 3 no. 8). Abrasives made of fine-grained sandstone for processing organic materials also feature in the assemblage (Fig. 6 no. 12). Among the category "divers" we would like to point out the presence of a chopper made on a quartzite pebble, a notched flake and a large cutting tool made of siliceous slate. This category also includes some unfinished bifaces and tool fragments.

Bone tools and ornaments

There are only three bone and antler tools in the assemblage: a blank of a projectile (?), rectangular in its cross-section and made of reindeer antler (Fig. 6 no. 1); a point, and an awl, both produced by diagonally cutting a working edge. The point was made from a horse (?) rib, while the small and fragile awl is composed of hare bone (Fig. 6 no. 3). Furthermore, there are five worked pieces of reindeer antler, cut from either one or both sides (Fig. 6 no. 2). There are several retouchers on the fragments of thick tubular bones.

Personal ornaments are represented by the pendants in oval forms which were cared from the shell of freshwater *Unio pictorum* mollusks (Fig. 6 nos 4-5) and beads made of fossil *Crinoidea* (Fig. 6 nos 7-8). One pendant has two one side drilled holes located near its center. The second pendant is represented by a large fragment with a partly preserved hole. In the assemblage there is also a blank of such a pendant (Fig. 6 no. 9). The assemblage also contains a fragment of a roundish thin bone plate with two cut-through holes (Fig. 6 no. 6). Pendants of this type have never been found in Upper Paleolithic sites before.

The find layer also yielded 11 fragments of mammoth tusks with traces of carving and two small fragments of polished ivory (Fig. 6 nos 10-11) and the fragment of a horse (?) rib dyed with ochre (Fig. 6 no. 13).

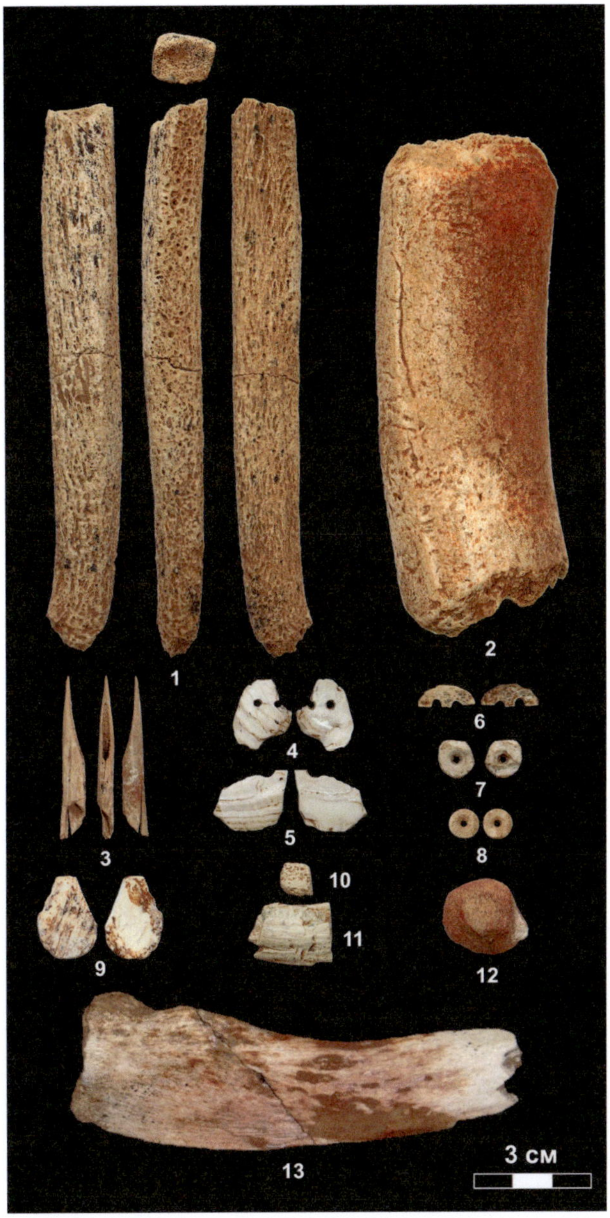

Figure 6. Zaozer'e. Bone and antler inventory and personal ornaments. 1: projectile blank of reindeer antler; 2: worked reindeer antler; 3: an awl; 4-5: pendants made of *Unio* shell; 6: fragment of an ornament made of bone; 7-8: beads made of fossil *Crinoidea*; 9: blank of the pendant; 10: blank of polished bead made of mammoth ivory; 11: worked fragment of mammoth tusk; 12: abrasive of fine-grained sandstone 13: horse rib dyed with ochre.

Discussion

Based on the technological and typological characteristics of the age of the stone artifact assemblage and the presence of bone/antler tools and personal ornaments we propose to assign the Zaozer'e site to the beginning of the Upper Palaeolithic. The main features of Zaozer'e assemblages and the spatial organization of the site itself evidently could be interpreted as requisites of a modern cultural model (Benazzi *et al.* 2020).

The surprisingly early presence of modern humans in the European subarctic has to be studied against the environmental background of the western foothills of the Ural Mountains. It is clear enough that the northward migration routes of the early communities depended heavily on physiographic characteristics of the region and specific features of the relief-forming processes. Here, in a rather small territory, extensive plains, foothills, and low mountains crossed by valleys of the large rivers are combined. This mosaic of landscapes created an especially attractive environment for various species of herbivores. The abundant fauna, the variety and availability of raw materials for stone tool manufacture, as well as abundant water resources, created the necessary conditions for Paleolithic hunter-gatherers (Svendsen *et al.* 2010).

The geographical location of Zaozer'e, practically on the border between Europe and Asia (Fig. 1), renders it necessary to carry out comparison of it with sites of two main neighboring areas where sites of the beginning of the Upper Palaeolithic are known – Eastern European and North Asian.

Sites from the beginning of Upper Palaeolithic (41-36 ka) belong to the ancient chronological group of the Kostenki region in the centre of East European plain (Fig. 1). These are assemblages of Kostenki XVII (layer II) and Kostenki XIV (layers of IVb-IVw and LVA), Kostenki I (layer III) resembling "Protoaurignacien" and early Aurignacien, and local Streletskian assemblages – Kostenki I, (layer V), Kostenki VI, Kostenki XII, (layer Ia and III) (Dinnis *et al.* 2019: 34-35). Zaozer'e's assemblage has practically no similarity with the latter group of assemblages mentioned here.

The site of Zaozer'e has similar traits with regards to the structure of the assemblage and some technical and typological characteristics of stone implements with the sites of the "Aurignacoid" group of Kostenki sites and especially with the assemblages of layer IVb of Kostenki XIV (41-39 ka). But these assemblages differ from Zaozer'e regarding the technology of primary knapping because of the absence of edge-faceted cores and the presence of bipolar splitting technology in Kostenki sites.

Another common and important feature is the presence of personal ornaments in the assemblages of these sites. Personal ornaments of Zaozer'e and sites of the beginning of the Upper Palaeolithic of the Kostenki area also have undoubted similarity. However, unlike its stone assemblage, the personal ornaments at Zaozer'e have significant common traits with ornaments from the second layer of Kostenki XVII. Pendants with drilled holes made of belemnite, flattened pebbles, fossil corals and canines of polar foxes are found in this layer (Stepanova *et al.* 2022).

Sites of the early stages of the Upper Palaeolithic are also known in a neighbouring Asian region – in the Altai region of Russia. Most representative Initial Upper Palaeolithic (IUP) assemblages are known from the Denisova cave (50-40 ka) and Kara-Bom (46-43 kyr BP) (Derevianko *et al.* 2020; Belousova 2018) (Fig. 1). In the industry of the Initial Upper Palaeolithic in the Denisova cave, subprismatic and edge-faceted cores" were created, often intended for the production large blades. Methods of radial and Levallois splitting also were used. Upper Palaeolithic categories are shown in the elongated points, end scrapers, burins, awls, pieces esquilles, and retouched prismatic blades (Derevianko *et al.* 2020).

In the primary knapping of early Kara-Bom assemblages the technology of bipolar splitting absolutely prevails. Characteristic of tools of this tradition are retouched points on blades with a ventral trimming of the base, the simple retouched points on blades and microblades, truncated-facetted implements and the pieces esquilles on blades could be mentioned (Belousova 2018).

An obvious similarity between Zaozer'e and the Altai sites could be traced in personal ornaments. In layers of the beginning of the Upper Palaeolithic in the Denisova cave, the representative assemblage of beads and pendants made of bone, stone and fossil shells were collected (Derevianko *et al.* 2020).

Another aspect that is noteworthy is the obvious similarity of Zaozer'e lithic technology with Western European (for instance Chatelperonian) assemblages of the beginning of Upper Palaeolithic (such as the use of double-platform subprismatic, edge-faceted and radial cores and the leading type of blank, a large blade) and in some categories of the toolset (Djakovic *et al.* 2023).

Of particular interest to scholars is the presence of crescent-shaped and arch-backed tools in the Zaozer'e assemblage. These categories are one of the most distinctive typological features of the units of the beginning of the Upper Palaeolithic or Transitional industries such as Chatelperonian in Central and Southern France and Northern Spain, and Uluzzian in Italy and Greece (Djakovic et al. 2023; Villa et al. 2018). A few assemblages with arch-backed points have been reported from Central and Eastern

Europe. Chronologically, these units are sandwiched between Mousterian and Aurignacian in Western and Southern Europe (Stefanski 2018, 67). Zaozer'e crescent-shaped tools resemble Uluzzian lunates (Villa *et al.* 2018) and arch-backed points like those of Kraków Zwierzyniec 1 (Stefanski 2018: 65-66).

In addition, it should be noted that personal ornaments of the Zaozer'e site, according to classification by M. Vanhaeren and F. D'Errico (2006), refer to the southern group of European Early Upper Palaeolithic ornaments in which use of shells of sea mollusks and fossils is characteristic.

Thus, Zaozer'e has some visible similarity in both lithic technology and the structure of tool kits within industries of the beginning of Upper Palaeolithic in Europe and Northern Asia. The essential similarity is also expressed in the categories of personal ornaments: i.e. the use of shells and fossils for their production.

At the same time with assemblages of the Initial Upper Palaeolithic (IUP) in Europe there is a considerable difference – the presence of a noticeable proportion of implements of Middle Palaeolithic morphology in the Zaozer'e assemblage practically absent in IUP assemblage. The presence of Middle Palaeolithic component also characteristic of the Altai sites of the beginning of Upper Palaeolithic, for example, for the Denisova cave (Derevianko *et al.* 2020). But Middle Palaeolithic components in Eastern European assemblages are represented, as usual, by plano-convex bifaces which are *fossil directeur* of the East European KMG industries (Chabay 2004), on the contrary, in contemporaneous Siberian assemblages Middle Palaeolithic component represented by Levallois forms typical for local Middle Palaeolithic (Derevianko et al. 2020: 15-30). (But recently late KMG assemblages were surprisingly discovered in Altai (Chagyrskaya and Okladnikov caves) and even further to the east in Northern China (Jinsitai Cave) (Kolobova *et al.* 2020; Li *et al.* 2018).

We suggest that Zaozer'e belongs to the one of initial waves of the anatomically modern human occupation of the European continent which began ca. 55 ka (Slimak 2023). Most likely Zaozer'e belongs to the last episode of penetration of the modern human to West Eurasia ca. 45-40 ka. According to Slimak these three phases, from the beginning of the Upper Palaeolithic, can be interpreted as three distinct migratory waves of biologically modern populations (Slimak 2023: 13). Data from the analysis of DNA of the first migrants show that the wide spread of the anatomically modern human in Europe was followed by continuous contacts with the native Neanderthal population (Fu *et al.* 2016, 203-204; Hajdinjak *et al.* 2021: 256; Slimak 2023: 14-15). This fact could serve as a reasonable explanation of the presence of Middle Palaeolithic components in some regional East European industries of the first half of the Upper Palaeolithic, including at Zaozer'e.

Conclusions

On basis of the technological and typological characteristics of the stone artifact assemblage its age and the presence of bone/antler tools and personal ornaments we propose to assign the Zaozer'e site to the beginning of the Upper Palaeolithic. The main features of Zaozer'e assemblages and spatial organization of the site itself evidently could be interpreted as requisites of a modern cultural model. Zaozer'e probably belongs to the one of pioneer waves of modern humans moving into Eastern Europe, expanding from the south of the European continent. The Zaozer'e materials evidently demonstrate that modern humans reached the Northeast of the East European plain and the Urals (i.e., Subarctic zone of the Eastern Europe) at least ca. 40 ka.

Acknowledgements

The 2003 and 2005 excavations of the Zaozer'e site were conducted together with Wil Roebroeks with financial support of the Wenner Gren Foundation for Anthropological research Inc. (New York), ICRG 41 *The Colonization of the Northern World*. I am hearty grateful to Wil Roebroeks for joint fruitful work, constant help, and valuable advice. I am indebted to A. Khazov for his help in the preparation of the illustrations.

References

Belousova N.E.. 2018. *Kamennye industrii nachala verkhnego paleolita Gornogo Altaya (Stone assemblages of the beginning of the Upper Palaeolithic in Altai mountains)*. PhD thesis. Novosibirsk: AEA (In Russ.).

Benazzi S., S. Arrighi, F. Badino, E. Bortolini, C. Figus, F. Lugli, G. Marciani, G. Oxilia, M. Romandini, S. Silvestrini, P. Boscato, A. Cipriani, A. Moroni, F. Negrino, M. Peresani, R. Pini, C. Ravazzi, A. Ronchitelli and E. Spinapolice. 2020. A focus on the Middle to Upper Palaeolithic transition in the Mediterranean area. *Quaternary International* 551: 1-6.

Chabay V. P.. 2004. *Srednij paleolit Kryma. (The Middle Palaeolithic of Crimea).* Simferopol: Shliakh. 2004 (In Russ.).

Derevianko A. P., M. V. Shunkov and M. B. Kozlikin M.B. 2020.. Who were the Denisovians? *Archaeology, Ethnology and Anthropology of Eurasia* 48: 3-32.

Djakovic I., M. Roussel and M. Soressi. 2023. *Prospects and pitfalls of an intrusive model for the Châtelperronian stone tool industry during the Middle to Upper Palaeolithic transition in France and northern Spain.* Preprint. DOI:10.1101/2023.07.14.549013

Dinnis R., A. Bessudnov , N. Reynolds, T. Deviese, A. Pate, M. Sablin, A Sinitsyn and T. Higham. 2019. New data for the Early Upper Palaeolithic of Kostenki (Russia). *Journal of Human Evolution* 127: 21-40.

Fu Q., C. Posth, M. Hajdinjak, M. Petr M, S. Mallick, D. Fernandes, A. Furtwängler, W. Haak, M. Meyer, A. Mittni, B. Nickel, F. Peltzer, R. Rohland, V. Slon, S. Talamo, I. Lazaridis, M. Lipson, I. Mathieson., S. Schiffels, P. Skoglund, A. Derevianko, N. Drozdov, V. Slavinsky, A. Tsybankov R.-G. Cremonesi, F. Mallegni, B. Gély, E. Vacca E, M. González Morales, L.-G. Straus, C. Neugebauer-Maresch, M. Teschler-Nicola, S. Constantin, O.-T. Moldovan, S. Benazzi, M. Peresani, D. Coppola, M. Lari, S. Ricci, A. Ronchitelli, F. Valentin, C. Thevenet, K. Wehrberger , D. Grigorescu, H. Rougier, I. Crevecoeur, D. Flas, P. Semal, M. A. Mannino, C. Cupillard, H. Bocherens, N. Conard, R. Harvati, V. Moiseyev, D.G. Drucker, J. Svoboda, M. P. Richards, D. Caramelli, R. Pinhasi, J. Kelso, N. Patterson, J. Krause, S. Pääbo and D. Reich. 2016. The genetic history of Ice Age Europe. *Nature* 534: 200-205.

Gribchenko Yu. N.. 2017. Environments of the northeastern East European Plain at the time of the initial human colonization. In: V.M.Kotliakov, A.A.Velichko and S.A.Vasil'ev (eds.) *Human Colonization of the Arctic: The Interaction Between Early Migration and the Paleoenvironment*: 105-136. Cambridge: Academic Press.

Hajdinjak M., F. Mafessoni, L. Skov, B. Vernot, A. Hübner, Q. Fu, E. Esse, S. Nagel, B. Nickel, J. Richter, O.-T. Moldovan, S. Constantin, E. Endarova, N. Zahariev, R. Spasov, F. Welker, G.M. Smith, V. Sinet-Mathiot, L. Paskulin, H. Fewlass, S. Talamo, Z. Rezek, S. Sirakova, N. Sirakov , Sh. McPherron, T. Tsanova, J.-J. Hublin, B. Peter, M. Meyer, P. Skoglund, J. Kelso and S. Pääbo. 2021. Initial Upper Palaeolithic humans in Europe had recent Neanderthal ancestry. *Nature* 592: 253-257.

Hoffecker J.. 2011. The early Upper Paleolithic of Eastern Europe reconsidered. *Evolutionary Anthropology* 20: 24-39.

Hublin J.-J. 2015. The modern human colonization of western Eurasia: when and where? *Quaternary Science Reviews* 118: 194-210.

Kanivets V. I.. *Paleolit krainego severo-vostoka Evropy (The Palaeolithic of North-Eastern Europe)* Moscow: Nauka. (In Russ.).

Kolobova K., R. Roberts, V. Chabai, Z. Jacobs, M. Krajcar, A. Shalagina, A. Krivoshapkin, Li Bo, T. Uthmeier, S. Markin, M. Morley, K. O'Gorman, N. Rudaya, S. Talamo, B. Viola and A. Derevianko. 2020. Archaeological evidence for two separate dispersals of Neanderthals into southern Siberia. *PNAS* 117: 2879-2885.

Li F., S. Kuhn, F. Chen, Y. Wang, J. Southon, F. Peng, M. Shan, Ch. Wang, J. Ge, X. Wang, T. Yun and X. Gao. 2018. The easternmost Middle Paleolithic (Mousterian) from Jinsitai Cave, North China. *Journal of Human Evolution* 114: 76-84.

Pavlov P. Yu.. 2009. Zaozer'e: an early Upper Paleolithic site in the European Northeast. *Rossijskaya arkheologia (Russian Archaeology)*: 5-17 (in Russ.)

Pavlov P.. 2021. On the human occupation of the northeast of East European plain and the Urals at the beginning of Upper Palaeolithic (MIS3). *Academia Letters*. Article 2077.

Pavlov P.Yu..2022. The Main Stages and Specific Features of the Initial Peopling of the Northeast of East European Plain and the Urals at the Beginning of the Upper Palaeolithic. *Stratum Plus*: 129-149 (In Russ.)

Pavlov P., W. Roebroeks and J.-I. Svendsen. 2004. The Pleistocene colonization of Northeastern Europe: a report on recent research. *Journal of Human Evolution* 47: 3-17.

Silaev V.I., S.N. Shanina , I. V. Smoleva, D. V. Kiseliova, N. V. Cheredinchenko N.V., A. F. Khazov, E. A. Tumanova and P. Yu. Pavlov. 2019. Opyt ispolzovania mineralogo-geokhimicheskikh svoistv kostnukh ostatkov dlia rekonstruktsii sredy obitania na paleoliticheskoj stoyanke Zaoze'e (Srednij Ural). (Testing the use of mineralogical and geochemical characteristics of fossil bone remains as a tool for reconstructing palaeolandscapes at the Paleolithic site of Zaozer'e (Middle Urals). *Pervobytnaya arkheoogia. Journal mezhdisciplinarnukh issledovanij (Prehistoric Archaeology. Journal of Interdisciplinary Research)* 1: 35-77. (In Russ.)

Slimak L. 2023. The three waves: rethinking the structure of the first Upper Paleolithic in Western Eurasia. *PLoS ONE* 18(5): e0277444.

Slimak L., Svendsen J-I., Mangerud J., Plisson H., Heggen H., Brugere A., Pavlov and P.Yu. 2011. "Late Mousterian Persistence near the Arctic Circle. *Science* 332: 841-845.

Stefanski D.. 2018. The Early Upper Palaeolithic assemblage of arch-backed points from Kraków-Zwierzyniec site 1. In: Valde-Nowak P., Sobczyk K., Nowak M., Źrałka J. (eds.), *Multas per gentes et multa per saecula. Amici magistro et collegae suo Ioanni Christopho Kozłowski dedicant*: 63-70. Krakow: Jagellonian University Press.

Stepanova K.N., A. A. Maliutina, A. A. Bessudnov and E. Yu. Girya . 2022. Personal ornaments from Kostenki 17, Layer II: Manufacturing, usage and cultural context within the Initial Upper Palaeolithic of Eastern Europe. *Stratum Plus* (1): 193-220. (In Russ.)

Svendsen J-I., H. P. Heggen, A.-K. Hufthammer, J. Mangerud, P. Pavlov and W. Roebroeks. 2010. Geo-archaeological investigations of Palaeolithic sites along the Ural Mountains – on the northern presence of humans during the last Ice Age. *Quaternary Science Reviews:* 29: 3138-3156.

Vanhaeren M. and F. D'Errico. 2006. Aurignacian ethno-linguistic geography of Europe revealed by personal ornaments. *Journal of Archaeological Science* 34: 1-24.

Villa P., L. Pollarolo, J. Conforti, F. Marra, C. Biagioni, I. Degano, J. J. Lucejko, C. Tozzi, M. Pennacchioni., G. Zanchetta, C. Nicosia, M. Martini, F. Sibilia and L. Panzeri. 2018. From Neandertals to modern humans: new data on the Uluzzian. *PLoS ONE* 13: e0196786.

In Dusseldorp, G.L., Chu, W., Bakels, C.C. and M.A. Soressi (eds) 2025. Intent on the Paleolithic. Papers in honour of Prof. Dr. Wil Roebroeks. *Analecta Praehistorica Leidensia* 55. Leiden: Sidestone Press, pp. 101-114.

L'apport de l'analyse lithologique et technologique des bifaces dans l'analyse des comportements des derniers neandertaliens

A. Turq, J. Ph. Faivre

Abstract

Les progrès réalisés ces dernières années dans les domaines de la pétro-archéologie et de la technologie lithique ont permis à la recherche préhistorique de progresser dans la connaissance des relations complexes établies entre les stratégies d'acquisition des géo-ressources et les traditions techniques à l'œuvre dans le cadre de la conception et de la gestion des outillages lithiques. Parmi ces outils figurent notamment les bifaces et outils bifaciaux du Paléolithique moyen récent dont le caractère polymorphe au sein des industries moustériennes relève d'une segmentation plus ou moins forte des étapes de production et de gestion mettant tout autant en lumière leur fort potentiel de réduction, d'utilisation et de mobilité.

Portant sur plus de 2300 bifaces, provenant tant de fouilles (moins de 5%) que de récoltes de surface et issus d'un espace géographique de 10 000 km^2 aux géo-ressources bien identifiées et caractérisées, cette étude confirme la complexité des stratégies techno-économiques appliquées à la gestion des outils bifaciaux dans le cadre de la mobilité humaine, mettant une nouvelle fois en lumière la parfaite connaissance des ressources minérales et des contraintes environnementales par les néandertaliens.

Keywords: Paléolithique moyen; Néandertal; géo-ressources; bifaces; techno-économie; Bassin aquitain; mobilité humaine; gestions de l'outillage et gestion des territoires

A. Turq
Chercheur associé à l'UMR 5199 – PACEA, Université de Bordeaux, conservateur du Musée de Sauveterre-la-Lémance, Lot-et-Garonne, France.

J. Ph. Faivre
Chargé de recherche au CNRS, UMR 5199 – PACEA, Université de Bordeaux, Bâtiment B2, allée Geoffroy St-Hilaire, 33615 PESSAC CEDEX.

Introduction

Depuis son identification comme outil taillé par la main de l'homme[1], le biface, anciennement qualifié de hache antédiluvienne (Boucher de Perthes 1857), de coup de poing, etc., n'a cessé d'intriguer et d'être questionné. À la suite de premiers essais classificatoires (Lartet and Christy 1875; Commont 1906), F. Bordes dressa une typologie des bifaces et proposa d'utiliser cette catégorisation comme critère de distinction chrono-culturelle entre le Paléolithique inférieur et moyen (Bordes 1953). Parallèlement, la taille expérimentale (Coutier 1929;

1 John Frere, dans un article de deux pages publié en 1800 dans la société des antiquaires de Londres, décrit des bifaces découverts, aux côtés d'ossements d'animaux, dans une carrière de Hoxne comme des objets« fabriqués et utilisés par des hommes qui n'avaient pas usage du métal ».

Bordes 1947; Tixier 2012) a ouvert la voie à l'approche technologique de ces outils en participant activement à la reconnaissance et à la définition des techniques et des méthodes impliquées dans leur confection. Plus récemment se sont développées les approches fonctionnelles (Anderson-Gerfaud 1981; Anderson-Gerfaud et Helmer 1987; Beyries 1984; Claud 2008, 2012; Brenet *et al.* 2015; Sorensen *et al.* 2018) documentant les modes d'utilisation ; ou encore lithologiques (Dibble 1984; Geneste 1985, 1990; Boëda 1986, 2013; Boëda *et al.* 1990; Bourguignon *et al.* 2006; Geneste *et al.* 1990; Turq 2000, 2001; Soressi 2002; Gravina et Discamps 2015), éclairant les stratégies de gestion techno-économiques développées sur ces outils, et notamment leur inscription dans la dynamique de mobilité humaine au cours du Paléolithique moyen récent (Moustérien) en Europe de l'ouest.

Pour cette chronologie et cette géographie, les populations néandertaliennes sont décrites comme extrêmement mobiles, déplaçant leurs campements suivant des modalités variées (Binford 1980 ; Gamble et Roebroeks 1999) en lien avec différents registres d'activités et traduisant, par là même, leur formidable capacité d'adaptation à des environnements distincts et des ressources diverses (Roebroeks *et al.* 1988; Féblot-Augustins 1993, 1997, 1999; Gamble 1993, 1999). Ces schémas de mobilités s'inscrivent sur des distances plus ou moins importantes dont les amplitudes et les trajectoires sont perçues à travers l'origine et la circulation des ressources minérales impliquées dans la composition des industries lithiques.

Au Paléolithique moyen, la tendance largement établie est à l'approvisionnement local en matières premières lithiques (90 % et plus), c'est-à-dire à moins de 5 km des lieux d'occupations où les produits issus de leur exploitation sont abandonnés (e.g. Geneste 1985; Turq 2000). Cependant, ces composantes principales coexistent le plus souvent avec des échantillons de matériaux plus distants constituant de véritables traceurs de la mobilité humaine (e.g. op. cit.). Qu'il s'agisse des matériaux locaux ou des matériaux exogènes, leur mode de gestion relève le plus souvent d'une segmentation spatiale et temporelle plus ou moins importante des chaînes opératoires de production (Turq *et al.* 2013a). Cette tendance techno-économique est globale et touche aux différents registres techniques de la production lithique au Paléolithique moyen, qu'ils s'agissent des méthodes de débitage (Turq *et al.* 2013a, 2013b) ou de façonnage (Faivre 2006).

Pour ces dernières, le biface, outil emblématique de la Préhistoire ancienne, figure parmi les objets dont le potentiel informatif, concernant les relations des groupes humains aux territoires qu'ils ont occupés et exploités, est reconnu de longue date (Bindford 1969, 1979; Geneste 1985). Jusqu'alors, ce potentiel fut principalement abordé à travers l'étude des seules pièces issues de fouilles programmées ou préventives qui ne constituent qu'un échantillon des ressources documentaires existantes en comparaison des corpus, bien plus nombreux, de bifaces provenant de ramassages effectués sur des stations de surface non fouillées et conservés dans des collections publiques ou privées. Cette masse de données s'avère ainsi méconnue (voire inconnue), et donc largement ignorée par la recherche préhistorique.

Ce constat est, sans doute, étroitement lié aux orientations méthodologiques de l'archéologie préhistorique accordant aux questions de contextualisation et d'intégrité un caractère central, les séries issues de ramassages « hors contextes » ne réunissant dès lors pas les niveaux de résolution archéologique requis. Difficile de ne pas souscrire à ce constat. Tout en restant critique et lucide quant à la nature des séries issues de ramassages, il est pourtant possible d'en cibler certains registres de données. Parmi eux figure en priorité celui des matières premières lithiques et de leurs origines qui sont susceptibles de documenter des tendances techno-économiques à des échelles géographiques jusqu'alors plus ou moins perçues, voire non perçues, à partir des séries issues des sites fouillés. Ce sont autant de données offrant une vision beaucoup plus large de la diversité des espaces lithologiques fréquentés, des degrés d'exploitation et de sélection de leurs géo-ressources, et bien sûr de leur diffusion via la mobilité des groupes humains.

Partant de ce constat, notre démarche d'étude s'est focalisée sur le nord-est du Bassin aquitain (Fig. 1), région marquée par l'existence de plusieurs espaces lithologiques aux géo-ressources diversifiées. Sa réalisation a impliqué, au préalable, un long et fastidieux travail de recherche et d'examen de collections de surfaces livrant des bifaces, tant dans les musées que chez les collectionneurs privés.

Corpus et méthode

Au total, près de 8000 bifaces de tout type (Paléolithique inférieur et Paléolithique moyen) ont été examinés et leurs lieux de découverte cartographiés (Fig. 2). Au sein de ce large corpus,

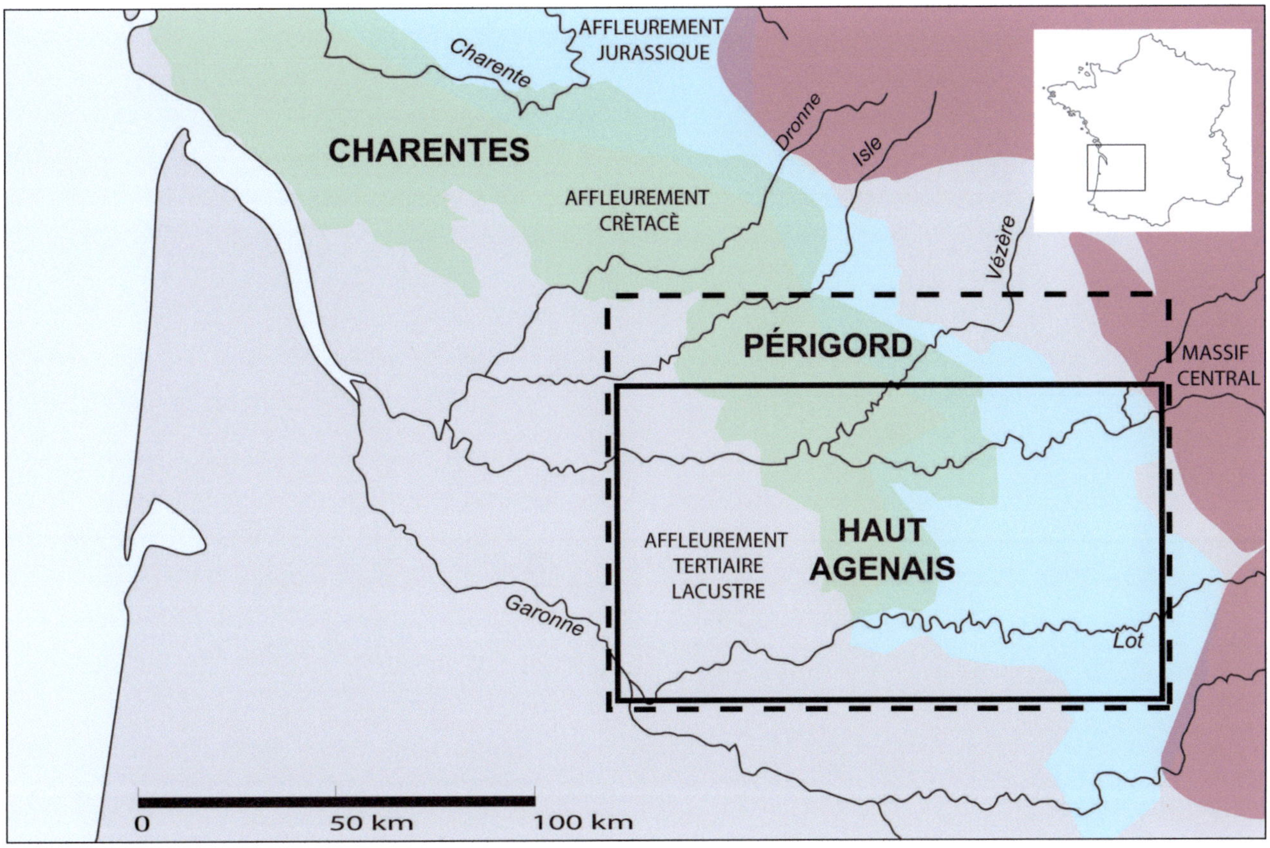

Figure 1. Géographie de la zone d'étude; les pointillés indiquent les limites des figures 2, 3 et 4.

un tri techno-typologique a ciblé les pièces caractéristiques du Moustérien récent (ou MTA; Boëda 2001, 2013; Soressi, 2002; Boëda *et al.* 1990; Faivre *et al.* 2017a, 2017b; Turq *et al.* 2016): des outils de sections biconvexes ou plano-convexes, plus ou moins réaffûtés, pouvant avoir été recyclés en nucléus, avoir servi de support à différents types d'outils (racloirs, becs, etc.; Boëda 2001; Turq 2001; Soressi 2002), ou encore avoir servi de briquet (Sorensen *et al.* 2014) ou de tablette de travail (Turq *et al.* 2016: 67), tout cela sans que ces utilisations n'altèrent la fonctionnalité des bords.

Un échantillon de 2361 bifaces « moustériens » a ainsi été constitué et a fait l'objet d'une caractérisation fine des matières premières permettant d'isoler 755 objets confectionnés à partir de matériaux dont les origines pétrographiques sont exogènes de l'espace lithologique environnant leur zone ou lieu de découverte. C'est ce second volet que nous aborderons ici.

Le nord-est aquitain: une zone d'étude privilégiée

Le nord-est du Bassin aquitain constitue une zone particulièrement riche en sites archéologiques attribués au Paléolithique moyen (Lartet Christy 1875; Peyrony 1930, 1934; Bordes 1948, 1954, 1955, 1972, 1984; Bourgon 1957; Guichard 1976; Le Tensorer 1979; Geneste 1985; Rigaud 1982, 1988; Le Brun Ricalens 1988; Dibble et Lenoir 1995; Turq 2000: Turq *et al.* 2008, 2011; Bourguignon *et al.* 2004; Delagnes *et al.* 2007; Jaubert 2011; Faivre *et al.* 2017a-b; Sebastianutti 2020; Turq *et al.* 2022), ayant fait l'objet de nombreuses fouilles étalées sur plus d'un siècle, et dont les ressources lithologiques sont bien connues à travers les nombreux travaux engagés depuis les années 1980 (Demars 1980; Chadelle 1983; Morala 1983; Larick 1983; Geneste 1985; Seronie-Vivien 1987; Lebrun Ricalens 1988; Turq 2000; Fernandez 2012; Turq *et al.* 2013a). Cette richesse documentaire en fait donc une zone privilégiée pour

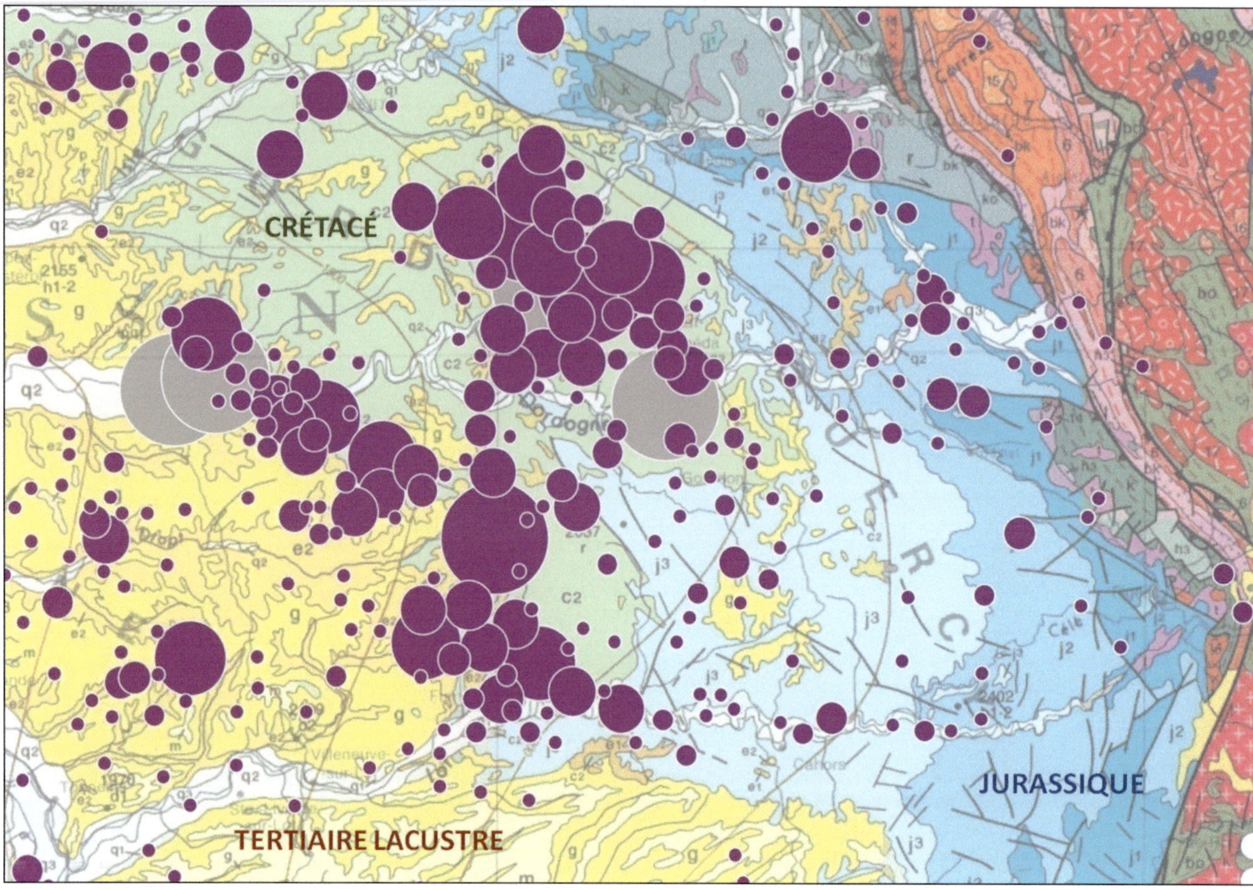

Figure 2. Carte de répartition des bifaces établie à partir d'un corpus de près de 8 000 bifaces de tout type découverts dans le nord-est du Bassin Aquitain: la dimension des cercles est proportionnelle au nombre de biface découvert sur le territoire d'une commune (de 1 à 200).

l'étude des stratégies techno-économiques développées par l'homme de Néandertal.

Ces dernières années deux synthèses ont été publiées sur les espaces lithologiques nord-est aquitains. Basés sur l'étude de plusieurs contextes archéologiques (en grotte, abri ou plein air) ayant livré des séries lithiques attribuées aux techno-complexes Levallois, Quina et Discoïde (Turq *et al.* 2016, 2013b), ces travaux ont montré que (Fig. 3):

- les affleurements des calcaires crétacés sont riches en grottes, abris, sources de matières premières lithiques, ressources en eau, et concentrent la plupart des grands sites moustériens autour de deux pôles, l'un charentais, l'autre périgourdin ;
- pour ce dernier pôle, le territoire au Paléolithique moyen récent s'organise autour d'une « zone préférentielle de résidence », définie par la concentration des principales archéo-séquences régionales (Jaubert 2011) et des sites ayant livré des squelettes néandertaliens sub-complets (Maureille *et al.* 2016). Elle recouvre une partie d'un vaste espace lithologique, que nous qualifierons de « Sénonien du Périgord », caractérisé par la présence de formations livrant des silex gris-noirs, blonds et de type Belvès. En périphérie de cet ensemble se trouve l'espace lithologique du bergeracois, celui des molasses tertiaires qui livrent des silcrètes-meulières[2], celui du sud Périgord-Haut-Agenais qui livre essentiellement des silex blonds mais aussi le silex du Fumélois et de Gavaudun, l'espace lithologique Quercynois (silex

2 Que nous appelions par le passé improprement silex calcédonieux ou lacustre tertiaire.

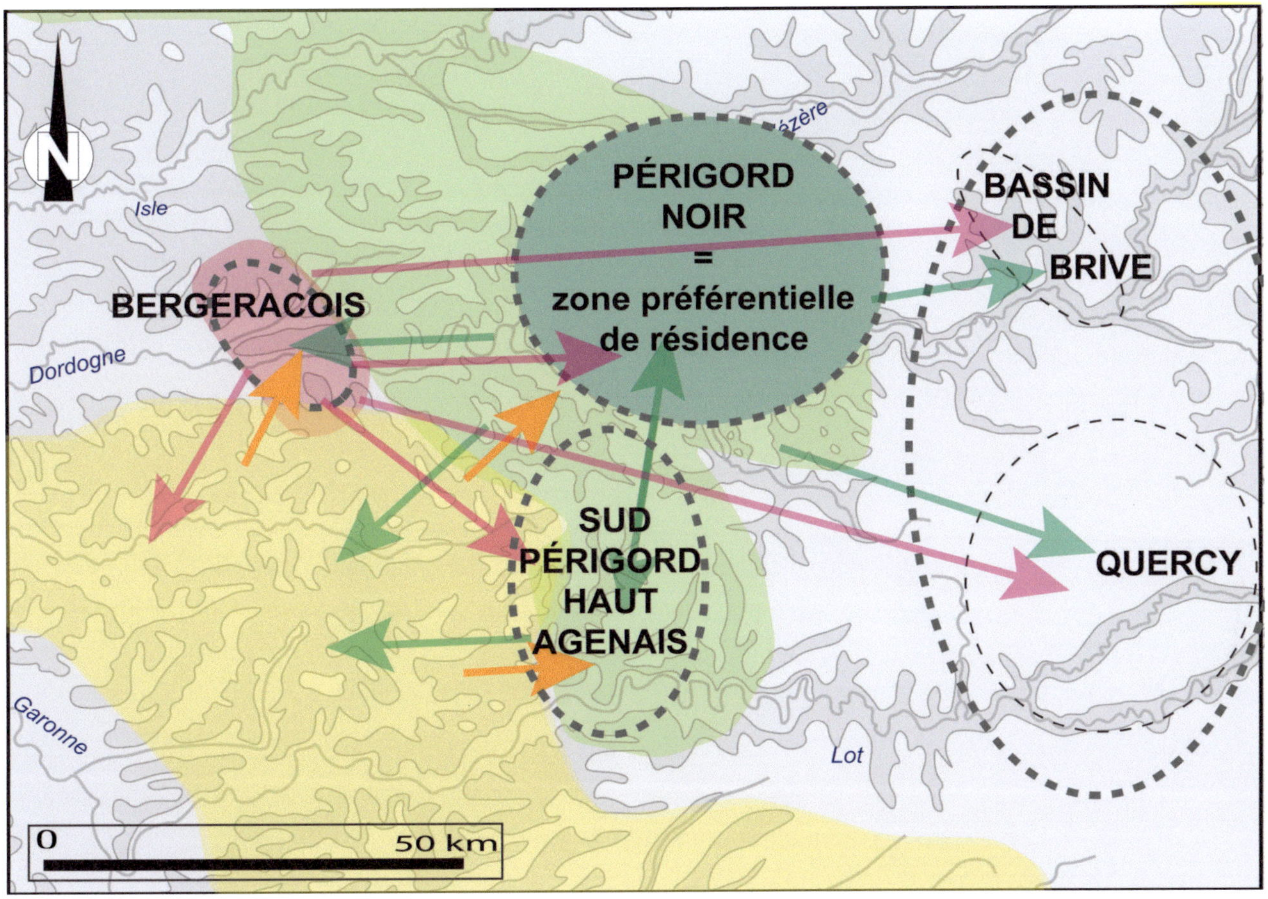

Figure 3. Contacts mis en évidence entre les divers espaces lithologiques du nord-est du Bassin aquitain (d'après Turq *et al.* 2016, 2017).

jurassiques), et enfin celui du Bassin de Brive (silex jaspéroïde);

- pour chacun des espaces lithologiques, l'approvisionnement en matières premières lithiques est toujours local (entre 95 à 100%), quelles que soient leurs qualités, la dimension des blocs ou leur abondance. Là où les affleurements de silex sont plus rares, comme c'est le cas en Quercy ou dans le Bassin de Brive, d'autres matériaux locaux, comme les métaquartzites, sont principalement exploités;
- les relations entre les espaces lithologiques reposent donc sur un nombre assez limité d'objets (de 1% à 15% des assemblages lithiques), mais permettent malgré tout de proposer une hypoth**èse d'organisation territoriale**: de l'ouest au sud de la « zone préférentielle de résidence » existent des mouvements de matériaux plus ou moins importants entre les différents espaces, ceci à l'exception de l'axe Haut-Agenais/Bergeracois. Les principaux mouvements de matériaux concernent le Bergeracois, zone fortement occupée, puis de manière décroissante le sud-Périgord et les terrains molassiques. Si tous ces espaces témoignent de relations géographiques s'établissant à partir de mouvements bidirectionnels des matériaux, le Quercy et le Bassin de Brive témoignent quant à eux de mouvements unidirectionnels, d'ouest en est, signant une absence de transfert des géoressources depuis ces espaces orientaux.

Le biface: traceur de mobilité humaine

Les résultats obtenus (Fig. 4; Sebastianutti 2020; Turq *et al.* 2022) n'ont fait que conforter les observations découlant de l'étude des séries régionales attribuées à différents techno-complexes lithiques (Levallois, Quina et Discoïde (cf. Figure 3): les mouvements de bifaces mettent en relation le Périgord avec les différents espaces

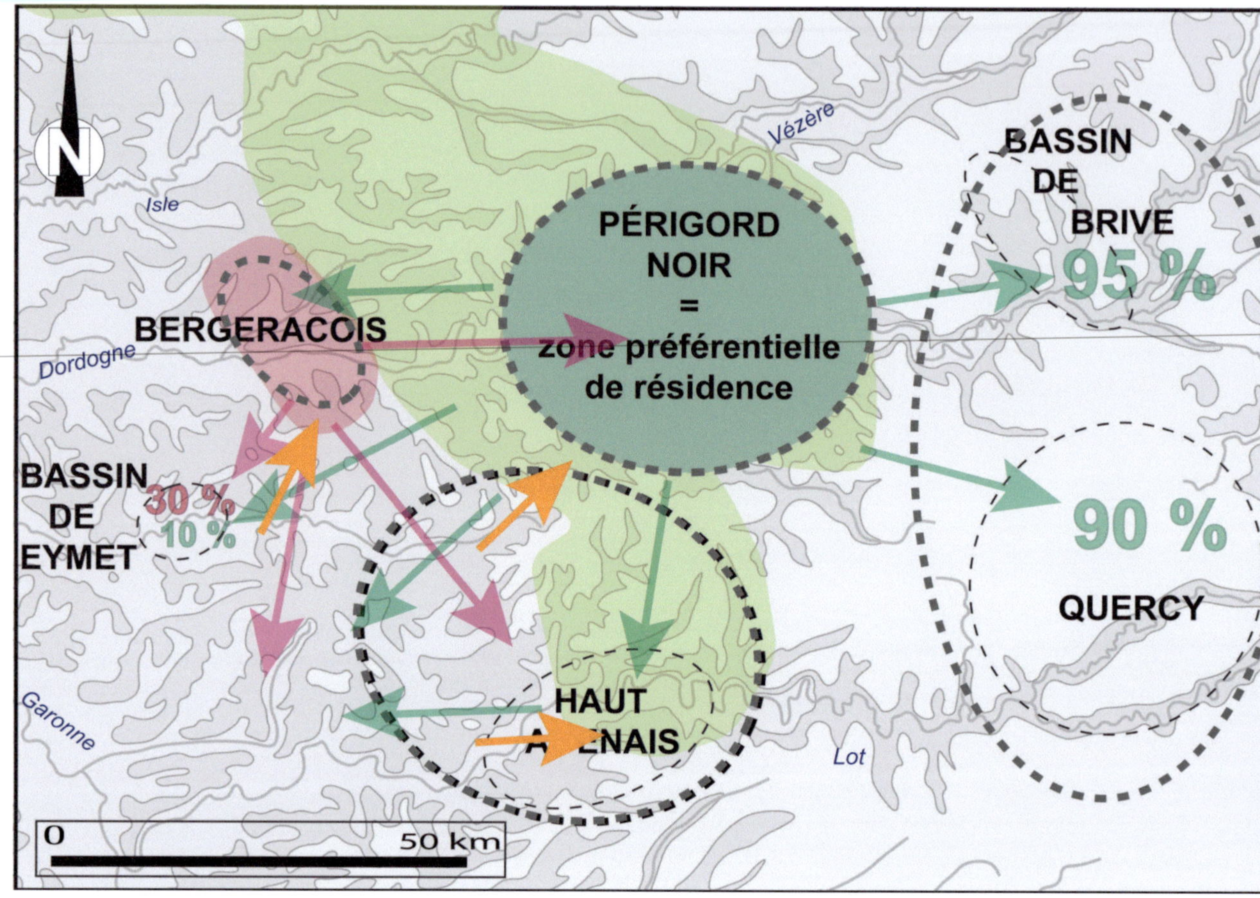

Figure 4. Mouvement des bifaces entre les différents espaces lithologiques du nord-est du Bassin aquitain (d'après Turq *et al.* 2016, 2017).

lithologiques l'environnant. Les trajectoires de matériaux sont bidirectionnelles à l'exception, une nouvelle fois, du Quercy qui ne constitue pas un espace lithologique émetteur de matériaux (trajectoire unidirectionnelle depuis le Périgord).

Les bifaces: quels états d'abandon?

La poursuite des recherches dans les domaines de la caractérisation des matières premières lithiques, des analyses technologiques et de l'enrichissement du corpus nous ont amenés à nous poser la question suivante: sous quelles formes les bifaces sont-ils abandonnés?

Pour tenter d'y répondre, notre attention s'est focalisée sur:

- trois espaces lithologiques: les molasses tertiaires, les formations sénoniennes du Périgord et celles du Haut-Agenais qui regroupent les plus fortes concentrations de bifaces (Fig. 5). Ce choix exclu les espaces lithologiques de la partie est qui n'ont enregistré que des déplacements ouest-est;
- les bifaces du moustérien récent trouvés en dehors de la zone d'affleurement du matériau dans lequel ils ont été fabriqués (tableau 1);
- les matériaux les mieux définis que sont les silex du Bergeracois, de « Belvès », les silex gris et noirs du Sénonien (qui n'existent qu'au nord de la ride anticlinale de Beaumont du Périgord). N'ayant pas les moyens de distinguer les silex blonds du Périgord de ceux du Haut-Agenais, ce matériau n'a pas été retenu.

Ainsi, sur un effectif de 2361 bifaces examinés et regroupant des caractéristiques techno-typologiques de facture « moustérienne », un échantillon de 755 pièces a été retenu (tableau 1).

ZONE	Bergeracois	GN du S	Belvés	TOTAL
A	13/275 (4.8)	10/275 (3.6%)	1/275 (0.4%)	24 /275 (8.7%)
B	9/252 (3.6%)	local	4/252 (1.6%)	13/252 (5.2%)
C	2/195 (1%)	5/195 (2.6%)	7/195 (3.6%)	14/195 (7.2%)
Total des zones	24/722 (3.3%)	15/722 (2.2%)	12/722 (1.7%)	51/722 (7.1%)
Hors zone mais figurés	6	11	16	33

Tableau 1. Nombre de bifaces par zone et par matériau.

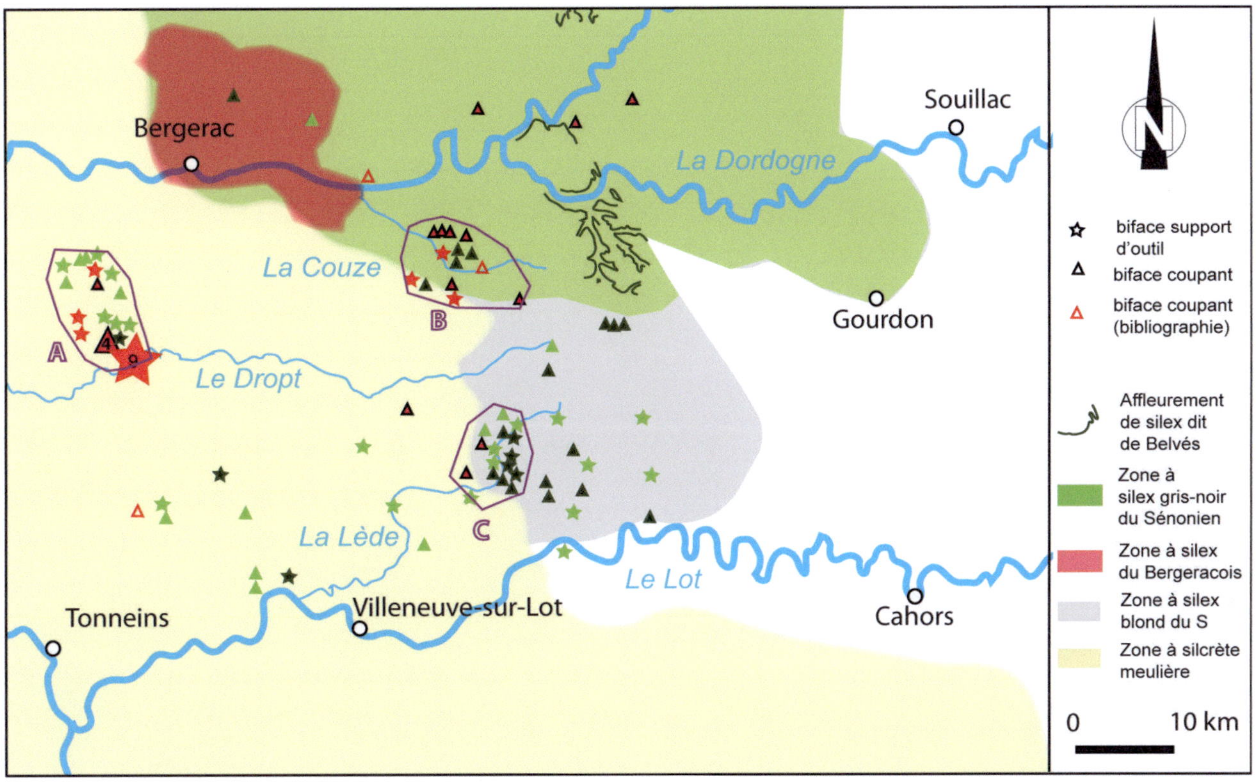

Figure 5. Carte de répartition des bifaces en matériaux exogènes. Les zones cernées de violet sont: A, la région d'Eymet; B, la moyenne vallée de la Couze ; C la vallée de la Lède entre Lacapelle et Salle-sur-Lède.

Résultats

Au sein de l'espace lithologique du Sénonien du Périgord, quelques rares bifaces (0,2 %) sont en matériaux exogènes (silex du Bergeracois). Entre Dordogne et Lot, sur les terrains crétacés et tertiaire, la proportion de bifaces exogènes (confectionnés sur des silex du Bergeracois, des silex gris et noirs du Périgord et de Belvès) est bien plus importante (7,1 %), mais très faible au regard des 80 % de bifaces exogènes du Quercy.

La figure 5 met également en lumière trois zones de concentration qui correspondent exclusivement à des ramassages de plein air, à des pièces isolées ou issues de stations riches en matériel: sur le Dropt et la région au nord d'Eymet (A); sur la Couze, entre Nojals-et-Clottes et Saint Avit Sénieur (B) et sur la Lède, entre Salle et Lacapelle-Biron (C). Deux des concentrations sont situées à la limite des espaces lithologiques du Sénonien du Périgord et du Haut-Agenais (concentrations B et C).

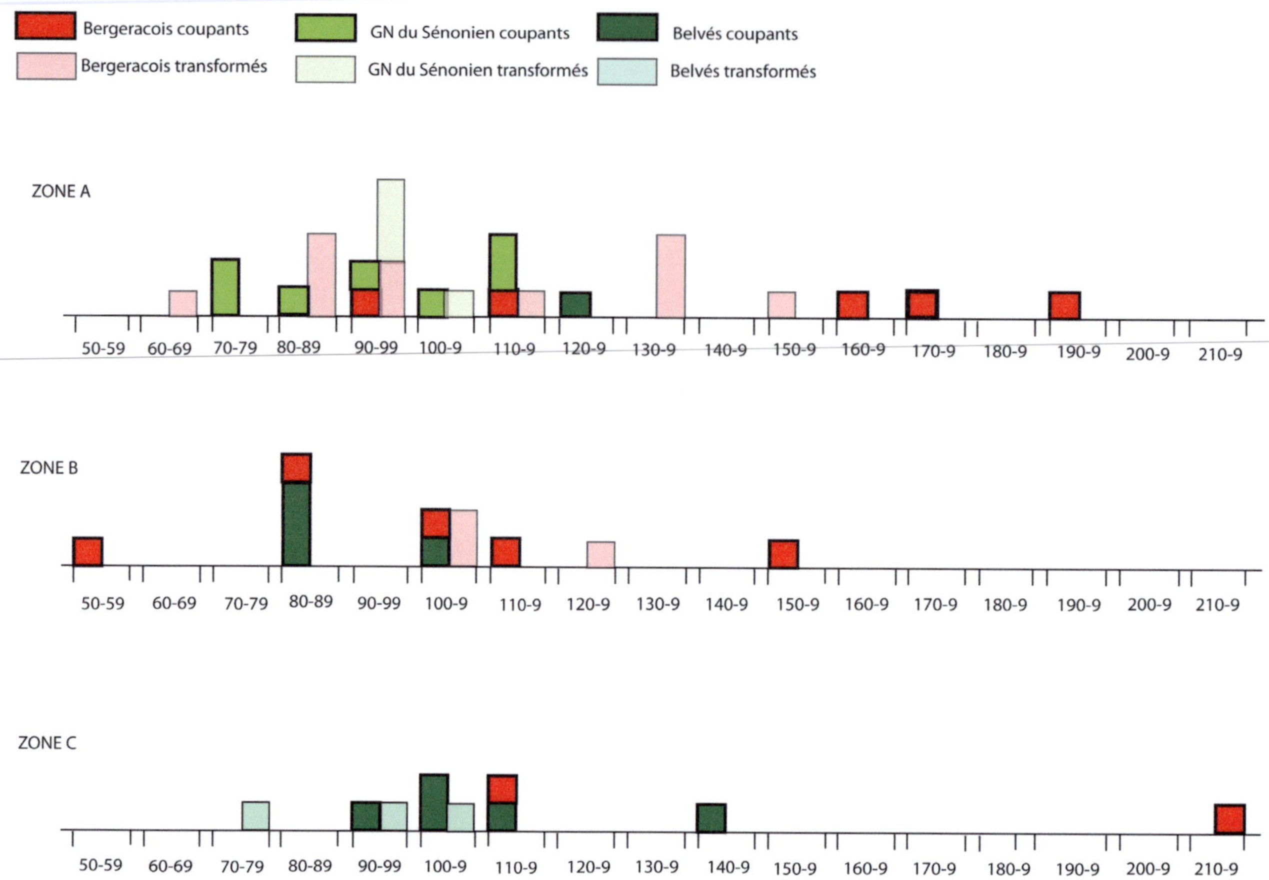

Figure 6. Classes de longueur et état des bifaces exogènes lors de leur abandon dans les trois zones étudiées.

Ces concentrations se caractérisent par la présence de bifaces faits aux dépens d'au moins deux matériaux extérieurs au litho-espace et qui représentent moins de 10% de l'ensemble des outils (A = 8.7%, B = 5.2%, C=7.2%).

La zone A

Elle fait partie du litho-espace des terrains molassiques qui livrent des silcrètes-meulières parfois abondantes mais de qualité variable. Elle correspond au bassin de l'Escourou et à celui du Dropt autour d'Eymet. Ici il n'y a pas ou peu d'abris mais des fonds de vallée à tendance marécageuse qui constituaient sans doute, à certains moments, des zones de chasse particulièrement favorables. Les chasseurs-cueilleurs néandertaliens ont majoritairement utilisé les silcrètes-meulières locales matérialisées par des bifaces de dimensions variables mais le plus souvent abandonnés lorsqu'ils n'étaient plus coupants (non réaffûtés). Quelques pièces (cinq au total dont trois de grandes dimensions ayant encore un net potentiel de réaffutage) se distinguent par leur caractère à la fois coupant et peu réduit. Des bifaces en silex du Bergeracois et des formations sénoniennes du Périgord (dont un en silex de Belvès) ont également été introduits et sont majoritairement coupants mais de taille modeste.[3] Trois pièces confectionnés à partir de ces matériaux ont été recyclés en nucléus ou en support d'outils retouchés (racloir, denticulé, etc.).

La zone B

Elle est principalement centrée sur les formations crétacées de la moyenne vallée de la Couze, en aval de Combe-Capelle (Landesque 1887; Peyrony 1943:

3 Cette dimension moindre pour ces bifaces peut s'expliquer par la taille des blocs d'origine (plus petits que ceux du Bergeracois) mais aussi par la production de bifaces sur des éclats de seconde génération.

247-257; Dibble et Lenoir 1995) et en amont de La Cavaille (Lacorre 1935), et déborde aussi sur les terrains molassiques tertiaires. Le premier litho-espace est riche en grottes et abris. Cette zone offre d'intéressantes ressources lithiques largement exploitées: les silex gris-noirs ou blonds du Sénonien et les silcrètes-meulières. Ici ont été apportés des bifaces en silex du bergeracois et en silex de la région de Belvès. Les premiers sont le plus souvent abandonnés alors qu'ils étaient encore tranchants et de taille importante, les seconds sont tous recyclés (devenus des supports d'outils retouchés).

La zone C

Il s'agit de la haute vallée de la Lède qui, entre Lacapelle-Biron et Salle-sur-Lède, est bordée de falaises. Elle coule ici dans le litho-espace sénonien du Haut-Agenais qui ne livre que des silex blonds du Sénonien dont le silex dit de Gavaudun. Juste à proximité se développe les terrains tertiaires lacustres assez pauvres en silcrètes-meulières. Pour des groupes humains venant de l'ouest, il s'agit de la première vallée offrant des structures naturelles permettant de s'abriter des vents océaniques chargés de limons qui balayent les plateaux. Dans cette zone protégée se trouve le seul gîte connu du silex dit de Gavaudun. Son affleurement est extrêmement limité (village de Gavaudun) et le matériau n'est représenté que dans les occupations en grotte (Moulin du Milieu et la Grotte Percée[4]; Turq *et al.* 2022). En plein air, ce sont les silex santoniens blonds qui prédominent, ceci malgré une qualité assez variable. En ce qui concerne les bifaces exogènes, ce sont les pièces peu ou pas réaffûtées et d'assez grande taille qui sont abandonnées et de rares pièces plus petites, réaffutées ou recyclées.

Essai Interpretatif

Le premier point concerne l'état des bifaces exogènes lors de leur abandon. Au regard du nombre de pièces retrouvées dans les trois zones, et celui observé en dehors de ces concentrations, il ne peut pas s'agir de perte accidentelle. Pour la moitié des bifaces retrouvés, l'abandon est tout à fait compréhensible, pour les autres il ne l'est pas à première vue. Pourquoi les laisser? On peut envisager qu'il s'agisse de caches ou d'abandons après avoir renouvelé l'outillage à partir des ressources locales. Cette hypothèse peut s'appliquer pour les zones A et B et dans une moindre mesure pour la C. Ici, on peut envisager un abandon qui, que l'on vienne de l'est ou de l'ouest, interviendrait avant d'entrer dans un nouvel espace lithologique. Dans tous les cas, on peut considérer que le *personalgear* (Binford and Binford 1969, 1979) intégrant des bifaces est renouvelé lors des changements d'espaces lithologiques, ce qui implique une parfaite connaissance des ressources et le rôle de matière première relai joué par chacune d'elle (Porraz 2005; Vaissié 2021).

Nous venons de voir la situation dans des zones riches en silex. Confrontons ces observations avec celles faites dans la zone préférentielle de résidence et un espace lithologique pauvre en silex, le Quercy.

À l'est du Périgord Noir (zone préférentielle de résidence) ou au cœur des causses lotois, y compris autour des rares gîtes de silex, les zones de concentration du type de celles décrites ci-dessus n'existent pas. Il n'y a pas de concentration de bifaces aussi importante que celles observées dans les autres espaces lithologiques du nord-est aquitain. Ici, les importations d'objets en silex sénoniens sont importantes et ces derniers sont généralement abandonnés après une très forte utilisation. Les gîtes à silex jurassiques, mais aussi les formations à galets de métaquartzite (utilisés pour la production de supports mais aussi pour le façonnage de biface), constituent les principales sources d'approvisionnement. Comme indiqué précédemment, il n'existe pas d'indice concret de mouvement de matériaux lithiques quercinois vers l'ouest.

Si des expéditions aller-retour entre les litho-espaces sénonien et quercynois semblent bien avoir eu lieu, les chasseurs-cueilleurs néandertaliens ne semblent donc pas avoir constitué de *personalgear* pour leur retour. Cette absence pourrait s'expliquer par la présence des riches ressources en silicites du Sénonien du Périgord, faiblement distantes du Quercy, offrant l'assurance d'un renouvellement des outillages.

Conclusions

Ces données sur les bifaces viennent s'ajouter et compléter les observations faites sur les autres vestiges lithiques inscrits dans la mobilité des groupes du Paléolithique moyen récent (produits de débitage, nucléus ou objets retouchés), et renforcent notre perception de la parfaite connaissance des ressources minérales par ces hommes, une véritable anticipation des besoins et une parfaite adaptation aux contraintes environnementales.

---

4 En dehors de ces gisements il n'est représenté en plein air que sur des sites distants de moins de 2 km et par 3 bifaces et quelques rares autres pièces.

L'analyse des nombreux ramassages de surface faits par ces sauveteurs du patrimoine que sont les amateurs a permis de confirmer les premières observations et de mettre en évidence des zones qui constituent probablement des relais dans l'exploitation des territoires. Malgré cela, il reste à rechercher les autres indices du déplacement des bifaces que sont les éclats de réaffutage qui existent dans bien des sites y compris ceux n'ayant pas livré de bifaces (Faivre, 2006). Pour compléter l'inventaire des pièces exogènes et l'estimation de leurs degrés de consommation, les investigations doivent être poursuivies.

Pour aller encore plus loin, il est indispensable de progresser dans la connaissance des ressources minérales, ce qui passe d'abord par une meilleure caractérisation des divers types de silex, en particulier les silexites du Sénonien et les silcrètes-meulières tertiaires qui constituent les deux plus importantes ressources en matières premières lithiques dans le nord-est aquitain.

References

Anderson-Gerfaud, P.. 1981. *Contribution méthodologique à l'analyse des microtraces d'utilisation sur des outils préhistoriques.* Thèse 3ème cycle: Géologie du Quaternaire et Préhistoire. Bordeaux: Université de Bordeaux I, 1607.

Anderson-Gerfaud, P. and D. Helmer. 1987. L'emmanchement au Moustérien. *MOM Éditions* 15: 37-54.

Beyries, S.. 1984. *Approche fonctionnelle de la variabilité des faciès du Moustérien.* Thèse de 3ème cycle. Paris: Université de Paris X, 1984, 306 p., ill. (Thèse de 3ème cycle).

Binford, L.-R.. 1979. Organization and formation processes: looking at curated technology. *Journal of Anthropological Research* 35: 255-273.

Binford, L.R.. 1980. Willow smoke and dogs' tails: hunter-gatherer settlement systems and archaeological site formation. *American Antiquity* 45: 4-20.

Binford, S.-L. and L.-R. Binford. 1969. Stone tools and human behaviour. *Scientific American* 224: 70-84.

Boëda, E.. 1986. *Approche technologique du concept Levallois et évaluation de son champ d'application: étude de trois gisements saaliens et weichséliens de la France septentrionale.* Thèse N.D. Lettres. Paris: université Paris X-Nanterre.

Boëda, E.. 2001. Détermination des unités techno-fonctionnelles de pièces bifaciales provenant de la couche acheuléenne C'3 base du site de Barbas I. In: D. Cliquet (ed.), *Les industries à outils bifaciaux du Paléolithique moyen d'Europe occidentale, Actes de la table ronde organisée à Caen (Basse-Normandie – France), 14 et 15 octobre 1999*: 51-75. Liège: Université de Liège (ERAUL 98).

Boëda, E.. 2013. *Techno-logique et technologie. Une paléo-histoire des objets lithiques tranchants. Collection préhistoire au présent*. Archéo éditions.

Boëda, E., J.-M. Geneste and L. Meignen. 1990. Identification de chaînes opératoires lithiques du Paléolithique ancien et moyen. *Paléo* 2: 43-80.

Bordes, F.1947. Etude comparative des différentes techniques de taille du silex et des roches dures. *L'Anthropologie*. 51 (1-2): 1-29.

Bordes, F.. 1948. Les couches moustériennes du gisement du Moustier (Dordogne): typologie et technique de taille. *Bulletin de la Société Préhistorique Française* 1948: 113-115..

Bordes, F., 1954. Les gisements du Pech de l'Azé (Dordogne): I. Le Moustérien de tradition acheuléenne. *L'Anthropologie* 58: 401-432.

Bordes, F.. 1955. Les gisements du Pech de l'Azé (Dordogne): I. Le Moustérien de tradition acheuléenne. *L'Anthropologie* 59: 1-32.

Bordes, F.. 1972. *A Tale of Two Caves.* New York: Harper and Row.

Bordes F.. 1984. *Leçons sur le Paléolithique T. II: le Paléolithique en Europe*. Paris: C.N.R.S. (Cahiers du Quaternaire 7).

Boucher-de-Perthes, J. 1857. *Antiquités celtiques et antédiluviennes: mémoire sur l'industrie primitive et les arts à leur origine*. Paris: Treuttel et Wurtz..

Bourgon, M.1957. *Les industries moustériennes et pré-moustériennes du Périgord*. Paris: Masson. (Archives de l'Institut de Paléontologie Humaine 27).

Bourguignon, L., A. Turq and J.-Ph. Faivre. 2004. Ramification des chaînes opératoires: Spécificité du Moustérien? *Paléo* 16: 37-48.

Bourguignon, L., A. Delagnes and L. Meignen, L.. 2006a. Systèmes de production lithique, gestion des outillages et territoires au Paléolithique moyen: où se trouve la complexité? In*:* L. Astruc, F. Bon, V. Lea, P.-Y. Milcent and Philibert (eds.), *Normes techniques et pratiques sociales. De la simplicité des outillages pré et protohistoriques*: 75-86. XXVIe rencontres internationales d'archéologie et d'histoire d'Antibes. Antibes: Editions APDCA .

Brenet, M., J. P. Chadelle, É. Claud, D. Colonge, A. Delagnes M. Deschamps, M. Folgado, B. Gravina and E. Ihuel. 2015. The function and role of bifaces in the Late Middle Paleolithic of southwestern France: Examples from the Charente and Dordogne to the Basque Country. *Quaternary International* 428: 151-169.

Chadelle, J.-P.. 1983. *Technologie et utilisation du silex au Périgordien supérieur: l'exemple de la couche VII du Flageolet I.* Toulouse: Ecole des Hautes études en Sciences sociales (mémoire).

Claud, E., 2008. *The function of recent Middle Palaeolithic handaxes of Southwestern France. Use-wear analysis of tool-kits from La Graulet, La Conne de Bergerac, Combe Brune 2, Fonseigner and Chez-Pinaud / Jonzac.* Université Sciences et Technologies – Bordeaux I.

Claud, E.. 2012. Les bifaces: des outils polyfonctionnels? Étude tracéologique intégrée de bifaces du Paléolithique moyen récent du Sud-Ouest de la France. *Bulletin de la Société préhistorique Française* 109: 413-439.

Commont, V.. 1906. Les découvertes récentes à Saint-Acheul – L'Acheuléen. *Revue de l'École d'Anthropologie.* XVI: 228-241.

Coutier, L., 1929. Expériences de taille pour rechercher les anciennes techniques paléolithiques, *Bulletin de la Société préhistorique de France* 3, 172-174.

Delagnes, A., J. Jaubert and L. Meignen. 2007. Les technocomplexes du Paléolithique moyen en Europe occidentale dans leur cadre diachronique et géographique. In: B. Vandermeersch and B. Maureille (eds.), *Les Néandertaliens: Biologie et culture:* 213-229. Paris: Editions du Comité des Travaux Historiques et Scientifiques.

Demars, P.-Y.. 1980. *Les matières premières siliceuses utilisées au Paléolithique supérieur dans le bassin de Brive.* Thèse 3ème cycle: Géologie du Quaternaire et préhistoire. Bordeaux: Université de Bordeaux I, 1566.

Dibble, H., L.. 1984. Interpreting Typological Variation of Middle Paleolithic Scrapers: Function, style, or sequence of reduction. *Journal of Field Archaeology* 11: 431-436.

Dibble, H and M. Lenoir. 1995. *The Middle Paleolithic Site of Combe-Capelle Bas (France).* University Museum Press, University of Pennsylvania. XVIII.

Faivre, J.-Ph.. 2006. L'industrie moustérienne du niveau Ks (locus 1) des Fieux (Miers, Lot): Mobilité humaine et diversité des compétences techniques, *Bulletin de la Société Préhistorique Française* 103: 17-32.

Faivre, J.-Ph. 2011. Organisation techno-economique des systèmes de production dans le Paléolithique moyen récent du nord-est aquitain. Oxford. (BAR International Series 2280).

Faivre, J.-Ph., B. Gravina, L. Bourguignon, E. Discamps and A. Turq. 2017a. Late Middle Palaeolithic lithic technocomplexes (MIS 5-3) in the northeastern Aquitaine Basin: Advances and challenges in the Aquitaine Basin (southwestern France). *Quaternary International.* 433 B: 116-132.

Faivre, J.-Ph., E. Discamps, B. Gravina, A. Turq and L., Bourguignon. 2017b. Cleaning up a messy Mousterian: how to describe and interpret Middle Palaeolithic chrono-cultural variability in Atlantic Europe. *Quaternary International* 433 B: 1-3.

Féblot-Augustins, J.. 1993. Mobility strategies in the late Middle Palaeolithic of central Europe and western Europe: elements of stability and variability. *Journal of Anthropological Archaeology* 12: 211-265.

Féblot-Augustins, J.. 1997. Middle and Upper Palaeolithic raw material transfers in western and central Europe: assessing the pace of change. *Journal of Middle Atlantic Archaeology* 13: 57-90.

Féblot-Augustins, J.. 1999. Raw material transport patterns and settlement systems in the European Lower and Middle Palaeolithic: continuity, change and variability. In: Roebroeks, W. and C, Gamble (eds.), *The Middle Palaeolithic Occupation of Europe:* 193-214. Leiden: Leiden University.

Fernandes, P.. 2012. *Itinéraires et transformations du silex: une pétroarchéologie refondée, application au Paléolithique moyen.* PHD Thesis, Université Bordeaux 1, ED Sciences et Environnement, Spécialité Préhistoire 1 and 2.

Gamble, C.S.. 1993. Exchange, foraging and local hominid networks. In: C. Scarre and F. Healy (eds.), *Trade and Exchange in Prehistoric Europe: Proceedings of a Conference held at the University of Birmingham, April 1992*: 35-44. Oxford: Oxbow Books in association with the Prehistoric Society and the Société Préhistorique Française.

Gamble, C.S.. 1999. *The Palaeolithic Societies of Europe.* Cambridge: Cambridge University Press.

Gamble, C.S and W. Roebroeks. 1999. The Middle Palaeolithic: a point of inflection. In: W. Roebroeks and C. Gamble (eds.), *The Middle Palaeolithic Occupation of Europe*: 3-21. Leiden: University of Leiden.

Geneste, J.-M.. 1985. *Analyse lithique d'industries moustériennes du Périgord: une approche technologique du comportement des groupes humains au Paléolithique moyen.* Thèse N. D. Sc.. Bordeaux: Université de Bordeaux I.

Geneste, J.-M.1990. Développement des systèmes de production lithique au cours du Paléolithique moyen en Aquitaine septentrionale. In: C. Farizy (ed.) *Paléolithique moyen et récent et Paléolithique supérieur ancien en Europe: Ruptures et transitions: examen critique des documents archéologiques.* Actes du colloque international de Nemours, 9-10 mai 1988: 203-213. Nemours: APRAIF (Mémoires du Musée de Préhistoire d'Ile-de-France 3).

Gravina, B and E. Discamps. 2015. MTA-B or not to be? Recycled bifaces and shifting hunting strategies at Le Moustier and their implication for the late Middle Palaeolithic in southwestern France. *Journal of Human Evolution* 84: 83-98.

Guichard, J.. 1976. Les civilisations du Paléolithique moyen en Périgord. In: H. Lumley (de) (dir.). *La Préhistoire Française: Les civilisations paléolithiques et mésolithiques de la France*: 1053-1069. Paris: C.N.R.S..

Jaubert, J.. 2011. Les archéo-séquences du Paléolithique Moyen du sud-ouest de la France. Quel bilan un quart de siècle après F. Bordes? In: Delpech F and J. Jaubert (eds.) *François Bordes et la Préhistoire, colloque international François Bordes, Bordeaux, 22-23 avril 2009*: 235-253. Paris: Editions du comité des Travaux Historiques et Scientifiques.

Larick, R.-R. 1983. *The circulation of Solutrean foliate point cherts: residential mobility in the Perigord.* Binghamton . Dept. of Anthropology. (Unpublished Ph. D. Dissertation).

Lacorre, F. 1935. *La grotte et l'abri de la Cavaille (Dordogne*), Congrès préhistorique de France. Périgueux 1934: 455-465.

Landesque M. A. 1887. *Excursion à la station préhistorique de Combe-Capelle », Bulletin de la Société Géologique de France*: 15 *:* 866-869.

Lartet, E. and H. Christy. 1875. *Reliquiae Aquitanicae: Being Contributions to the Archaeology and Palaeontology of Périgord and the Adjoining Provinces of Southern France*; 1865-75. Text (Vol. 1). Williams and Norgate.

Le Brun-Ricalens, F., 1988. *Contributions à l'étude du Paléolithique du Pays des serres du bas Quercy et de l'Agenais entre les vallées du Lot et de la Garonne.* Mémoire de D.E.A., Université de Toulouse II le Mirail, 452.

Le Tensorer, J.-M.. 1979. *Recherches sur le Quaternaire en Lot-et-Garonne: stratigraphie, paléoclimatologie et préhistoire paléolithique.* Thèse Bordeaux: Université de Bordeaux I. IV-812 p., ill. Thèse Sc. Bordeaux: Université de Bordeaux I, 87.

Maureille, B, P. Bayle and A. Balzeau. 2016. Neandertal et ses morts. In: Turq A., J. Ph. Faivre, B. Maureille, Ch. Lahaye and P. Bayle (eds.), *Néandertal à la loupe*: 120-130. Les Eyzies.

Morala, A.. 1983. A propos des matières premières lithiques en Haut-Agenais. *Bulletin de le Société Préhistorique Française* 80: 169.

Peyrony, D.. 1930. Le Moustier. Ses gisements, ses industries, ses couches géologiques. *Revue Anthropologique* XL: 48-76 and 155-176.

Peyrony, D.. 1934. La Ferrassie: Moustérien, Périgordien, Aurignaciens. *Préhistoire* III.

Peyrony, D.. 1943. Combe-Capelle. *Bulletin de la Société préhistorique de France* 40 (10-12): 243-257.

Porraz, G.. 2005. *En marge du milieu alpin-dynamiques de formations des ensembles lithiques et modes d'occupations des territoires au Paléolithique moyen* (doctoral dissertation Aix en Provence).

Rigaud, J.-Ph.. 1982. *Le Paléolithique en Périgord: les données du Sud-Ouest sarladais et leurs implications.* Thèse Sc.. Bordeaux: Université de Bordeaux I, 737.

Rigaud, J.-Ph. (ed). 1988. *La grotte Vaufrey, paléoenvironnement, chronologie, activités humaines.* Chalons-sur-Marne: Paquez et fils. (Mémoires de la Société Préhistorique Française LXIX).

Roebroeks, W., J. Kolen and E. Rensink. 1988. Planning depth, anticipation and the organization of Middle Palaeolithic technology: the 'archaic natives' meet Eve's descendants. *Helinium* 28: 17-34.

Sebastianutti, D.. 2020. *Sites préhistoriques de plein air en Périgord noir.* Dolmen editions.

Seronie Vivien, M.-R.. 1987. Les silex du Mésozoïque nord-aquitain: approche géologique de l'étude du silex pour servir à la recherche. *Bulletin de la Société Linéenne de Bordeaux* suppl. au t. XV.

Sorensen, A.C., E. Claud and M. Soressi. 2018. Neandertal fire-making technology inferred from microwear analysis. *Scientific Reports* **8**: 10065.

Sorensen, A., W. Roebroeks and A. Van Gijn. 2014. Fire production in the deep past? The expedient strike-a-light model. *Journal of Archaeological Science* 42: 476-486.

Soressi, M.. 2002. *Le Moustérien de tradition acheuléenne du Sud-ouest de la France. Discussion sur la signification du faciès à partir de l'étude comparée de quatre sites: Pech de l'Azé I, le Moustier, La Rochette et la Grotte XVI.* Thèse de doctorat. Bordeaux: Université de Bordeaux I..

Tixier, J.. 2012, *Méthode pour l'étude des outillages lithiques.* Luxembourg : Musée National d'Histoire et d'Art, Luxembourg, Centre National de Recherche Archéologique du Luxembourg. (ArchéoLogiques 4).

Turq, A.. 2000. Paléolithique inférieur et moyen entre Dordogne et Lot, *Paléo* Supplément 2.

Turq, A.. 2001. Réflexions sur le biface dans quelques sites du Paléolithique ancien-moyen en grotte et abri du Nord-est du bassin Aquitain. In: D. Cliquet (ed.), *Les industries à outils bifaciaux du Paléolithique moyen d'Europe occidentale, Actes de la table ronde organisée à Caen (Basse-Normandie – France), 14 et 15 octobre 1999*: 141-148. Liège: Université de Liège. (ERAUL 98).

Turq A., H. L. Dibble, J.-Ph. Faivre J.-Ph., P. Goldberg, S. McPherron and D. Sandgathe. 2008. Le Moustérien du Périgord Noir: Quoi de Neuf ? In: J. Jaubert and J.-G. Bordes (eds.*), Les Sociétés du Paléolithique dans un Grand Sud-Ouest de la France: Nouveau Gisements, Nouveau Résultats, Nouvelles Méthodes:* 83-93. Bordeaux (Mémoire XLVII de la Société Préhistorique Française).

Turq, A., H. L. Dibble, P. Goldberg, S. McPherron, D. Sandgathe, H. Jones, K. Maddison, B. Maureille, S. Mentzer, J. Rink and A. Steenhuysen 2011. Les Fouilles Récentes du Pech de l'Azé IV (Dordogne). *Gallia Préhistoire* 53:1-70.

Turq, A., A. Morala and M.-R. Seronie-Vivien. 2013a. Inventaire des silicifications du Quercy, de ses marges et des marqueurs lithologiques du nord-est aquitain: synthèse des données. In: M. Jarry J.-Ph. Brugal and C. Ferrier, C. (dirs), Modalité d'occupation et d'exploitation des milieux dans le Sud-Ouest de la France: l'exemple du Quercy, *Paléo, supplément* 4: 59-180.

Turq, A., W. Roebroeks, L. Bourguignon and J.-Ph Faivre. 2013b. The fragmented character of Middle Palalaeolithic stone tool technology. *Journal of Human Evolution* 65: 641-655.

Turq, A., J. Ph. Faivre, B. Maureille Ch. Lahaye,and P.. 2016. *Néandertal à la loupe*. Les Eyzies.

Turq, A, P. Billiant, P. Blot, P. Gouyou and C. Ribouillault. 2022. *Préhistoire à fleur de terre: les chasseurs cueilleurs de la haute vallée de la Lède*. Toulouse: Colibri.

Vaissié, E.. 2021. *Géographie culturelle du Paléolithique Moyen récent dans le Massif Central et ses marges: Territoires, mobilités et systèmes techniques lithiques* (doctoral dissertation, Bordeaux).

In Dusseldorp, G.L., Chu, W., Bakels, C.C. and M.A. Soressi (eds) 2025. Intent on the Paleolithic. Papers in honour of Prof. Dr. Wil Roebroeks. *Analecta Praehistorica Leidensia* 55. Leiden: Sidestone Press, pp. 115-122.

Reflections on the Eemian hornbeam (*Carpinus*) forest at Neumark-Nord 2

Corrie Bakels

Abstract

The second half of the Eemian in Central Europe was characterised by a vegetation dominated by hornbeam (*Carpinus*) trees. The surroundings of the small lake Neumark-Nord 2, Germany, were no exception. Hornbeam forest covered the entire landscape. Was this a vast stretch of closed-canopy forest? Considering the presence of large browsing mammals, first and foremost the huge straight-tusked elephant, this was most probably not the case. In this paper, a reconstruction is proposed in which the forest is broken up into islands, separated by elephant paths, bordered by a kind of rough coppice due to recurrent browsing. In some places intensive browsing, followed by grazing, may have led to the development of open spaces devoid of trees.

Keywords: Carpinus forest; Eemian; Neumark-Nord 2; straight-tusked elephant

To Wil Roebroeks, who lured me to Neumark-Nord 2 and loves forests

Introduction

The pollen diagram from the infill of a small lake or pond, situated at Neumark-Nord 2, Germany, contains a pollen association zone (PAZ), which is characterised by the dominance of hornbeam (*Carpinus*) pollen (Bakels 2012; 2014) (Fig. 1). The infill is dated to the Eemian and the zone, PAZ V, belongs to the second half of this interglacial era (Fig. 2 above). Pollen percentages of herbs are much lower than those of trees, indicating that the main vegetation type might be considered 'forest'. Contemporaneous infills in similar small lakes in the same region, for instance at Gröbern, show even lower herb pollen percentages (Fig. 2 below). Hornbeam was not the only tree present. In fact, in the second, later phase of PAZ V fir (*Abies*) became important as well, but this phase is absent in the infill at Neumark-Nord 2 and in the paper presented here only the first phase, PAZ Va, will be considered. In this first phase hornbeam was accompanied by deciduous oak (*Quercus*), alder (*Alnus*), elm (*Ulmus*), ash (*Fraxinus*), hazel (*Corylus*) and yew (*Taxus*). The coniferous tree pollen may be considered as having been transported by wind from a distance. However, hornbeam dominates the record and the forest can be labelled a 'hornbeam forest'. The question is: what did this forest look like?

Corrie Bakels
Faculty of Archaeology, Leiden University, Leiden, The Netherlands
c.c.bakels@arch.leidenuniv.nl

Hornbeam and hornbeam forests

The common hornbeam (*Carpinus betulus* L.) is a medium-sized tree reaching heights of 20-25 m (Sikkema *et al.* 2016). It grows in temperate climates on all kinds

Figure 1. The situation of Neumark-Nord and Gröbern.

of soils except acid ones. The tree thrives in sunny conditions as well as in the shade. It is actually one of the strongly shade-tolerant native trees of Europe. Common hornbeam is also characterised by a good regeneration capacity.
It regenerates faster than deciduous oak species and ash (Sikkema *et al.* 2016), for example. A second European hornbeam species is the oriental hornbeam (*Carpinus orientalis* Mill.), a small tree or shrub, reaching 1-5 m, rarely 15 m. It is a drought-resistant species preferring calcareous soils (Sikkema *et al.* 2016). But pollen grains from oriental hornbeam differ from those of continental hornbeam (Beug 2004) and the grains retrieved from the Neumark-Nord 2 deposits are definitively those of the common hornbeam.

At present, common hornbeam is found in mixed deciduous forests, together with deciduous oaks and, to a lesser extent, ash (*Fraxinus excelsior* L.), small-leaved lime (*Tilia cordata* Mill.), wild cherry (*Prunus avium* L.), smooth-leaved elm (*Ulmus minor* Mill.), hazel (*Corylus avellana* L.) and field maple (*Acer campestre* L.). Pure stands, or stands with a dominance of hornbeam, are rare (Sikkema *et al.* 2016; Novák *et al.* 2020) (Fig. 3). The forest is commonly described as an oak-hornbeam forest in which oak reaches the greater height and hornbeam constitutes the understory (Weeda *et al.* 1985). Another characteristic is the presence of a well-developed shrub and herb layer. On rich soils suitable for beech (*Fagus sylvatica* L.), the oak-hornbeam forest is absent (Stortelder *et al.* 1999: 318). The co-occurrence of light-demanding oaks, shrubs and herbs, and shade-tolerant tree species such as hornbeam is usually ascribed to former management by humans including coppicing, livestock grazing and litter raking (Novák *et al.* 2020).

The Eemian hornbeam forest

The question is whether the Eemian hornbeam forest resembled the Holocene oak-hornbeam forest. A first point to stress is that hornbeam did not suffer from competition with beech, because beech was absent during the Eemian (see for instance Turner 2000). This implies that hornbeam could flourish in more regions and to a greater extent than during the Holocene.
A second observation is that in the pollen diagrams considered here, oak is only present in low pollen percentages. This cannot be attributed to an important difference in pollen production, although hornbeam is a slightly greater producer and disperser of pollen than oak (Matthias *et al.* 2012). Therefore, the Eemian forest was not an oak-hornbeam forest but a true hornbeam forest with a moderate admixture of other deciduous trees like oak, ash and elm. This forest has no parallel in the present day. What did it look like?

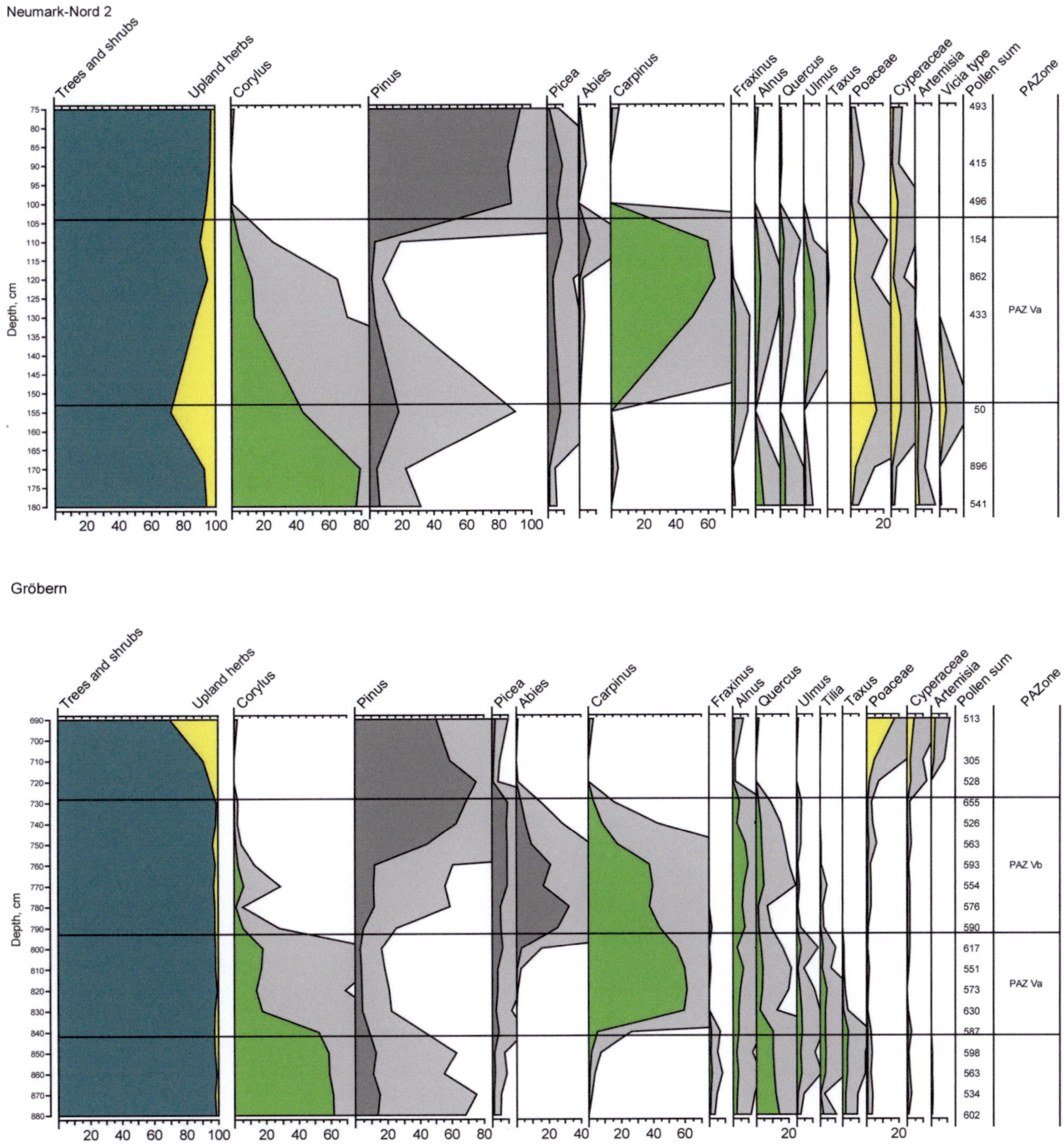

Figure 2. Above: the pollen association zone PAZ V part of the Neumark-Nord 2 pollen diagram, selection of curves, exaggeration 5 x. Upland pollen sum. Lime green: trees growing in the local forest, yellow: herbs. Source Bakels 2012; 2014. Below: the pollen association zone PAZ V part of the Gröbern diagram, curves as above. Source Litt 1990; 1994.

Figure 3. A hornbeam forest. Source: The Original Garden.

If it is assumed that hornbeam trees grew to the same height as today, it was not a forest of awe-inspiring tall trees. Moreover, full-grown hornbeams are shade-tolerant, but the germinating plants are light-demanding (Weeda *et al.* 1985). The forest can, therefore, hardly have been very dark everywhere. The impact of browsing animals also cannot be ignored.

The impact of animals

In 1975, Turner wrote: "It is suggested that large animals were exerting a considerable local effect on the vegetation by means of their grazing, browsing and trampling activities". Vera (2000) came to the same conclusion. During the investigations at the site Neumark-Nord 1, a small lake very close to Neumark-Nord 2, remains of straight-tusked elephants (*Palaeoloxidon antiquus*), red deer (*Cervus elaphus*), fallow deer (*Dama dama*) and bovids were found in profile unit 7, which consists of sediments deposited during the period of hornbeam dominance (Mania and Thomae 2013: Abb. 53). The largest animal is undeniably the straight-tusked elephant, the largest terrestrial mammal of the Pleistocene, with a shoulder height of up to 4 m for bulls and more than 3 m for cows. They were two times larger than the largest elephant still living today, the African elephant (*Loxodonta africana*) (Larramedi *et al.* 2017). Elephants are known to shape the environment and are considered to function as ecosystem engineers (Inogwabini *et al.* 2013 and the literature cited therein). Most investigations are focussed on African elephants in savannah ecosystems. The study of the impact of elephants on forests is limited, but some information is available concerning the effect of forest elephants (*Loxodonta cyclotis*) and Asian elephants (*Elephas maximus*).

Vanleeuwe and Gauthier-Hion (1998) describe an area in Odzala National Park (Republic of Congo) with Marantaceae forest, a forest with clearings and a network of wide paths, called 'boulevards', as well as smaller paths, all created due to the activities of forest elephants.

The boulevards are used for long distance travel and the others for foraging. Patterns of well-defined elephant paths are also reported from other African areas, for instance in the Lac Tumba region (Democratic Republic of Congo). The animals feed on fruits, leaves and saplings along the paths (Inogwabini *et al.* 2013 and the literature cited therein). One of the results is an increase in mean tree diameter perpendicular to the path (Inogwabini *et al.* 2013: fig. 4). Matsubayashi *et al.* (2006) studied Asian elephants in the Deramakot Forest Reserve, Sabah, Malaysia. This study concerns one of the largest populations of these elephants on Borneo. They eat the shoots of *Macaranga* tree species, and in doing so they may fatally damage the trees. The percentage of severely damaged trees increased in proportion to their size. The authors offer as an explanation for this effect the claim that larger trees are more likely to be broken when the animals reach for the shoots (Matsubayashi *et al.* 2006). Also, Asian elephants follow specific paths in the forest and the relative survival of young trees may lead to a gradient from young to old trees perpendicular to the path.

The straight-tusked elephant: a browser in a forest?

The forest elephant is a fruit-consumer and browser. On Borneo the Asian elephant is a browser. Both species affect trees by their feeding behaviour. But was the straight-tusked elephant also a browser? Fortunately, the fossilised stomach contents of two Neumark-Nord 1 elephants were discovered *in situ* together with their skeletons. They were found to contain fragments of leaves and twigs, though no truly lignified parts (Grube *et al.* 2010). As far as the leaf-fragments could be identified, they belonged to grasses (Poaceae), sedges (Cyperaceae) and mistletoe (*Viscum* L. cf *album*). Some buds could be ascribed to either hornbeam, hawthorn (*Crataegus* L. sp.) or sloe (*Prunus spinosa* L.) (Grube *et al.* 2010). The stomach contents also revealed a few pollen grains. Half of these were from hazel, maple and oak. Other trees, for instance hornbeam, were present as well. Herb pollen was scanty. The pollen grains must be considered to have been adhering to consumed plant matter or ingested with drinking water. They do not necessarily give information on the diet. Nevertheless, it is strange that the proportion of grass pollen is almost negligible. Concerning sedges their absence is more understandable as their pollen grains are far from robust. All in all, the straight-tusked elephants seem to have been browsers, but possibly grazers as well.

A second way of getting insight into the diet is to look for dental microwear, *i.e.* scratches on tooth enamel. Analysis of twenty Neumark-Nord 1 elephants' molars revealed that the straight-tusked elephants were mixed feeders. As phrased by Grube *et al.* (2010), they could change their dietary behaviour depending on either the most abundant or the most pleasing food. The same conclusion was reached elsewhere (Saarinen and Lister 2016; Rivals *et al.* 2019).

A third approach is offered by the study of stable isotope values obtained from dentine and bone. The outcome from these analyses is that $\delta^{13}C$ values points towards elephants feeding in a temperate wooded grassland. The $\delta^{15}N$ values suggest that the animals were mainly grazers in a wooded environment (Grube *et al.* 2010). What these studies do not mention is in which pollen association zone the analysed remains were found. In phases prior to the establishment of the hornbeam forest the land was more open and would match a lightly wooded country with stretches of grassy steppe (Pop and Bakels 2015). Nevertheless, the observation made by Grube *et al.* (2010) that straight-tusked elephants could change their dietary behaviour depending on either the most abundant or the most pleasing food makes it possible to consider these elephants as potential browsers.

The most abundant food in the hornbeam forest was, of course, hornbeam. But was it also a pleasing food? The forage quality of leaf-fodder was investigated by Hejcmanova *et al.* (2014). Tree species with low nutritive value are hornbeam and oak. Hazel is of intermediate value and species with high forage quality are elm, ash and maple. Other information is provided by the leaf-fodder that was and in some regions still is collected by traditional farmers in Europe to feed their livestock. They prefer elm, ash, maple and lime (Dreslerová 2012). Dreslerová does not mention hornbeam, but Sikkema *et al.* (2016) write that hornbeam is a good fodder tree. Further data can be found in Čermak *et al.* (2011). They analysed browsing damage in ten localities with a various abundance of game in the Czech Republic. One of their results is that maple, hornbeam and ash rank among the highly attractive tree species as food for animals. Oak and lime were less so. This implies that, although according to Hejcmanova *et al.* (2014) hornbeam has low nutritive value, the tree is attractive to browsers. It must be admitted that the animals concerned in these studies are ruminants and elephants are non-ruminants. Does this make a difference? This must be guessed at, but at present it will be assumed that the elephants might have preferred ash, elm, maple and lime, but have not avoided hornbeam as a food source.

The impact of elephant browsing in the hornbeam forest

Another guess is that the straight-tusked elephant may have resembled its more or less close relative, the forest elephant of Africa (Palkopoulou *et al.* 2018). Perhaps not exactly in diet, because the forest elephant consumes a considerable amount of fruit, but rather in the way it searches for preferred trees that are far from dominant in the forest. As mentioned before, Vanleeuwe and Gautier-Hion (1998) studied forest elephant paths and discovered two kinds of paths not connected with travelling: 'foraging paths' and 'clearing alleys'. The foraging paths are sinuous and wind in every direction. Clearing alleys are narrow paths forming a network around clearings. Similar kinds of paths may have been made by straight-tusked elephants, and the search for preferred trees may have led to winding foraging paths. Where small concentrations of such trees were present, frequent visits may have damaged the trees to such an extent that they died, and subsequent visits of elephants to the spot may have removed saplings resulting in clearings. Alongside the paths the animals may have eaten young twigs of hornbeam, resulting in a belt of damaged trees with increasing thickness and age perpendicular to the axis of the path. Hornbeam, as mentioned before, is characterised by a good regeneration capacity. Its regrowth is faster than that of other tree species. Elephant paths may have been lined by hornbeams resembling a coppice, with younger trees close to the path, older and more varied at some distance.

What did the Eemian hornbeam forest look like?

As described before, the Holocene hornbeam forest is a forest with tall oak trees and a hornbeam understory. Substitution of humans by elephants may result in a different picture, caused by a difference in their attitude towards oak trees. Humans value and spare oak trees, whilst elephants may have been less reticent in destroying tall oak trees. The network of winding foraging paths with their belt of damaged hornbeams and other woody vegetation may have resulted in islands of relatively intact trees surrounded by a kind of coppice. Frequently visited spots may have developed into open spaces with a dense herb vegetation in which regrowth of trees might have been difficult. Such clearings did offer opportunities to graze (Fig. 4). A 'by-product' of this elephant engineering work is that the vegetation would also have suited the red deer and fallow deer that are found in the same time horizon.

How dense the network of paths was, how many clearings there were, and to which extent the trees were 'coppiced' by animals depends on the number of animals and this remains unknown. Is this 'reconstruction' of the forest reflected in the pollen diagrams? The answer is yes. The low percentage of herb pollen in, for instance, the Gröbern diagram, suggests the presence of a forest (Litt 1990; 1994). The Neumark-Nord 2 diagram offers more open space, but this might be attributed entirely or partly to a clearing bordering the pond.

Can the pollen percentages in a pollen diagram tell us more? In the text above reference was already made to differences in pollen production by plants. Pollen productivity estimates (PPEs) are indispensable prerequisites for proper vegetation reconstructions. Matthias *et al.* (2012) investigated pollen-vegetation relationships in north-eastern Germany by sampling small lakes with as narrow a peat margin as possible. They did this in forested areas with known tree cover and vegetation structure. The kind of lakes they sampled is similar to those which might have dotted the Eemian landscape around Neumark-Nord 2. The present vegetation is, of course, quite different from that of the past. Regional herb vegetation consists mainly of crops and anthropogenic grassland, and beech, a tree absent during the Eemian, is a common tree. Nevertheless, Matthias *et al.*'s results offer some inkling of differences in PPEs. One outcome is that hornbeam is overrepresented by its pollen, especially so when stands border a lake. Therefore, Eemian hornbeam forests may not have been the closed canopy forests as suggested by the pollen diagrams. On the other hand, a study in the Białowieża Forest, Poland, showed that hornbeam seedlings can stay for years in the groundlayer, but, if not suppressed, can suddenly grow up very fast and overgrow both light-demanding and shade tolerant low vegetation. In a later stage, due to self-thinning, the density of saplings decreases, but

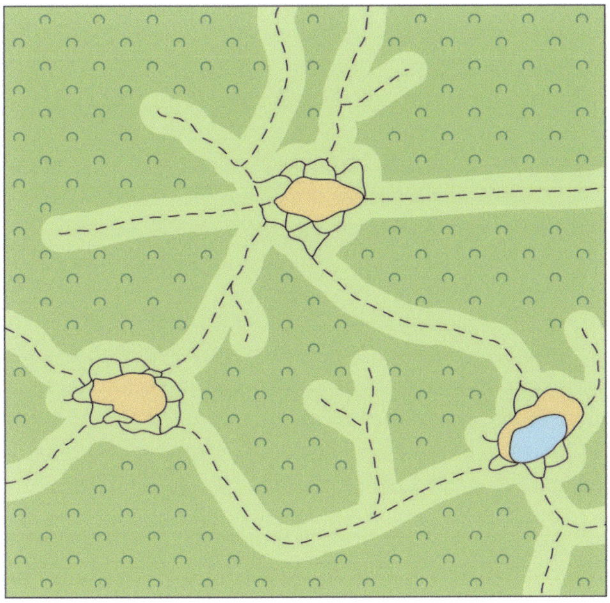

Figure 4. Reconstruction of the vegetation. Dark green: hornbeam forest with old and mature trees; light green: coppice-like vegetation; yellow: tree-less space; blue: lake; broken line: foraging path; drawn line: intensively used path around clearings. Drawing inspired by Vanleeuwe and Gautier-Hion 1998, fig. 2. Not to scale. Drawing T. Rooijakkers.

much of the lower vegetation has vanished by then (Kwiatowska *et al.* 1997). In this stage hornbeams do not flower yet. Bunting *et al.* (2016) reviewed the effect of coppicing on pollen. One outcome is that the coppicing of hazel reduces its pollen production for only one year, but that hornbeam might take 30 years to recover (Bunting *et al.* 2016 and the literature cited therein). Likewise, hornbeam browsed by animals will be underrepresented by its pollen. It remains, therefore, difficult to estimate the real openness of the Eemian hornbeam forest. It depends very much on the density of large mammals, especially elephants, which is very difficult to reconstruct.

Conclusion

The Eemian hornbeam forest was a mosaic of patches of older trees surrounded by belts of younger and coppice-like woody vegetation, interspersed by clearings covered in turf. The 'engineer' of this landscape was the straight-tusked elephant.

Acknowledgements

I am very grateful to Eduard Pop for providing me with Figure 1 and to Tineke Rooijakkers for the realisation of Figure 4. Thanks are also due to Oda Nuij for the references concerning PPEs.

References

Bakels, C.. 2012. Non-pollen palynomorphs from the Eemian pool Neumark-Nord 2: determining water quality and the source of high pollen percentages of herbaceous taxa. *Review of Palaeobotany and Palynology* 186: 58-61.

Bakels, C.C.. 2014. A reconstruction of the vegetation in and around the Neumark-Nord 2 basin, based on a pollen diagram from the key section HP7 supplemented by section HP10. In: S. Gaudzinski-Windheuser and W. Roebroeks (eds.), *Multidisciplinary studies of the Middle Palaeolithic record from Neumark-Nord (Germany), volume 1*: 97-107. Halle an der Saale: Landesmuseum für Vorgeschichte.

Beug, H.-J. 2004. *Leitfaden der Pollenbestimmung für Mitteleuropa und angrenzende Gebiete*. München: Verlag Dr. Friedrich Pfeil.

Bunting, M.J., M.J. Grant and M. Waller. 2016. Approaches to quantitative reconstruction of woody vegetation in managed woodlands from pollen records. *Review of Palaeobotany and Palynology* 22: 53-66.

Čermak, P., P. Beranová, J. Orálková, P. Horsák and J. Plšek. 2011. Relationships between browsing damage and the species dominance by the highly food-attractive and less food-attractive trees. *Acta Universitatis Agriculturae et Silviculturae Mendelianae Brunensis* LIX: 29-36.

Dreslerová, D. 2012. Forest in the prehistoric landscape II. *Archeologiské Rozhledi* 64: 199-236.

Grube, R, M.R. Palombo, P. Iacumin and A. Di Matteo. 2010. What did the fossil elephants from Neumark-Nord eat ? In: H. Meller (ed.), *Elefantenreich-Eine Fossilwelt in Europa, Katalog zur Sonderausstellung im Landesmuseum für Vorgeschichte Halle*: 253-272. Halle: Landesmuseum für Vorgeschichte Halle.

Hejcmanova, P., M. Stejskalová and M. Hejcman. 2014. Forage quality of leaf-fodder from the main broad-leaved woody species and its possible consequences for the Holocene development of forest vegetation in Central Europe. *Vegetation History and Archaeobotany* 23: 607-613.

Inogwabini, B.-I., M. Ngama-Nkosi, L. Wema-Wema and M. Longwango. 2013. Elephant effect on forest physical structure and plant species composition in Salonga and Malebo (Lac Tumba landscape),

Democratic Republic of Congo. *Pachyderm* 53: 28-37.

Kwiatowska, A.J., K. Spalik, E. Michalak, A. Palińska and D. Panufnik. 1997. Influence of the size and density of *Carpinus betulus* on the spatial distribution and rate of deletion of forest-floor species in thermophylous oak Forest *Plant Ecology* 129: 1-10.

Larramedi, A., M.R. Palombo and F. Maran. 2017. Reconstructing the life appearance of a Pleistocene giant: size, shape, sexual dimorphism and ontogeny of *Palaeoloxodon antiquus* (Proboscidea: Elephantidae) from Neumark-Nord 1 (Germany). *Bollettino della Società Paleontologica Italiana* 56: 299-317.

Litt, T.. 1990. Stratigraphie und Ökologie des eeminterglazialen Waldelefanten-Schlachtplatz von Gröbern, Kreis. Gräfenhainichen. In: D. Mania, M. Thomae, T. Litt and T. Weber (eds.), *Neumark-Gröbern, Beiträge zur Jagd des mittelpaläolithischen Menschen*: 193-208. Berlin. (Veröffenlichungen des Landesmuseum für Vorgeschichte in Halle 43).

Litt, T. 1994. Paläoökologie, Paläobotanik und Stratigraphie des Jungquartärs im nordmitteleuropäischen Tiefland (Unter besonderer Berücksichtigung des Elbe-Saale-Gebietes). *Dissertationes Botanicae* 227: 1-185.

Mania, D. and M. Thomae. 2013. *Zur Geologie und Stratigraphie der pleistozänen Becken von Neumark Nord (Geiseltal).* Halle and der Saale. (Veröffentichungen des Landesamtes für Archäologie Sachsen-Anhalt – Landesmuseum für Vorgeschichte 67).

Matsubashi, H., P. Laga and J.R.A. Sukor. 2006. Utilization of *Macaranga* trees by the Asian elephants (*Elephas maximus*) in Borneo. *Mammal Study* 31: 115-118.

Matthias, I., A.B. Nielsen and Th. Giesecke. 2012. Evaluating the effect of flowering age and forest structure on pollen productivity estimates. *Vegetation History and Archaeobotany* 21: 471-484.

Novák, P., W. Willner, D. Zukal, J. Kollár, J. Roleček, K. Świerkosz, J. Ewald, Th. Wohlgemut, J. Csiky, V. Onyshenko and M. Chytry. 2020. Oak-hornbeam forests of Central Europe: a formalized classification and syntaxonomic revision. *Preslia* 92: 1-34.

Palkopoulou, E., M. Lipson, S. Mallick and 34 others. 2018. A comprehensive genomic history of extinct and living elephants. *PNAS* 115: E2566-E2574.

Pop. E. and C. Bakels. 2015. Semi-open environmental conditions during phases of hominin occupation at the Eemian Interglacial basin site Neumark-Nord 2 and its wider environment. *Quaternary Science Reviews* 117: 72-81.

Rivals, F., G.M. Semprebon and A.M. Lister. 2019. Feeding traits and dietary variation in Pleistocene proboscideans: a tooth microwear review. *Quaternary Science Reviews* 219: 145-153.

Saarinen, J. and A.M. Lister. 2016. Dental microwear reflects local vegetation and niche separation in Pleistocene proboscideans from Britain. *Journal of Quaternary Science* 32: 799-808.

Sikkema, R., G. Caudullo and D. de Rigo. 2016. *Carpinus betulus* in Europe: distribution, habitat, usage and threats. In: J. San-Miguel-Ayanz, D. de Rigo, G. Caudullo, T. Houston Durrant and A. Mauri (eds.), *Europa.eu.media: The European Atlas of Forest Tree Species*: 73-75. Luxemburg: Publication office of the European Union.

Stortelder, A.H.F., J.H.J. Schaminée and M. Hermy. 1999. Querco-Fagetea. In: A.F.H. Stortelder, J.H.J Schaminée and P.W.F.M. Hommel (eds.), *De Vegetatie van Nederland* 5: 287-331. Uppsala, Leiden: Opulus Press.

Turner, C.. 1975. Der Einfluß großer Mammalier auf die interglaziale Vegetation. *Quartärpaläontologie* 1: 13-19.

Turner, C. 2000, The Eemian interglacial in the North European plain and adjacent areas. *Netherlands Journal of Geosciences* 79: 217-231.

Vanleeuwe, H and A. Gauthier-Hion. 1998. Forest elephant paths and movements at the Odzala National Park, Congo: the role of clearings and Marantaceae forests. *African Journal of Ecology* 36: 174-182.

Vera F.W.M.. 2000. *Grazing Ecology and Forest History.* Oxford: CAB International.

Weeda, E.J., R. Westra, Ch. Westra, T. Westra. 1985. *Nederlandse oecologische flora* 1. Hilversum: IVN, VARA and VEWIN.

In Dusseldorp, G.L., Chu, W., Bakels, C.C. and M.A. Soressi (eds) 2025. Intent on the Paleolithic. Papers in honour of Prof. Dr. Wil Roebroeks. *Analecta Praehistorica Leidensia* 55. Leiden: Sidestone Press, pp. 123-136.

Fire in the hole: A prototype agent-based model for studying hunter-gatherers' fire-driven hunting

Anastasia Nikulina, Fulco Scherjon

Abstract

This article presents a conceptual agent-based model (ABM) designed to explore the dynamics and consequences of hunter-gatherers' fire-driven hunting of big mammals around Neumark-Nord. Based on the recently developed HUMan impact on LANDscapes (HUMLAND) ABM, we created the HUMLAND-Fire-in-the-hole special. To illustrate the potential of this model, we used mock datasets and tested two scenarios: 1) elephants influencing vegetation without humans; 2) humans affecting vegetation through the use of fire while hunting, with elephants also influencing vegetation. The initial findings show that the model output mainly depends on three parameters: vegetation openness required for animal plant consumption, number of hunter-gatherer groups, and foragers' accessible areas around campsites. Although movement of campsites does not directly impact the primary model observations, foragers' local actions determine animal behaviour during the course of the simulation run. Overall, our results show that hunter-gatherer groups reducing elephant impact are necessary to approach pollen-based reconstructions, and that elephant presence can cause substantial vegetation change, in accordance with the observed behaviour of these animals. The developed ABM can be used in related research. If Last Interglacial local scale input data becomes accessible, the opportunity to re-run this model with such data will be compelling.

Keywords: Neanderthals; elephants; hunting; agent-based modelling; landscape burning

Anastasia Nikulina
Faculty of Archaeology, Leiden University, Leiden, The Netherlands
a.nikulina@arch.leidenuniv.nl;
nikulina1302@gmail.com

Fulco Scherjon
MONREPOS Archaeological Research Centre and Museum for Human Behavioural Evolution, Römisch-Germanisches Zentralmuseum, Schloss Monrepos, 56567 Neuwied, Germany
Fulco.Scherjon@leiza.de

Introduction

Throughout history, humans have used the power of fire for various purposes (Wrangham 2009; Roebroeks and Villa, 2011; Gowlett and Wrangham 2013). Among many ancient practices, the use of fire by hunter-gatherer societies has long fascinated researchers seeking to understand the dynamics of human interactions with their environment. The use of fire represents one of the key aspects of foragers' subsistence strategies which could have facilitated (un)intentional landscape transformations driven by the use of fire for hunting purposes, among other factors (Nikulina *et al.* 2022). The Late Pleistocene witnessed faunal extinctions wherein hunting often played a role (Smith *et al.* 2018; Andermann *et al.* 2020), and these events were linked to reduced vegetation structural diversity (Bakker *et al.* 2016; Berti and Svenning 2020) as well as alterations in natural fire regimes and other ecological processes (Nikulina *et al.* 2022). Along with natural and human-induced fires, herbivory pressure is often considered as one of the reasons for the creation of vegetation openness in the past (Vera 2000; Feurdean *et al.* 2018; Pearce *et al.* 2023). Furthermore, the impact of herbivores extended beyond direct interactions, as they influenced the fire regimes of inhabited areas through the reduction of litter accumulation (Bond and Wilgen 1996; Pringle *et al.* 2023).

Ethnographic observations show that driving small and big game is the most frequent way in which fire is used by hunter-gatherers to obtain food (Scherjon *et al.* 2015). Archaeological evidence also supports the use of fire by hunter-gatherers around their campsites, particularly in Mesolithic contexts (e.g. Bos and Urz 2003; Gumiński and Michniewicz 2003; Hörnberg *et al.* 2006; Rayan and Blackford 2010; Woldring *et al.* 2012) when numerous large animal species, such as the straight-tusked elephant (*Palaeoloxodon antiquus*), had already gone extinct (Elias and Schreve 2013). However, one of the earliest occurrences of human-induced vegetation burning was discovered at Neumark-Nord (NN) in Germany, dating back to the Last Interglacial (Eemian, ~130,000-116,000 BP) (Roebroeks *et al.* 2021). The archaeological site NN is located in a former lignite mine in Saxony-Anhalt, Germany (51°19′28″N, 11°53′56″E). Here, compelling evidence also suggests that hominins systematically and regularly targeted adult elephants in their hunts (Gaudzinski-Windheuser *et al.* 2023). Selective harvest, such as through hunting, impacts target group-living species by altering their social networks and reducing the collective effect on the environment of the group (Wiśniewska 2002).

The straight-tusked elephant assemblage from NN is genetically closest related to current day African forest elephants (Meyer *et al.* 2017). Forest elephants (*Loxodonta cyclotis*) are smaller than savanna elephants (*Loxodonta africana*) and their tusks curve inwards. The forest elephant is a species of African elephant that is found primarily in the dense forests of Central and West Africa. The forest elephant is considered to be a separate species from the savanna elephant based on genetic and morphological differences. Elephants, when not opposed, will populate an area against or just over the carrying capacity (He *et al.* 2020). Individuals tend to migrate only within a certain limited area (Beirne *et al.* 2021).

Hunting elephants with fire is a method employed by various cultures and tribes throughout history (Agam and Barkai 2018). This technique involved fire used strategically to drive and trap elephants, making them easier targets for hunters. The use of fire creates a sense of panic and confusion among elephants. This allowed hunters to take advantage of the situation and capitalize on their vulnerability. The smoke and flames from the fire would impair the elephants' vision and respiratory system. As the fire spread, the elephants would be forced to charge through the flames, often sustaining burns and becoming easy targets, being blinded by the smoke. Additionally, the use of fire in elephant hunting allowed the coordination of many hunters in a single game drive. By strategically positioning themselves around the elephants and setting fire to the surrounding grass, the hunters could create a ring of fire that would encircle the elephants. This prevented the elephants from escaping and forced them towards the waiting hunters, who would spear them. The use of fire also allowed for the simultaneous hunting of multiple elephants, resulting in a higher success rate. The use of fire at such large scales would have altered the landscape significantly if used often, although elephants themselves also cause landscape changes (Parker 2017).

To shed light on the complexities of the effects of this hunting technique, we present an agent-based model (ABM) designed to explore the dynamics and consequences of hunter-gatherers' fire-driven hunting of elephants. ABM is a bottom-up approach to model the decisions made by individual agents to explain system-wide behaviours. This ABM serves as a prototype, a first step towards a deeper understanding of these hunting strategies and their ecological consequences. A useful approach to answering such open questions is to implement a modelling tool which allows one to explore different scenarios and hypotheses (Gilbert and Troitzsch 2005; Romanowska *et al.* 2021). Such a modelling tool could be used to explore the impact of hunting techniques on the success rate of a hunting community (for example, see Ko *et al.* (2023)).

Materials and Methods

We used the recently developed HUMan impact on LANDscapes (HUMLAND) model code (Nikulina *et al.*, 2024) as a starting point and modified the code according to our needs, thus creating the HUMLAND-Fire-in-the-hole special (Fig. 1). To gain insights into the operation of the model and its inherent interactions, we performed sensitivity analysis for two model observations: 1) mean vegetation openness of the study area calculated for grid cells with both climate-based and pollen-based estimates for vegetation cover;

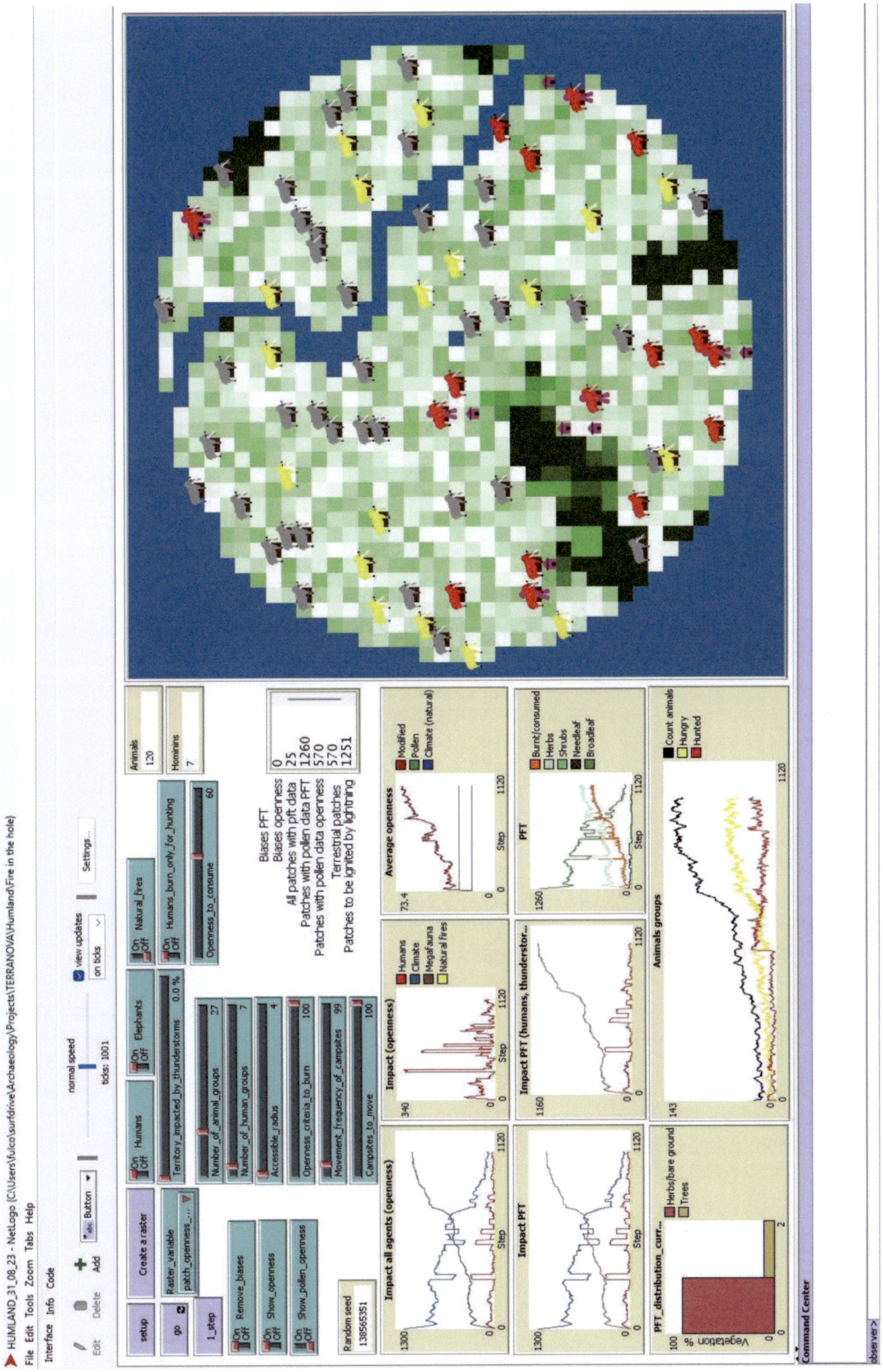

Figure 1. Screenshot of HUMLAND-Fire-in-the-hole. Grey elephants are thriving populations, yellow elephants represent hungry herds, and red elephants are being hunted.

2) the number of animal groups. This analysis was performed using the nlrx R package (Salecker *et al.* 2019). We used the Latin Hypercube Sampling (LHS) sensitivity analysis technique and the Partial Rank Correlation Coefficient (PRCC), in accordance with the approach undertaken by Nikulina *et al.* (2024). The LHS set up had ten random seeds and collected 240 samples for each run.

Afterwards, we varied the most influential parameters between their minimal, intermediate and maximal values to explore the following scenarios: 1) no hunting on animals, only elephants impact vegetation; 2) humans use fire to hunt elephants, both humans and animals impact vegetation. Each combination of parameters was executed on 10 runs with a different random seed. The outputs were analyzed in R with the ggplot2 (Wickham 2016) and tidyverse (Wickham *et al.* 2019) packages.

Before discussing the results, we present below a summarized Overview, Design concepts, and Details (ODD) document (after Nikulina *et al.* (2024)) for this special version of HUMLAND. The ODD protocol is a structured approach to document the design and implementation of an ABM (Grimm *et al.* 2020). Those sections of the ODD that are irrelevant for our version of HUMLAND have been omitted.

ODD HUMLAND-Fire-in-the-hole (after Nikulina et al., 2024)

Purpose

Humans started transforming their environment long before the emergence of agriculture and industrialization. Landscape dynamics are complex and include variable components such as climatic fluctuations, megafauna impact, natural fires, and anthropogenic activities. Thus, there is a need for further research which can allow us to assess different possible scenarios for anthropogenic impact which plays a role in landscape change, with specific focus on burning organized by hunter-gatherers. The model has been implemented in NetLogo (version 6.2.2) and is accessible via the ComSES model repository (https.//www.comses.net/, search for HUMLAND) or from the authors.

Entities, state variables, and scales

The following entities are included in the model: agents representing hominin groups (one agent is one group) and their campsites (one campsite per group), agents representing elephant groups, and grid cells. The model is two-dimensional, and its spatial extent is a rectangle with 43 × 42 grid cells. The world wraps horizontally and vertically. This prototype version of the model imports mock datasets, taken for one time window (9200-8700 BP) for illustrative purposes. One simulation step equals one year, and the current simulation does not account for seasonal variability. One run is 1000 time steps.

Process overview and scheduling

A simulation starts with the setup where input datasets are read, entities are created, and their state variables are set. The process overview of a simulation run is shown in Fig. 2. In HUMLAND it is possible to run simulations with natural fires caused by thunderstorms. In the simulations that we analyzed in this paper these are not included.

Each time step starts with vegetation regeneration of grid cells which were previously burnt or consumed. Afterwards, animals move to one of the grid cells with vegetation openness equal to or lower the value of the *Openness_to_consume* variable. The chosen grid cell is within three grid cells' (30 km) radius around the grid cell where they originated. After animal movements, humans randomly walk to one of the grid cells near their campsite within an accessible area defined by the *Accessible_radius* parameter. If humans see an animal within their accessible radius, they move directly to this grid cell, burn it and decrease the number of animals within a hunted group by one. Vegetation openness of a grid cell where an animal was hunted is set to 100% (completely open) after this fire event. The first dominant Plant Functional Type (PFT) for these areas is replaced by bare ground.

Once humans finished hunting, and group sizes of animals were reduced, the surviving animal groups impacted vegetation on grid cells where they were present. Vegetation openness of these grid cells is set to 100%, and first dominant PFT is replaced by bare ground, due to megafauna impact. It is assumed that elephants consume, strip bark, and tramp vegetation of occupied grid cells.

Design concepts

In accordance with ethnographic observations, driving game is the most frequent way in which fire is used by hunter-gatherers to obtain food. Besides ethnographic data, evidence from an archaeological context shows that fire was an important part of the technological repertoire of *Homo* since at least the Middle Palaeolithic. Notably, one of the earliest pieces of evidence of such activities was discovered at Neumark-Nord in Germany, dated to the Last Interglacial. In addition to that, foraging groups were suggested as one of the primary drivers of vegetation openness in Europe during the Last Glacial Maximum. Human-induced vegetation burning conducted by hunter-gatherers during the Early-Middle Holocene (~11,700-6,000 BP) is often accepted. Despite the presence of the case studies for anthropogenic burning of past landscapes by hunter-gatherers, it is still difficult to establish whether these local-scale impacts caused changes on the regional and (sub-)continental levels (Nikulina *et al.* 2022; Verpoorte and Scherjon 2025). Furthermore, overall landscape dynamics not only depends on humans, but rather represents the synergy of natural and cultural processes on different spatio-temporal scales. Therefore, it is often challenging to distinguish different types of impact on landscapes in proxy-based reconstructions (e.g., palynological datasets), but this makes impacted landscapes very suitable for studies using ABM.

Emergence

The model's key results are the increase of average vegetation openness and the decrease of the percentage of grid cells dominated by trees. These results emerge from the sole impact of elephants and the combined impact of humans and elephants on vegetation.

Objectives

Hunting animals using fire is the objective for hominins. Once they are in a grid cell with an elephant group, the fire is started. If there are no animals in this grid cell, humans do not burn it.

Elephants have the following objectives: reduction of vegetation cover (via consumption, trampling, bark stripping, pathway construction, etc.) and reproduction. When a group of elephants has more than 15 individuals, the group splits into two groups.

Sensing

Humans are aware of the presence of animals within the accessible area around campsites. Elephants sense grid cells with sufficient resources for consumption (i.e., vegetation openness is equal to or lower than the *Openness_to_consume* value).

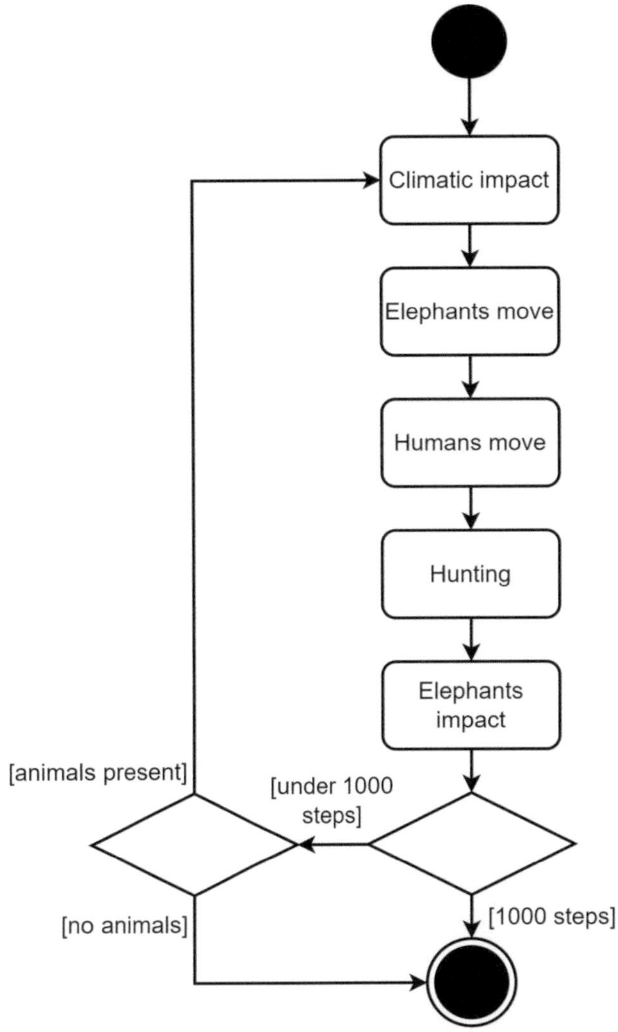

Figure 2. Activity diagram of HUMLAND-Fire-in-the-hole.

Interaction

Hominins directly affect grid cells and elephants. If a hominin decides to burn the vegetation in a grid cell, its state variables are modified. After every hunt, an elephant group loses one individual. Animals impact grid cells by reduction of vegetation cover.

Stochasticity

The model is initialized with a random distribution of hominins and elephants. Additionally, hominins randomly choose one of the grid cells around their campsite to move to. Campsites move to random locations during simulation runs. This happens with the frequency defined via the *Movement_frequency_of_campsites* variable. Elephants also have a stochasticity element in their behaviour because they randomly go to one of the grid cells with vegetation openness equal to or lower than the value of the *Openness_to_consume* value.

Observation

The primary model observations are the number of animals and mean vegetation openness for grid cells which have both pollen-based vegetation cover estimations and climate-based vegetation cover values (i.e., not all grid cells are considered).

Initialization

The environment is created during initialization. Topography, vegetation type and openness are read from input files. Once the environment is created, elephants, hominin groups and their campsites are randomly distributed on the land grid cells. The number of campsites and hominins is defined via the parameter *Number_of_human_groups*. Grid cells around campsites are defined as accessible areas. The *Accessible_radius* parameter defines the size of this area in the number of grid cells around campsites. Hominins cannot move beyond their accessible areas, across water bodies (seas, large lakes, and main rivers) or over high mountains (elevation above 2500 m a.s.l.). Water bodies and the most elevated areas do not have vegetation cover, and, therefore, cannot be burned nor consumed.

Input data

The simulation uses several datasets (detailed below) for illustrative purposes in this version of HUMLAND. Grid cell sizes of all datasets were changed to 10 km × 10 km. For this prototype model only NN and the surrounding grid cells within a radius of 200 kilometres around the site were included.

The initial landscape is reconstructed with the following datasets: Global Topography 30 Arc-Second elevation data set (GTOPO30) (Gesch *et al.* 1999; Danielson and Gesch 2011), Water Information System for Europe (WISE) and two outputs of a dynamic vegetation model. GTOPO30 is a digital elevation model (DEM) which was used to represent elevation data in the current ABM. The WISE dataset was included to define the distribution of large rivers and lakes in the model as they represent natural barriers for fire spread.

In this research, the climate-based vegetation cover dataset represents theoretical potential natural vegetation (PNV) distribution driven by climatic conditions only. As this dataset represents climate-only forced vegetation, it is used in this ABM as the starting point (i.e., before any impact of humans and animals) of each simulation and as a target for vegetation regrowth. For illustrative purposes we use a subset of CARbon Assimilation in the Biosphere (CARAIB) dataset (Zapolska *et al.* 2023a, b).

Simulation outputs and climate-based vegetation cover were compared against pollen-based estimates of plant cover obtained from the pollen-based vegetation cover model. In the current ABM we used a mock subset of Regional Estimates of Vegetation Abundance from Large Sites (REVEALS) dataset (Serge *et al.* 2023) as a target vegetation cover for simulations.

The following datasets are integrated in the HUMLAND-Fire-in-the-hole special:

- GTOPO30, Raster 1 km, Digital elevation model, in meters (www.usgs.gov);
- WISE, Vector 1:10000000, Distribution of large rivers and lakes (https.//water.europa.eu/);
- CARAIB first dominant PFT, Raster ~26 km (0.25°), PNV PFT (http.//www.umccb.ulg.ac.be/Sci/m_car_e.html);
- CARAIB vegetation openness, Raster ~26 km (0.25°), PNV openness, in percentage (http.//www.umccb.ulg.ac.be/Sci/m_car_e.html);
- REVEALS first dominant PFT, Vector ~100 km (1°), Pollen-based past first dominant PFT (Serge *et al.* 2023);
- REVEALS vegetation openness, Vector ~100 km (1°), Pollen-based past vegetation openness, in percentage (Serge *et al.* 2023);
- REVEALS vegetation openness standard errors, Vector ~100 km (1°), Standard errors for estimates of observed past vegetation openness (Serge *et al.* 2023).

Submodels

Climate impact

The regrowth of vegetation after the impact of thunderstorms, megafauna, and/or human burning is determined by climatic conditions. We used the climate-based vegetation cover mean number of years to recover to calculate the vegetation openness recovery rate and to define the step when natural PFT would reestablish in the grid cell after vegetation burning and/or consumption. PFT recovery on all impacted grid cells always begins with herbs, which replace bare ground after seven simulation steps. Subsequently, depending on the initial dominant PFT estimated from the climate-based vegetation cover after the required number of years, the herbs may be replaced by trees or shrubland.

After megafauna plant consumption and fires the rate of vegetation openness recovery (V_{or}) is calculated via the following formula: $V_{or} = O_i - O_c / \mu$.

O_i represents the vegetation openness after the impact caused by fire or megafauna, O_c refers to the climate-based vegetation cover estimates of vegetation openness, and μ – the mean number of years required for the recovery of the initial vegetation openness. During each simulation step, V_{or} is subtracted from the current simulation openness until it reaches the climate-based vegetation cover estimates of vegetation openness.

Anthropogenic impact

This submodel introduces changes to the vegetation through human-induced fires to hunt elephants. Each step, every human group will hunt elephants within their accessible radius. Hunter-gatherers do this by burning the patch on which the elephant group resides. People can hunt and burn more than one grid cell per turn. Fire can spread to neighbouring grid cells depending on their probability of ignition. Every hunt will result in the loss of one elephant per hunted group.

There are several parameters which define human behaviour and the intensity of their impact: *Number_of_human_groups*, *Accessible_radius*, *Campsites_to_move*, their *Movement_frequency_of_campsites*, *Openness_criteria_to_burn*, and the *Humans_burn_only_for_hunting* switch.

The first parameter defines the number of groups present in the study area during one simulation run. The accessible radius parameter defines the territory within which humans move and set fires around campsites.

There are two parameters associated with movements of campsites: *Movement_frequency_of_campsites* (the number of simulation steps after which groups can relocate a campsite) and *Campsites_to_move* (the percentage of hominin groups that relocate a campsite at a certain step). Due to the temporal resolution of the current simulation, hunter-gatherers' highest possible frequency of camp movements is every step (i.e., once per year). The search radius for the new grid cell to establish a site is twice the accessible radius. Any grid cell can be chosen for the new site, except the previously occupied grid cell. The newly established accessible area can overlap with the previous one.

Since the main focus of the HUMLAND-Fire-in-the-hole version is on fire use for hunting purposes, we added the *Humans_burn_only_for_hunting* switch. If its value is set to TRUE, humans burn only when they are hunting animals. In such cases, the *Openness_criteria_to_burn* parameter does not impact the intensity of hunter-gatherer vegetation burning. However, if you intend for humans to burn vegetation for reasons other than hunting (e.g., signalling or promoting certain plants) you should set the switch to FALSE and select the preferred value for the *Openness_criteria_to_burn* parameter. In this case humans will only burn grid cells dominated by trees or shrubs with vegetation openness lower than or equal to this criterion. A low value minimizes the number of positive decisions to start a fire, and higher values increase human-induced fires, because even relatively open areas can be burned by people in this case.

Natural fires

Based on the value of the parameter *Territory_impacted_by_thunderstorms*, the number of grid cells experiencing thunderstorms per simulation step is determined. This parameter is expressed as a percentage, and based on its value, the calculation determines how many grid cells will be affected by thunderstorms. These thunderstorms randomly occur in different grid cells within the study area. Thunderstorms can occur over rivers, lakes, and high mountains, but these areas are not susceptible to burning.

Following the occurrence of thunderstorms, fires may initiate depending on the probability of ignition for the affected grid cells. Fire can spread to neighbouring grid cells, influenced by their probability of ignition.

Megafauna impact

Megafauna, that is elephant groups, will move each turn to a grid cell that has sufficient vegetation to consume and that is located within their foraging range (a constant value set to 30 km from the center). They will subsequently remove all the vegetation on a grid cell. If they cannot find sufficient food, the elephant herd will become hungry. Hungry herds will lose one member per time step until extinction. What is sufficient for a herd is configurable by the user via *Openness_to_consume*. A higher value indicates that needs can be easier fulfilled because animals can

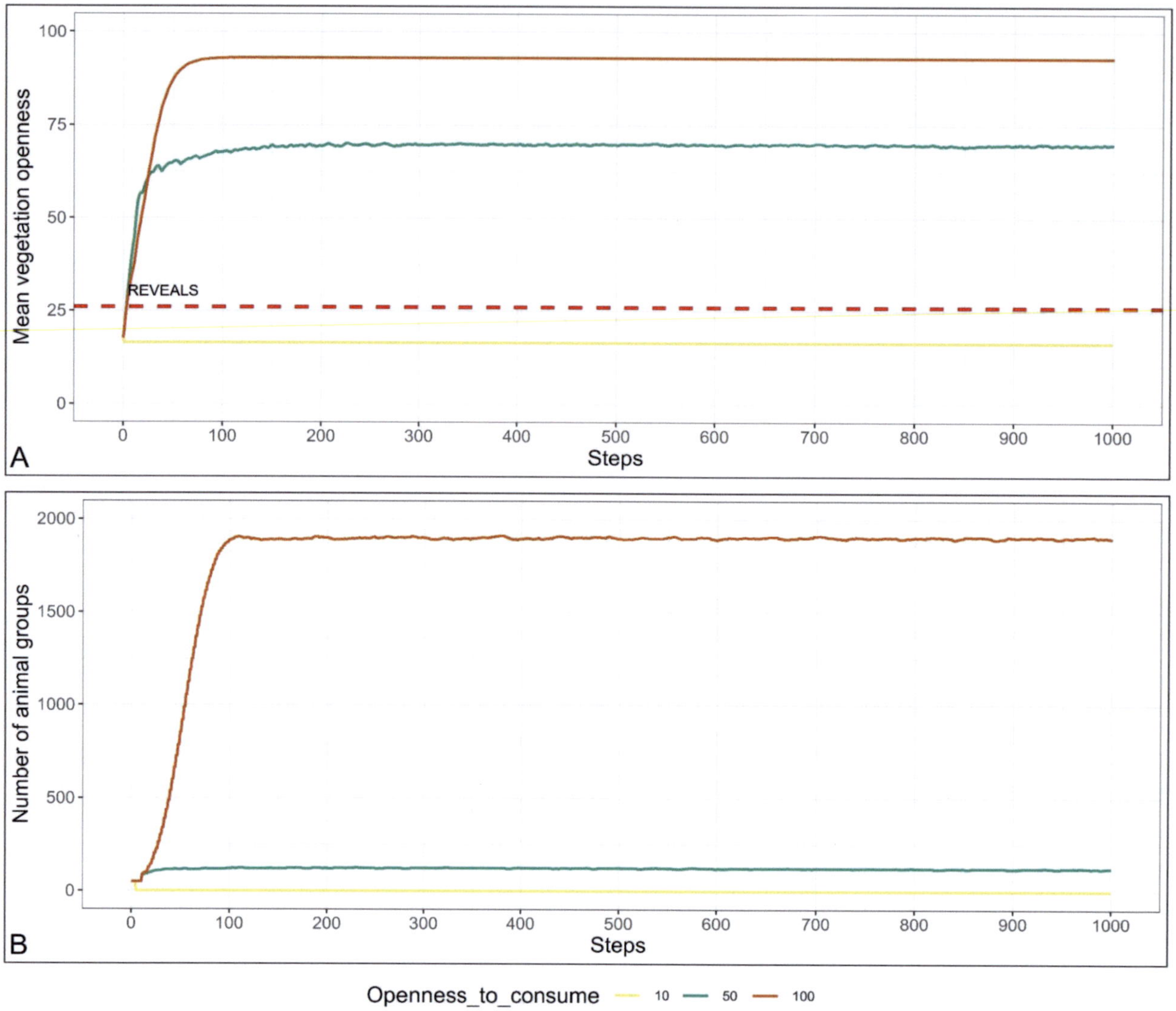

Figure 3. Results of simulation runs for the first scenario (no hunting, only elephants impact vegetation). A – mean vegetation openness; B – the number of animal groups. Each line depicted on the experiment output graph represents the mean of 10 simulation runs. The horizontal dashed red line indicates pollen-based estimates.

consume all grid cells even if the vegetation is rather open (i.e., 80%-90%).

Results

Sensitivity analysis shows that values for the *Accessible_radius* parameter have the most influence on the mean vegetation openness of the study area (PRCC = -0.71 and p-value = 0). Following this, *Openness_to_consume* (PRCC = 0.48) and *Number_of_human_groups* (-0.48) emerge as the second most substantial factors affecting the mean vegetation openness with p-values = 0. The number of animal groups mainly depends on the *Openness_to_consume* parameter (PRCC = 0.48, p-value = 0). Notably, what captures the attention is that both *Accessible_radius* and *Number_of_human_groups* have more impact on vegetation openness than on the number of elephant groups (PRCC = -0.51, p-value = 0 and PRCC = -0.32, p-value = 0 accordingly). Taken as a whole, the model's output predominantly hinges on these parameters: *Openness_to_consume*, *Accessible_radius*, and *Number_of_human_groups*.

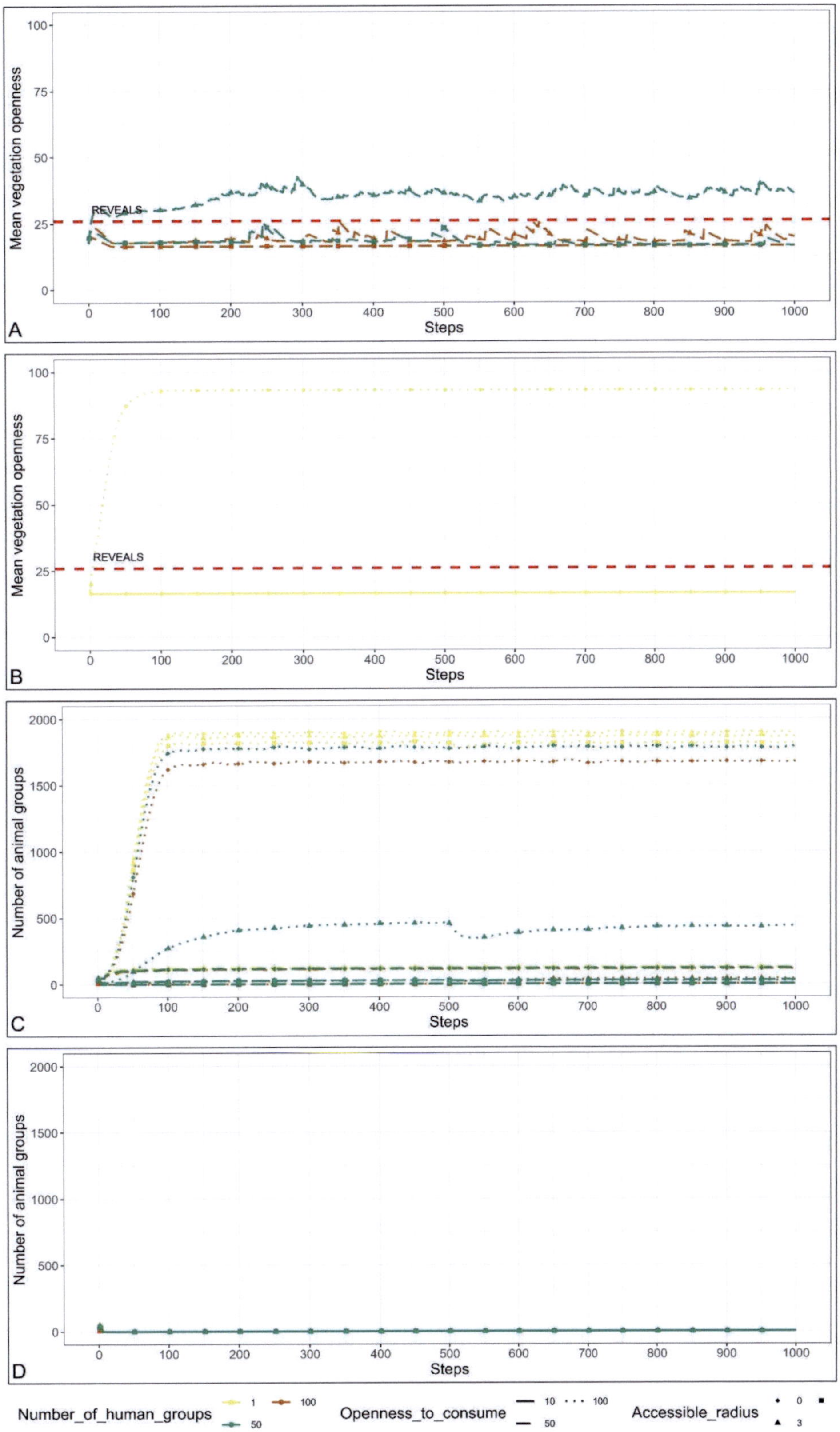

Figure 4. Results of simulation runs for the second scenario (anthropogenic and animal impacts): A – relatively satisfactory results for mean vegetation openness; B – relatively unsatisfactory results for mean vegetation openness; C – relatively satisfactory results for the number of animal groups; D – relatively unsatisfactory results for the number of animal groups. Each line depicted on the experiment output graph represents the mean of 10 simulation runs. The horizontal dashed red line indicates pollen-based estimations.

We manually varied the values for these three parameters to test two scenarios. During the first scenario elephants were only present in the study area without humans. Consequently, the effects on vegetation were solely attributed to megafauna impact. During these runs, we exclusively varied the *Openness_to_consume* parameter, utilizing values of 10, 50, and 100, while maintaining the *Number_of_animal_groups* at a constant value of 50.

Fig. 3 shows results for the first scenario. When the varied parameter is set at its minimal value (10), all animals shortly disappear after the start of the simulation, as evidenced by the yellow line in Fig. 3B. In this scenario, the influence of elephants on altering vegetation openness within the study area remains negligible, as indicated by the yellow line in Fig. 3A. Conversely, employing maximal values for the *Openness_to_consume* parameter results in excessive vegetation consumption and a consequent exponential growth in animal population size. To achieve a balance that aligns with pollen-based estimations while preventing an abrupt surge in the elephant population without human presence, it seems advisable to choose a relatively low value for the *Openness_to_consume* parameter, surpassing 10 yet remaining below 50.

In the second scenario humans and megafauna impacted landscapes. While the two parameters related to elephant impact had the values from the first scenario, we changed values for the important parameters related to anthropogenic impact. Specifically, *Number_of_human_groups* was adjusted to values of 1, 50, and 100; *Accessible_radius* was modulated across 0, 3, and 5. Additionally, *Movement_frequency_of_campsites* was set to 500, and *Campsites_to_move* was established at 50.

The results of this scenario are shown in Fig. 4. Regarding vegetation openness, the ABM shows unsatisfactory results when *Number_of_human_groups* and *Accessible_radius* are set to their minimal values (Fig. 4B). In such instances, the limited influence of humans on the elephant population contributes to its rapid expansion and subsequent excessive consumption, particularly noticeable when *Openness_to_consume* is set at 100.

Contrastingly, Fig. 4A shows relatively favourable outcomes in terms of vegetation openness when *Number_of_human_groups* is established at values of 50 and 100, while *Accessible_radius* is set to 3 and 5. In these configurations, a more balanced interplay emerges, leading to improved vegetation openness compared to the pollen-based estimates.

Finally, the outcomes derived from this scenario confirm the robust linkage between the number of animal groups and the assigned value of *Openness_to_consume*. Even within simulation runs incorporating both anthropogenic and animal influences, a low value attributed to this parameter leads to the rapid extinction of elephant groups (Fig. 4D). Conversely, elevated values of this parameter introduce a spectrum of model outcomes in terms of animal population size, wherein instances of greater *Openness_to_consume* values entail a reliance on their interaction with other parameters (Fig. 4C). In some cases, it is possible to notice declines in the number of animal groups indicative of shifts in campsite occupancy, evidenced by a reduction in elephants' population size around step 500 (the green dotted line in Fig. 4C).

Discussion and Conclusion

The initial findings obtained from the HUMLAND-Fire-in-the-hole model demonstrate that movement frequency of campsites and the percentage of moved sites does not impact primary observations (i.e., mean vegetation openness and the number of animal groups). However, these parameters impact animal behaviour while the model is running. The decline observed in Fig. 4C (dotted green line) may potentially signify shifts in elephant behaviour, indicating instances where these animals are prompted to leave areas where humans have initiated hunting activities after establishing a new campsite. Elephant presence can have a major impact on the landscape, causing a decline in woody vegetation (Bakker *et al.* 2016; Parker 2017). Our simulation results match these observations when elephant population growth is either unrestricted or subject to relatively fewer constraints from hunting. Note that high openness does not mean that there are no trees at all on grid cells. In some cases, trees and bare ground can be present on the same grid cells together which makes trees the first dominant PFT in a very open landscape.

The changes of values for the three most influential parameters confirmed the differences in their importance for primary observations (mean vegetation openness and the number of animal groups). Although all three have substantial impact on openness, the number of animals mainly depends on the required resources for consumption (the *Openness_to_consume* parameter). Hence, when this parameter exhibits relatively low values (i.e., allowing animals to inhabit and affect areas with relatively close vegetation), the reduction in resources due to overconsumption or fires would constrict the animal population size, in addition to a direct decrease resulting from hunting pressure.

We know that Neanderthals hunted elephants for many hundreds of years at the site of NN. Neanderthal diet in general was shown to include quite a large proportion of animal resources (Bocherens 2011). There is some evidence for communal hunting and for storage strategies for consumption in the (near) future (Rendu *et al* 2012). Therefore, we assume that the elephants were hunted mainly for consumption, and we have focussed on the hunting method and its effect on the environment. When local scale Eemian data becomes available for future research it is advisable to implement vegetation consumption by elephants based on intake estimates. In addition, Eemian climate-based and pollen-based estimates should be used in future applications of the presented ABM to reconstruct the landscape more realistically and find plausible parameter values.

We have written this modelling tool to address certain research questions involving Neanderthals, elephants, and NN. This tool is designed to help explore the potential environmental impact of hunting strategies using fire in the landscape and to gain insights into the effectiveness and feasibility of these strategies, the impact they have directly and indirectly on the environment, and specifically on its vegetation. This tool is based on the existing HUMLAND model, and the presented new version is a conceptual model that incorporates various factors such as environmental conditions, hunting techniques, and animal behaviour. This modelling tool can be used in related research, and we look forward to hearing feedback on its effectiveness and applicability.

Acknowledgements

We would like to thank Wil Roebroeks for his ongoing support, his humor and wit, his energy in whatever he embarks upon, and for taking us along for the ride!

The research is financed through the European Union's Horizon 2020 research and innovation programme within the TERRANOVA project, No 813904, and supported by the Liveable Planet programme of Leiden University. The paper reflects the views only of the authors, and the European Union cannot be held responsible for any use which may be made of the information contained therein. This work was performed using compute resources from the Academic Leiden Interdisciplinary Cluster Environment (ALICE) provided by Leiden University. We would like to thank Anhelina Zapolska (Vrije Universiteit Amsterdam, The Netherlands) and Maria Antonia Serge (CNRS, Université de Toulouse-Jean Jaurès, France).

References

Agam, A., and R. Barkai. 2018. Elephant and Mammoth Hunting during the Paleolithic: A Review of the Relevant Archaeological, Ethnographic and Ethno-Historical Records. *Quaternary* 1:3. https.//doi.org/10.3390/quat1010003.

Andermann, T., S. Faurby, S. T. Turvey, A. Antonelli, and D. Silvestro. 2020. The past and future human impact on mammalian diversity. *Science Advances* 6: eabb2313. https.//doi.org/10.1126/sciadv.abb2313.

Bakker, E. S., J. F. Pagès, R. Arthur, and T. Alcoverro. 2016. Assessing the role of large herbivores in the structuring and functioning of freshwater and marine angiosperm ecosystems. *Ecography* 39: 162-179. https.//doi.org/10.1111/ecog.01651.

Beirne, C., T. M. Houslay, P. Morkel, C. J. Clark, M. Fay, J. Okouyi, L. J. T. White, and J. R. Poulsen. 2021. African forest elephant movements depend on time scale and individual behavior. *Scientific Reports* 11: 12634. https.//doi.org/10.1038/s41598-021-91627-z.

Berti, E., and J.-C. Svenning. 2020. Megafauna extinctions have reduced biotic connectivity worldwide. *Global Ecology and Biogeog*raphy 29: 2131-2142. https.//doi.org/10.1111/geb.13182.

Bocherens, H.. 2011. Diet and ecology of Neanderthals: Implications from C and N Isotopes, insights from Bone and Tooth Biogeochemistry. In: Conard, N. J., and J. Richter (eds.), *Neanderthal Lifeways, Subsistence and Technology: One Hundred Fifty Years of Neanderthal Study*: 73-85. Berlin: Springer.

Bond, W. J., and B. W. Van Wilgen. 1996. *Fire and Plants*. Springer Science and Business Media.

Bos, J. A. A., and R. Urz. 2003. Late Glacial and early Holocene environment in the middle Lahn river valley (Hessen, central-west Germany) and the local impact of early Mesolithic people – pollen and macrofossil evidence. *Vegetation History and Archaeobotany* 12: 19-36. https.//doi.org/10.1007/s00334-003-0006-7.

Danielson, J. J., and D. B. Gesch. 2011. "Global multi-resolution terrain elevation data 2010 (GMTED2010). In: *Open-File Report*. https.//doi.org/10.3133/ofr20111073.

Elias, S., and D. Schreve. 2007. Late Pleistocene Megafaunal Extinctions. *Encyclopedia of Quaternary Science*. DOI:10.1016/B0-44-452747-8/00266-0.

Feurdean, A., E. Ruprecht, Z. Molnár, S. M. Hutchinson, and T. Hickler. 2018. Biodiversity-rich European grasslands: Ancient, forgotten ecosystems. *Biological Conservation* 228: 224-232. https.//doi.org/10.1016/j.biocon.2018.09.022.

Gaudzinski-Windheuser, S., L. Kindler, K. MacDonald, and W. Roebroeks. 2023. Hunting and processing of straight-tusked elephants 125,000 years ago: Implications for Neanderthal behavior. *Science Advances* 9: eadd8186. https.//doi.org/10.1126/sciadv.add8186.

Gesch, D. B., K. L. Verdin, and S. K. Greenlee. 1999. New land surface digital elevation model covers the Earth. *Eos, Transactions American Geophysical Union*, 80: 69-70. https.//doi.org/10.1029/99EO00050.

Gilbert, N., and K. Troitzsch. 2005. *Simulation for the Social Scientist*. New York: McGraw-Hill Education (UK).

Gowlett, J. A. J., and R. Wrangham. 2013. Earliest fire in Africa: towards the convergence of archaeological evidence and the cooking hypothesis. *Azania: Archaeological Research in Africa* 48: 5-30. https.//doi.org/10.1080/0067270X.2012.756754.

Grimm, V., S. F. Railsback, C. E. Vincenot, U. Berger, C. Gallagher, D. L. DeAngelis, B. Edmonds, J. Ge, J. Giske, J. Groeneveld, A. S. A. Johnston, A. Milles, J. Nabe-Nielsen, J. Gareth Polhill, V. Radchuk, M.-S. Rohwäder, R. A. Stillman, J. C. Thiele, and D. Ayllón. 2020. The ODD Protocol for Describing Agent-Based and Other Simulation Models: A Second Update to Improve Clarity, Replication, and Structural Realism. *Journal of Artificial Societies and Social Simulation* 23: 7. https.//doi.org/10.18564/jasss.4259.

Gumiński, W., and M. Michniewicz. 2003. Forest and Mobility. A Case from the Fishing Camp Dudka, Masuria, north-eastern Poland. In: L. Larsson (ed.), *Mesolithic on the Move. Papers presented at the Sixth International Conference on the Mesolithic in Europe*: 119-127. Oxford: Oxbow Books.

He, C., J. Du, D. Zhu, and L. Zhang. 2020. Population viability analysis of small population: a case study for Asian elephant in China. *Integrative Zoology* 15: 350- 362.

Hörnberg, G., E. Bohlin, E. Hellberg, I. Bergman, O. Zackrisson, A. Olofsson, J.-E. Wallin, and T. Påsse. 2006. Effects of Mesolithic hunter-gatherers on local vegetation in a non-uniform glacio-isostatic land uplift area, northern Sweden. *Vegetation History and Archaeobotany* 15: 13-26. https.//doi.org/10.1007/s00334-005-0006-x.

Ko, C., Cho W., Hwang B., Chang B., Kang W., and D. W. Ko. 2023. Simulating Hunting Effects on the Wild Boar Population and African Swine Fever Expansion Using Agent-Based Modelling. *Animals* 13: 298. https.//doi.org/10.3390/ani13020298.

Meyer, M., E. Palkopoulou, S. Baleka, M. Stiller, K. E. H. Penkman, K. W. Alt, Y. Ishida, D. Mania, S. Mallick, T. Meijer, H. Meller, S. Nagel, B. Nickel, S. Ostritz, N. Rohland, K. Schauer, T. Schüler, A. L. Roca, D. Reich, B. Shapiro, and M. Hofreiter. 2017. Palaeogenomes of Eurasian straight-tusked elephants challenge the current view of elephant evolution. *eLife* 6: e25413.

Nikulina, A., K. MacDonald, F. Scherjon, A. E. Pearce, M. Davoli, J. C. Svenning, E. Vella, M. J. Gaillard, A. Zapolska, F. Arthur, A. Martinez, K. Hatlestad, F. Mazier, M. A. Serge, K. J. Lindholm, R. Fyfe, H. Rensse, D. M. Roche, S. Kluiving, and W. Roebroeks. 2022. Tracking Hunter-Gatherer Impact on Vegetation in Last Interglacial and Holocene Europe: Proxies and Challenges. *Journal of Archaeological Method and Theory* 29: 989-1033. https.//doi.org/10.1007/s10816-021-09546-2.

Nikulina, A., K. MacDonald, A. Zapolska, M. A. Serge, D. M. Roche, F. Mazier, M. Davoli, J.-C. Svenning, D. van Wees, E. A. Pearce, R. Fyfe, W. Roebroeks, and F. Scherjon. 2024. Hunter-gatherer impact on European interglacial vegetation: a modelling approach. *Quaternary Science Reviews* 324: 108439. DOI:10.1016/j.quascirev.2023.108439.

Parker, I. S. C. July 2017-June 2018. Uganda: elephants, people and fire in Murchison Falls NationalPark and north Bunyoro district. *Pachyderm* 59: 91 – 96.

Pearce, E. A., F. Mazier, S. Normand, R. Fyfe, V. Andrieu, C. Bakels, Z. Balwierz, K. Bińka, S. Boreham, O. K. Borisova, A. Brostrom, J.-L. de Beaulieu, C. Gao, P. González-Sampériz, W. Granoszewski, A. Hrynowiecka, P. Kołaczek, P. Kuneš, D. Magri,... and J.-C. Svenning. 2023. "Substantial light woodland and open vegetation characterized the temperate forest biome before *Homo sapiens. Science Advances* 9: eadi9135. https.//doi.org/10.1126/sciadv.adi913.

Pringle, R. M., J. O. Abraham, T.M. Anderson, T. C. Coverdale, A. B. Davies, C. L. Dutton, A. Gaylard, J. R. Goheen, R. M. Holdo, M. C. Hutchinson, D. M. Kimuyu, R. A. Long, A. L. Subalusky, and M. P. Veldhuis. 2023. Impacts of large herbivores on terrestrial ecosystems. *Current Biology* 33: R584-R610. https.//doi.org/10.1016/j.cub.2023.04.024.

Rendu, W., S. Costamagno, L. Meignen, and M. C. Soulier. 2012. Monospecific faunal spectra in Mousterian contexts. Implications for social behavior. *Quaternary International* 247: 50-58.

Roebroeks, W., and P. Villa. 2011. On the earliest evidence for habitual use of fire in Europe. *PNAS* 108: 5209-5214. https.//doi.org/10.1073/pnas.1018116108.

Roebroeks, W., K. MacDonald, F. Scherjon, C. Bakels, L. Kindler, A. Nikulina, E. Pop, and S. Gaudzinski-Windheuser. 2021. Landscape modification by Last Interglacial Neanderthals. *Science Advances* 7: 1-14. https.//doi.org/10.1126/sciadv.abj5567.

Romanowska, I.; C. D. Wren, and S. A. Crabtree. 2021. *Agent-Based Modeling for Archaeology. Simulating the Complexity of Societies*. The Santa FE Institute Press. https.//doi.org/10.37911/9781947864382.

Ryan, P. A., and J. J. Blackford. 2010. Late Mesolithic environmental change at Black Heath, south Pennines, UK, a test of Mesolithic woodland management models using pollen, charcoal and non-pollen palynomorph data. *Vegetation History and Archaeobotany* 19: 545-558. https.//doi.org/10.1007/s00334-010-0263-1.

Salecker, J., Sciaini, M., Meyer, K. M., and K. Wiegand. 2019. The nlrx r package: A next-generation framework for reproducible NetLogo model analyses. *Methods in Ecology and Evolution* 10: 1854-1863. https.//doi.org/10.1111/2041-210X.13286.

Scherjon, F., C. C. Bakels, K. MacDonald, and W. Roebroeks. 2015. Burning the Land. An Ethnographic Study of Off-Site Fire Use by Current and Historically Documented Foragers and Implications for the Interpretation of Past Fire Practices in the Landscape., *Current Anthropology* 56: 299-326. https.//doi.org/10.1086/681561.

Scherjon, F. 2019. *Virtual Neanderthals: a study in agent-based modelling Late Pleistocene hominins in western Europe*. Leiden: Leiden University.

Serge, M. A., F. Mazier, R. Fyfe, M. J. Gaillard, T. Klein, A. Lagnoux, D. Galop, E. Githumbi, M. Mindrescu, A.B. Nielsen, A. K. Trondman, A. Poska, S. Sugita, J. Woodbridge, D. Abel-Schaad, C. Åkesson, T. Alenius, B. Ammann, S. T. Andersen, ... and V. P. Zernitskaya. 2023. Testing the Effect of Relative Pollen Productivity on the REVEALS Model: A Validated Reconstruction of Europe-Wide Holocene Vegetation. *Land* 12: 986. https.//doi.org/10.3390/land12050986.

Smith, F. A., R. E. E. Smith, S. K. Lyons, and J. L. Payne. 2018. Body size downgrading of mammals over the late Quaternary. *Science* 360: 310-313. DOI: 10.1126/science.aao5987.

Vera, F. W. M. 2000. *Grazing ecology and forest history*. CABI Publishing.

Verpoorte, A. and F. Scherjon. 2025. Inland/Forest. In: L. N. Stutz, R.P. Stjerna and M. Tõrv,(eds), *The Oxford Handbook of Mesolithic Europe*, Oxford Handbooks. DOI: 10.1093/oxfordhb/97801988536 57.001.0001.

Wickham, H. 2016. *ggplot2: Elegant Graphics for Data Analysis*. https.//ggplot2.tidyverse.org.

Wickham, H., Averick, M., Bryan, J., Chang, W., McGowan, L., François, R., Grolemund, G., Hayes, A., Henry, L., Hester, J., Kuhn, M., Pedersen, T., Miller, E., Bache, S., Müller, K., Ooms, J., Robinson, D., Seidel, D., Spinu, V., ... and H. Yutani. 2019. Welcome to the Tidyverse. *Journal of Open Source Software* 4: 1686. https.//doi.org/10.21105/joss.01686.

Wiśniewska M., I. Puga-Gonzalez, P. Lee, C. Moss, G. Russell, S. Garnier, and C. Sueur. 2022. Simulated poaching affects global connectivity and efficiency in social networks of African savanna elephants – An exemplar of how human disturbance impacts group-living species." *PLOS Computational Biology* 18: e1009792. https.//doi.org/10.1371/journal.pcbi.1009792.

Woldring, H., M. Schepers, J. Mendelts, and R. Fens. 2012. Camping and foraging in Boreal hazel woodland – The environmental impact of Mesolithic huntergatherers near Groningen, the Netherlands. In: M. J. L. T. N. Terberger, R. N. E. Barton, and M. Street (eds.), *A mind set on flint. Studies in honour of Dick Stapert*: 381-392. Groningen: Barkhuis Publishing.

Wrangham, R. 2009. *Catching fire. How cooking made us human*. New York: Basic Books.

Zapolska, A., M. A. Serge, F. Mazier, A. Quiquet, H. Renssen, M. Vrac, R. Fyfe, and D. M. Roche. 2023a. More than agriculture. Analysing time-cumulative human impact on European land-cover of second half of the Holocene. *Quaternary Science Reviews* 314: 108227. https.//doi.org/10.1016/j.quascirev.2023.108227.

Zapolska, A., M. Vrac, A. Quiquet, T. Extier, and F. Arthur. 2023b. Improving biome and climate modelling for a set of past climate conditions: evaluating bias correction using the CDF-t approach Improving biome and climate modelling for a set of past climate conditions : evaluating bias correction using the CDF-t approach*Environmental Research Climate* 2: 025004. https.//doi.org/10.1088/2752-5295/accbe2.

In Dusseldorp, G.L., Chu, W., Bakels, C.C. and M.A. Soressi (eds) 2025. Intent on the Paleolithic. Papers in honour of Prof. Dr. Wil Roebroeks. *Analecta Praehistorica Leidensia* 55. Leiden: Sidestone Press, pp. 137-146.

The power of the water vole

Thijs van Kolfschoten

Abstract

The vole clock, a method of dating archaeological strata using vole teeth, has proven invaluable in debates over the age of European Lower and Middle Palaeolithic sites. With voles abundant in the fossil record, the method relies on the unidirectional evolution observed in vole lineages. This study explores the evolutionary history of voles, particularly the water vole lineage, emphasizing its role in biostratigraphical correlations. Case studies, including debates on the age of sites like Boxgrove and Schöningen, underscore the importance of vole data. The vole clock's accuracy and limitations are discussed, highlighting the method's strengths in abundance and evolutionary trends, while noting constraints due to geographical distribution and preservation issues. In conclusion, the vole clock emerges as a powerful tool for relative dating, complementing broader chronological contexts in archaeological and paleontological studies.

Keywords: Arvicola; Mimomys; vole clock; biostratigraphy; dating Palaeolithic sites

Introduction

In the year 1982, at the Natural History Museum in Maastricht (the Netherlands) when I met Wil Roebroeks for the first time, he invited me to visit their archaeological excavation in the Maastricht-Belvédère quarry. He showed me mammalian fossils they excavated (including a large skull fragment of a giant deer) and pointed at molars of a vole and some small mammal bone fragments in one of the section they had recently cleaned. Soon after, Wil asked fundamental questions that often preoccupy archaeologists investigating Palaeolithic sites *"Can you say something about the age of the fauna? And what about the climate and environment?* Of particular interest to Wil was the comparison with the recently published study on fauna from the ice-pushed ridge near Rhenen (the Netherlands) (van Kolfschoten 1981).

For the reconstruction of the palaeoclimate and the palaeoenvironment, it was necessary to have an overview of the entire fossil record. Delving deeper into the question about the age of the fauna, it became clear: voles are the clue to the answer.

This paper starts with general information about voles and details about the evolutionary changes that can be observed in the different vole lineages, with a specific focus on the water vole lineage. These changes provide the foundation for a biostratigraphical framework, which has proven instrumental in dating Palaeolithic faunal assemblages. To demonstrate the potency of this knowledge, we present a selection of case studies that illustrate the biostratigraphical value of voles. In the Discussion and Conclusion section, the biostratigraphical use of voles is summarized and put into a broader context.

Thijs van Kolfschoten
Faculty of Archaeology, Leiden University, Leiden, The Netherlands; Joint International Research Laboratory of Environmental and Social Archaeology, Shandong University, Qingdao; Institute of Cultural Heritage, Qingdao, Shandong University, China.
t.van.kolfschoten@arch.leidenuniv.nl

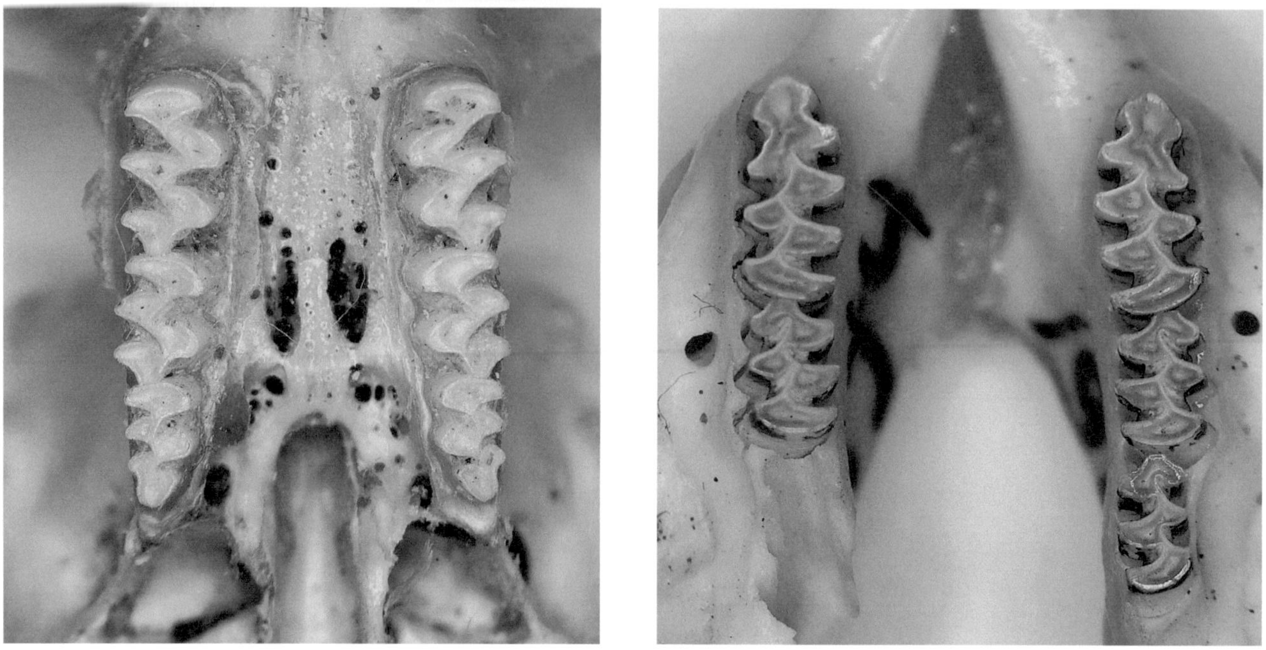

Figure 1. The upper (a) and lower (b) molars of the European water vole *Arvicola terrestris*.

Voles

Voles are small, burrowing rodents that belong to the subfamily Arvicolinae (Order Rodentia, Family Cricetidae). Voles have stout bodies, short legs, and short tails, while the dentition of voles is characterised by two pairs of continuously growing incisor teeth in the upper and lower jaws and a large gap, or diastema, between the incisors and the high-crowned molars. The occlusal surface of the worn molars shows a number of dentine fields that are covered on the outside with enamel (Fig. 1).

Voles are herbivorous rodents with a primary diet consisting of plant materials like grasses, roots, stems, leaves, and seeds. With over 150 recognized species, voles inhabit diverse ecosystems, including grasslands, meadows, and forests, depending on the species. Their geographical distribution covers North America and Eurasia.

The evolutionary history of voles is complex and remains an active area of investigation in paleontological and evolutionary biological research. The subfamily Arvicolinae, encompassing voles, lemmings, and other related rodents, constitutes a relatively recent branch within the broader family Cricetidae. It is thought to have originated during the Late Miocene period, roughly 11 to 5 million years ago, specifically in the southeastern regions of Eurasia (Alexey Tesakov pers. comm. September 29, 2023).

The fossil record reveals the presence of numerous vole lineages each with a (large) number of species. Notably, these lineages exhibit a sequential, albeit non-simultaneous, increase in crown height in their molars. The evolutionary history of many of these lineages is further characterized by distinctive features such as the development of crown cementum, the emergence of rootless molars, alterations in enamel structure and thickness differentiation, as well as the creation of additional enamel in the upper M3 and lower m1. These evolutionary shifts contribute to an enhanced complexity of the occlusal surface in both molars.

Vole remains are exceptionally abundant in the fossil record of Eurasia. Due to their relatively rapid evolution, especially in the Late Pliocene and Pleistocene, voles serve as a foundation for the biostratigraphical subdivision Villanyian-Biharian-Toringian spanning the past 3-4 million years. The Pliocene base of the Villanyian is marked by the appearance of *Mimomys* species such as *M. minor* and *M. hajnackensis*, species that are characterized by the presence of crown cementum (van der Meulen and van Kolfschoten 1986). The appearance of the genus *Microtus* in the European fossil record is regarded as the marker of the Villanyian/Biharian boundary (Fejfar and Heinrich 1990). The evolutionary changes

Figure 2. The European water vole *Arvicola terrestris*.

in the water vole lineage encompass the replacement of *Mimomys savini*, a species characterized by rooted molars, by *Arvicola cantiana*, a species with unrooted molars. This transition marks the Biharian/Toringian boundary (van Kolfschoten 1990).

Voles belonging to the genus *Microtus*, characterized by unrooted molars, are well-represented and often a dominant taxon in the micro-mammalian fauna since their Early Pleistocene appearance ca. 2 million years ago (Tesakov 1998). The most primitive *Microtus* species characterized by lower m1 molars exhibiting three triangular dentine fields that separate the posterior and anterior lobes, are assigned to the subgenus *Microtus (Allophaiomys)*. Notably, the Mid-Pleistocene, around 1 million years ago, marks a phase of rapid evolutionary changes within the *Microtus* lineage. This includes the increase of complexity of the occlusal pattern attributed to the formation of additional enamel ridges and an increase of the number of dental fields. This period also witnesses a radiation into various species classified under the subgenus *Microtus (Microtus)*. The transition between the two subgenera demarcates the boundary between the Early and Late Biharian Stages.

The water vole

Among current Eurasian mammalian fauna, two distinct water vole species are recognized: the European water vole *Arvicola terrestris* (also referred to as *Arvicola amphibius*) (Fig. 2) and the southwestern water vole, *A. sapidus*. *A. terrestris* exhibits a wide geographical distribution, while *A. sapidus* is confined to southwestern Europe (Fig. 3).

The evolutionary and biostratigraphical significance of the Eurasian water vole lineage has attracted considerable attention from various scholars, including M. Kretzoi and D. Janossy (Hungary), O. Fejfar (Czech Republic), G. Rabeder (Austria), J. Chaline (France), W.D. Heinrich and L. Maul (Germany), and A.J. van der Meulen (the Netherlands). Examining the earlier segment of the water vole lineage, spanning from *M. occitanus* via *M. polonicus* and *M. pliocaenicus* to *M. ostramosensis*, reveals notable features such as the emergence of cementum in the re-entrant angles of the molars and an increase in crown height (Neraudeau *et al.* 1995).

In the more recent phase of the lineage, the transition from *Mimomys savini*, characterized by rooted molars, to *Arvicola cantiana* with unrooted molars is observed. Additionally, there are noteworthy evolutionary changes within the succeeding *Arvicola* population.

The *Mimomys* – *Arvicola* transition, that marks the Biharian/Toringian boundary, took place during the second half of the early Middle Pleistocene ca. 0,5 million years ago. The transition has been documented in western Europe (Chaline 1986; van Kolfschoten 1990), central Europe (*e.g.* Fejfar and Heinrich 1990) and eastern Europe (Danukalova *et al.* 2016) and appeared to take place more or less simultaneously across these regions. While it is anticipated that there might be some asynchronicity in the regional shift from *Mimomys* to *Arvicola*, the current chronological resolution of dating methods does not permit precise age estimations with the required reliability to trace such regional differences.

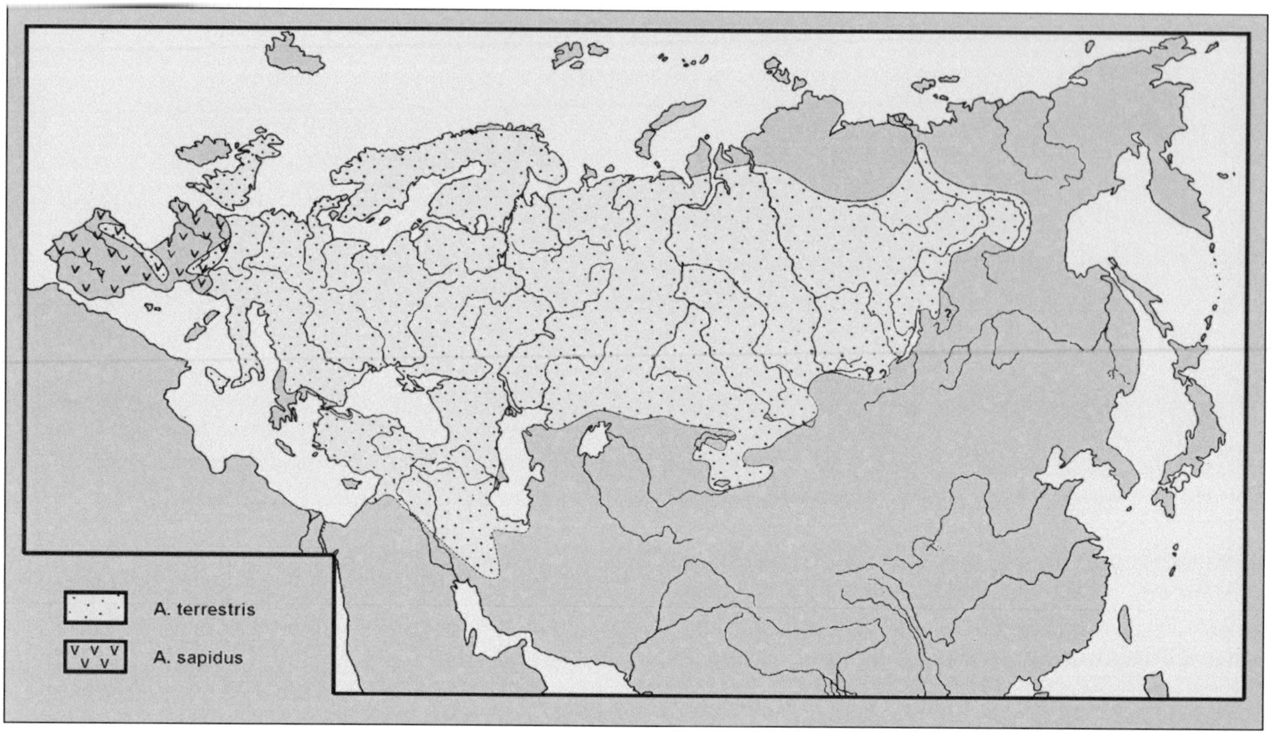

Figure 3. The geographical distribution of the two living water vole species: the European water vole *Arvicola terrestris* and the southwestern water vole, *A. sapidus*. After Panteleev et al., 1990.

Within the *Arvicola* lineage, notable evolutionary changes in the enamel of the molars have been observed. The enamel covering both the concave (leading) and convex (trailing) edges of the molars exhibits a three-dimensional structure (von Koenigswald 1980). In the Toringian water vole (*Arvicola*) record, there are discernible evolutionary changes leading to the thinning of the enamel specifically at the convex sides of the molars.

To quantitatively assess differences in enamel thickness, Heinrich (1978) proposed the calculation of the enamel thickness quotient, introducing the *Schmelzband-Differenzierungs-Quotient* (SDQ) values. This involves measuring enamel thickness on both sides of the salient angles of the m1, dividing the value of the trailing edge by that of the leading edge, and multiplying the quotient by 100. The SDQ of one molar is the mean of the SDQ values of all salient angles. Higher values indicate more primitive populations, while lower values suggest more advanced forms.

Heinrich applied this method to calculate SDQ values for numerous *Arvicola* assemblages from localities in central Europe and the Pannonian Basin. He observed a continuous decrease in SDQ values during the Middle and Upper Pleistocene. These data formed the foundation for his biostratigraphical framework, providing insights into the evolutionary changes within the *Arvicola* lineage (Heinrich 1982, 1987).

Establishing the evolutionary stage of fossil *Arvicola* remains has proven to be a valuable and widely utilized tool for biostratigraphic subdivision in the fossil record (Heinrich 1987; van Kolfschoten 1990, 1992; von Koenigswald and van Kolfschoten 1996; Parfitt 1998). This method has become one of the most extensively applied tools, serving to date and characterize various glacial and interglacial faunal assemblages. It is particularly instrumental in correlating with the stratigraphical subdivision of the late Middle Pleistocene, which includes three interglacial phases between the Elster and Saale (Drenthe) glaciations, as proposed by several authors (*e.g.* Urban 1995).

The Vole clock

The information provided indicates that vole fossils possess the potential to offer a relative, biostratigraphical age for a given faunal assemblage.

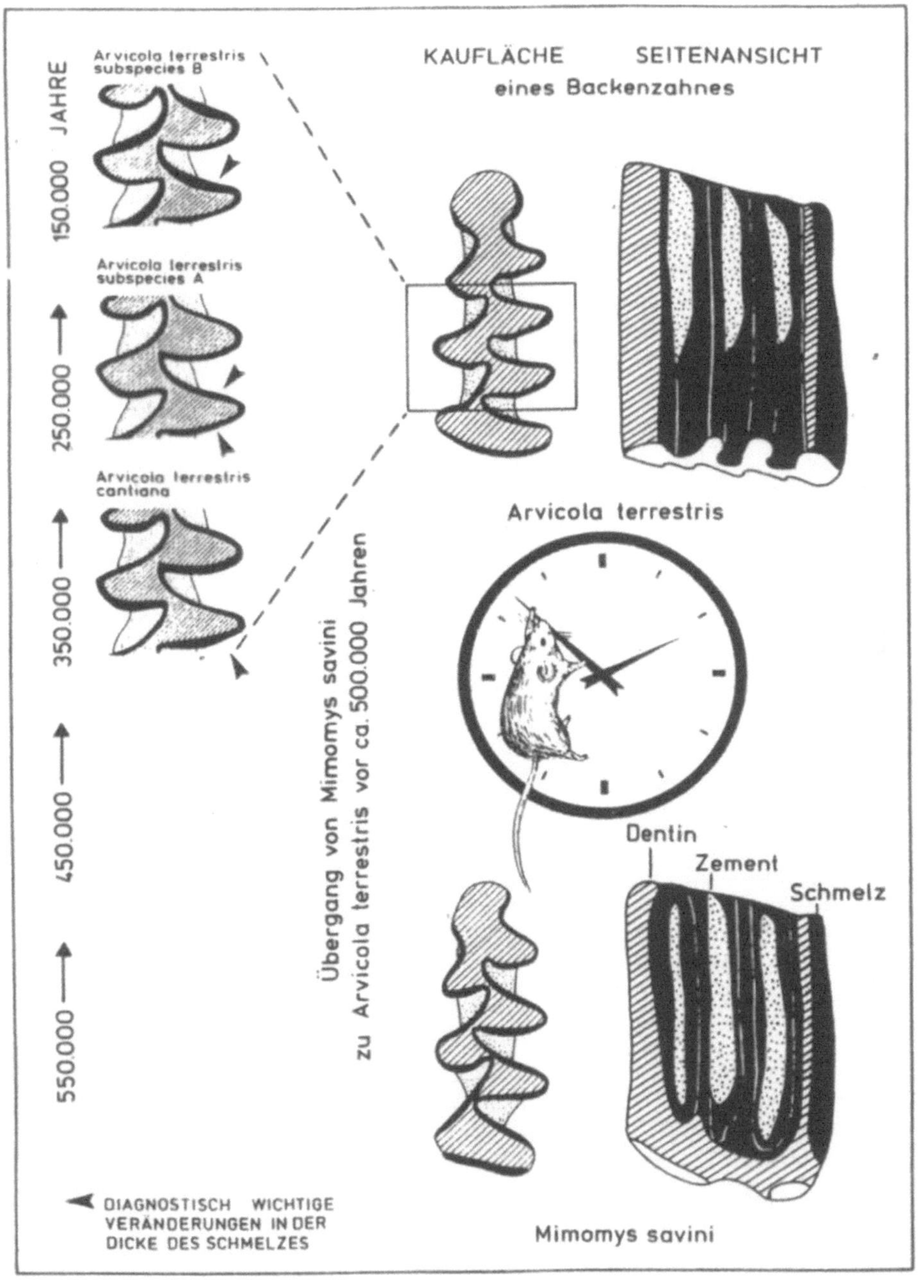

Figure 4. The vole clock. The illustration shows the transition from *Mimomys savini* to *Arvicola terrestris* ca. 500.000 years ago (Übergang von Mimomys savini zu Arvicola terrestris vor ca. 500.000 Jahren). The figure shows the occlusal view (Kaufläche) and lateral view (Seitenansicht) of the lower m1 molar (Backenzahn) of *Mimomys savini* (base) and *Arvicola terrestris* (top). The arrows indicate the important diagnostic changes in the enamel thickness. Dentin = dentine; Zement = cementum; Schmelz = enamel. After Bosinski *et al.* 1988: 23.

This age, when considered alongside other proxies, has the capacity to provide a rough estimate of the absolute age of the assemblage and its associated strata. The fossil record underscores the temporal changes in vole molars, with evolution following a unidirectional pattern, akin to the movement of the hands on a clock. Consequently, van Kolfschoten (1988) introduced the term "*Die Mäuseuhr*," commonly known as the vole clock (Fig. 4). This term is familiar particularly to scholars investigating the Lower and Middle Palaeolithic records of Europe.

On a Wikipedia page about the vole clock, it is stated that:

> *The vole clock is a method of dating archaeological strata using vole teeth. Investigations at sites across Europe have allowed construction of a detailed framework of how different vole species evolved*

> *over the last million years, and where and when specific species became extinct. .... Dr. Francis Wenban-Smith of Southampton University described the vole clock as "one of the wonders of modern science"... (http://en.wikipedia.org/wiki/Vole_Clock).*

Nevertheless, it is crucial to recognize that the pace of the vole clock is primarily contingent upon environmental and evolutionary pressures, exhibiting variation among distinct vole lineages. Within the *Clethrionomys* lineage, the clock ticks at a notably slow rate, while the early part of the *Microtus* lineage experienced a much swifter pace. Furthermore, a significant geographical gradient exists. Present-day European water voles in the northern regions, from an evolutionary perspective, demonstrate greater advancement, as indicated by their lower SDQ-values, compared to their counterparts in southern Europe (Röttger 1987).

Case studies

The vole clock dating method has been applied in numerous studies, especially in the context of debates regarding the age of European Lower and Middle Palaeolithic sites. This prominence is attributable to the absence of accurate alternative dating methods or their nascent development, rendering them less reliable. In these studies, the evolutionary changes within the water vole lineage, notably the transition from rooted to unrooted molars and subsequent alterations in enamel differentiation, have emerged as pivotal criteria.

An important study, wherein the evolutionary stage of the water vole played a crucial role, was the reassessment of both the artefactual and chronological evidence pertaining to the earliest occupation of Europe (Roebroeks and van Kolfschoten 1994). This study scrutinized the existing Palaeolithic record and revealed a notable absence of concrete evidence of hominin presence in Europe at sites predating the *Mimomys – Arvicola* transition approximately 500,000 years ago. This finding laid the foundation for the short chronology hypothesis, asserting that there is no unequivocal proof of human occupation in Europe before around 500,000 years ago (Roebroeks and van Kolfschoten 1994). This hypothesis sparked a fervent debate regarding the evidence from the examined sites.

Shortly after the publication of the short chronology hypotheses, new discoveries from localities such as the Spanish site Atapuerca (Carbonell *et al.* 1995) provided compelling evidence of hominin occupation, alongside a *Mimomys* vole assemblage, predating the *Mimomys – Arvicola* transition. These findings posed challenges to the original hypotheses, prompting a revised model known as the phased-colonization model. According to this revised model, the short chronology concept applies exclusively to Europe north of major mountain chains like the Alps and the Pyrenees, while the Mediterranean region, particularly Spain, was already inhabited by hominins during the late Early Pleistocene (Roebroeks 2001).

However, more recent discoveries have further complicated the short chronology hypotheses. Parfitt *et al.* (2010) presented evidence, including stone artifacts combined with mammalian fauna featuring at least two *Mimomys* species, from the Happisburgh locality in the United Kingdom. This evidence demonstrates the presence of Early Pleistocene hominins in northern Europe, challenging the applicability of the short chronology hypotheses as initially defined by Roebroeks and van Kolfschoten (1994), even in this region.

The role of voles in the ongoing discussion about the stratigraphical position of specific sites is notably evident in the debate surrounding the age of Boxgrove in the UK. The site is biostratigraphically dated to the last temperate stage of the Cromerian Complex, correlated with Marine Isotope Stage (MIS) 13 (Stringer *et al.* 1998). The correlation of the mammalian fauna with Marine Isotope Stage (MIS) 13 at Boxgrove leans heavily on the identification of the primitive water vole *Arvicola terrestris cantiana*, in conjunction with other fossils characteristic of the later stage of the Cromerian Complex. This correlation takes precedence over aminostratigraphy and nannofossil data, both of which point towards a younger age, specifically MIS 11 (Stringer *et al.* 1998).

Biostratigraphical data play a pivotal role in the ongoing discussion about the age of the Lower Palaeolithic finds from Schöningen, Germany, particularly concerning the age of the Schöningen Spear horizon, correlated with the locally defined Reinsdorf Interglacial (Thieme 1997). The Schöningen water voles, among the most well-represented species in the Spear Horizon faunal record, exhibit SDQ-values ranging from 125.4 (the base of the Reinsdorf sequence) to 115.2 (the Schöningen Spear horizon). This range indicates a Toringian fauna (van Kolfschoten 2012; 2014), which is younger than the small mammalian assemblages from British

MIS 11/Hoxnian (=Holsteinian) sites like Barnham and Beeches Pit, which exhibit higher SDQ-values (Parfitt, 1998).

Moreover, the rare occurrence of a *Mimomys*-fold in the Schöningen record suggests that water vole populations from Barnham and Beeches Pit, where *Mimomys*-fold frequencies are much higher than in Schöningen, are more primitive and, therefore, stratigraphically older. Conversely, numerous late Middle Pleistocene faunal assemblages, such as those from German localities like Weimar-Ehringsdorf, Ariendorf, Schweinskopf, and Maastricht-Belvédère (the Netherlands) (van Kolfschoten 1990) exhibit lower SDQ-values.

A more recent Mediterranean illustration of the significance of voles is evident in the determination of the stratigraphical age of two Palaeolithic sites situated in the Megalopolis Basin, Greece: Marathousa 1 (Doukas *et al.* 2018) and Kyparissia 4 (van Kolfschoten *et al.* in press). The faunal assemblage at Marathousa 1 encompasses unrooted *Arvicola* molars displaying a positive '*Mimomys*' enamel differentiation, with a mean SDQ-value of 122, indicative of a late Middle Pleistocene age for the lacustrine deposits (Doukas *et al.* 2018). In contrast, the limited rodent remains from Kyparissia 4 include at least three different voles: *Mimomys* sp. (cf. *Mimomys savini*) (large), *Mimomys* sp. (small), and *Microtus (Microtus)* sp. This combination of species suggests a Late Biharian age for the deposits, indicating an older age compared to the stratigraphical position of the Marathousa 1 site (van Kolfschoten *et al.* in press).

Other sites previously exposed in the Megalopolis Basin, such as Thoknia and the Choremiou section with four different fossiliferous layers, have yielded small mammal assemblages with vole remains. The vole record provides an avenue to explore the intricate stratigraphical position of the various Megalopolis Basin lignite seams including those exposed at the Palaeolithic sites .

But what about Wil's question about the age of the fauna from Maastricht-Belvédère? Did the voles from the excavated archaeological site yield any useful biostratigraphical information? Yes, they did. The voles excavated from this archaeological site indeed provided valuable biostratigraphical information. The mean SDQ-values of the *Arvicola* molars from Maastricht-Belvédère layers 3 and 4 are both 102. These values are markedly higher than the mean SDQ-value of the late Middle Pleistocene *Arvicola* molars from Rhenen and considerably lower than the mean SDQ-values of the *Arvicola* molars from Neede (the Netherlands), a site with a Holsteinian age (van Kolfschoten 1990), correlated with MIS 11.

The Maastricht-Belvédère faunas from layers 3 and 4, along with the small mammal faunal assemblage from Rhenen, predate the Saalian glaciation correlated with MIS 6. Consequently, both faunas are assigned to MIS 7. This designation is based on the *Arvicola* molars being evolutionarily much more advanced than those from the Schöningen Spear horizon, which is correlated with MIS 9.

Discussion and Conclusion

The limited case studies presented above underscore the pivotal role of information about fossil voles in debates on the age of fossil assemblages and facilitating long-distance correlations between sites. Numerous other examples abound, including Tesakov's analyses of voles from Tegelen (the Netherlands). The biostratigraphical correlation between the Tegelen data and sites in Eastern Europe, as explored by Tesakov (1998), led to a revised and higher age for the Tegelen fauna. This adjustment aligns more seamlessly with the presumed age of the hominin finds from Dmanisi (Georgia). These instances emphasize the significance of vole data not only in local stratigraphy but also in establishing broader chronological contexts for archaeological and paleontological sites.

But how good is the method and how accurate is the vole clock? The vole clock method, like any dating technique, comes with both strengths and limitations. One of the strengths lies in the abundance of voles in the fossil record. Their frequent presence offers a rich source of material for analysis. Another notable strength is the high rate of unidirectional evolution observed in several vole lineages. The clear evolutionary trends in vole characteristics contribute to the reliability of the method. The European fossil vole record, along with voles from specific regions in Asia, has also been extensively studied for decades by a large group of scholars. This wealth of data forms a robust and unique dataset that serves as the foundation for biostratigraphical correlations.

A significant constraint of the method is its applicability being limited to Eurasia and North America. Moreover, the Quaternary geographical distribution of the European water vole is confined to Eurasia (Fig. 3). Additionally, for an accurate correlation between sites, it is crucial to consider the observed geographical gradient in the evolutionary stage of water vole molars.

For the precise determination of the evolutionary stage of a specific vole population, it is imperative to assemble a collection of at least 20 molars, including several lower m1 molars from (semi-)adult individuals. A larger sample size provides a more representative picture of the population's characteristics and helps mitigate individual variations. Including molars from (semi-)adult individuals is crucial because the evolutionary changes in dental features that are examined in the vole clock method are often associated with age and maturity. Juvenile molars might not exhibit the same characteristics as those from mature individuals.

The quality of fossilized vole material is crucial. Suboptimal preservation can constrain both the quantity and accuracy of information extractable from the specimens. Equally vital is ensuring that the fossil vole remains have not undergone reworking from older deposits.

In essence, the success of the vole clock method relies not only on the quantity but also on the quality of the vole specimens. A meticulous approach to sampling, considering the age structure of the individuals, and accounting for preservation biases are integral aspects of ensuring accurate and meaningful results in biostratigraphical studies.

In summary, the vole clock method is a powerful proxy for relative dating, particularly in regions where voles are abundant and well-preserved. Its strengths lie in the abundance of vole fossils, the clear evolutionary trends, and the extensive dataset available. However, like any dating method, it has limitations and depends on various factors, including regional variations and the quality of fossil preservation.

Acknowledgements

This paper is a tribute to commemorate the retirement of Wil Roebroeks. Throughout the years, Wil has consistently remained a source of inspiration and a valued colleague. Wil, thank you for inviting me to join your research team and your important support during my scientific career. I also wish to extend my appreciation to the editors of the *Festschrift*, Marie Soressi, Gerrit Dusseldorp and Wei Chu, for the invitation to contribute to this special volume. Lastly, I would like to express my thanks to André Ramcharan (Faculty of Archaeology, Leiden University) for the pictures he made of the European water vole skull.

References

Bosinski, .G.. 1992. *Eiszeitjäger im Neuwieder Becken.* Koblenz, Archäologische Denkmalpflege.

Carbonell, E., J.-M. Bermúdez de Castro, J. L. Arsuaga, J. C. Diez, A. Rosas, G. Cuenca-Bescós, R. Sala, M. Mosquera and X.-P. Rodriguez. 1995. Lower Pleistocene hominids and artifacts from Atapuerca-TD6 (Spain). *Science* 269: 826-829.

Chaline, J.. 1986. Continental faunal units of the Plio-Pleistocene of France. *Memorie della Società Geologica Italiana* 31: 175-183.

Danukalova, G., A. Yakovlev, E. Osipova, R. G. Kurmanov, and Th. van Kolfschoten. 2016. Biostratigraphy of the early Middle Pleistocene of the Southern Fore-Urals. *Quaternary International* 420: 115-35.

Doukas, C., Th. van Kolfschoten, K. Papayianni, E. Panagopoulou and K. Harvati. 2018. The small mammal fauna from the Palaeolithic site Marathousa 1 (Greece). *Quaternary International* 497: 95-107.

Fejfar, O. and W.-D. Heinrich. 1990. Muroid rodent biochronology of the Neogene and Quaternary of Europe. In: *European Neogene Mammal Chronology*, E. H. Lindsay, V. Fahlbusch and P. Mein (eds.): 91-117. New York: Plenum Press.

Heinrich, W.-D..1978. Zur biometrischen Erfassung eines Evolutiontrends bei *Arvicola* (Rodentia, Mammalia) aus dem Pleistozän Thüringens. *Säugetierkundliche Informationen* 2: 3-21.

Heinrich, W.-D.. 1982. Zur Evolution und Biostratigraphie von *Arvicola* (Rodenlia, Mammalia) im Pleistozän Europas. *Zeitschrift für Geologische Wissenschaften* 10: 683-735.

Heinrich, W.-D.. 1987. Neue Ergebnisse zur Evolution und Biostratigraphie von *Arvicola* (Rodentia. Mammalia) im Quartär Europas. *Zeitschrift für Geologische Wissenschaften* 15: 389-406.

Néraudeau, D. L. Viriot, J. Chaline, B. Laurin and Th. van Kolfschoten. 1995. Discontinuity in the Plio-Pleistocene Eurasian water vole lineage. *Paleontology* 38, no. 1: 77-85.

Parfitt, S. A.. 1998. The interglacial mammalian fauna from Barnham. In: *Excavations at the Lower Palaeolithic Site at East Farm, Barnham, Suffolk, 1989-94*, Nick Ashton, Simon Lewis and Simon A. Parfitt (eds.): 111-47. London: Occasional Paper of the British Museum 125.

Parfitt, S. A., N. M. Ashton, S. G. Lewis, R. L. Abel, G. R. Coope, M. H. Field, R. Gale, P. G. Hoare, N. R. Larkin, M. D. Lewis, V. Karloukovski, B. A. Maher, S. M. Peglar, R. C. Preece, J. E. Whittaker and Ch. B.

Stringer. 2010. Early Pleistocene human occupation at the edge of the boreal zone in northwest Europe. *Nature* 466: 229-233.

Panteleev, P. A., A. N. Terekhina, A. A. Varshavskiĭ and V. Nikolaevich Bol'shakov. 1990. *Ėkogeograficheskaia͡ izmenchivost' gryzunov*. (Ecogeographical Variability of Rodents). Moscow: Nauka Press. (in Russian)

Roebroeks, W. and Th. van Kolfschoten. 1994. The earliest occupation of Europe: a short chronology. *Antiquity* 68: 489-503.

Roebroeks, W.. 2001. Hominid behaviour and the earliest occupation of Europe: an exploration. *Journal of Human Evolution* 41: 437-61.

Röttger, U.. 1987. Schmelzbandbreiten an Molaren von Schermäusen (*Arvicola* Lacépède, 1799). *Bonner Zoologische Beiträge* 38: 95-105.

Stringer, Ch., E. Trinkaus, M. B. Roberts, S. A. A. Parfitt and R. I. Macphail. 1998. The Middle Pleistocene human tibia from Boxgrove. *Journal of Human Evolution* 34: 509-47.

Tesakov, A. S.. 1998. Voles of the Tegelen fauna. *Mededelingen Nederlands Instituut voor Toegepaste Geowetenschappen TNO* 60: 71-134.

Thieme, H.. 1997. Lower Palaeolithic hunting spears from Germany. *Nature* 385: 807-810.

Urban, B.. 1995. Palynological evidence of younger Middle Pleistocene Interglacials (Holsteinian, Reinsdorf and Schöningen) in the Schöningen open cast lignite mine (eastern Lower Saxony, Germany). *Mededelingen Rijks Geologische Dienst* 52: 175-186.

van der Meulen, A. J. and Th. van Kolfschoten. 1986. Review of the Late Turolian to Early Biharian mammal faunas from Greece and Turkey. *Memorie della Società Geologica Italiana* 31: 201-11.

van Kolfschoten, Th.. 1981. On the Holsteinian? and Saalian mammal fauna from the icepushed ridge near Rhenen (The Netherlands). *Mededelingen Rijks Geologische Dienst* 35: 223251.

van Kolfschoten, Th.. 1988. Die Mäuseuhr. *Archäologie in Deutschland* 3: 18-19.

van Kolfschoten, Th.. 1990. The evolution of the mammal fauna in the Netherlands and the Middle Rhine area (Western Germany) during the late Middle Pleistocene. *Mededelingen Rijks Geologische Dienst* 43 (3): 1-69.

van Kolfschoten, Th.. 1992. Aspects of the migration of mammals to northwestern Europe during the Pleistocene, in particular the reimmigration of *Arvicola terrestris. Courier Forschungsinstitut Senckenberg* 153: 213-220.

van Kolfschoten, Th.. 2012. The Schöningen mammalian fauna in biostratigraphical perspective. In: *The chronological setting of the Palaeolithic sites of Schöningen*, K. Behre (ed.): 113-124. Mainz: Verlag des RGZM.

van Kolfschoten, Th.. 2014. The Palaeolithic locality Schöningen (Germany): A review of the mammalian record. *Quaternary International* 326-327: 469-80.

van Kolfschoten, Th., G. E. Konidaris, C. Doukas, A. Athanassiou, V. Tourloukis, E. Panagopoulou, P. Karkanas and K. Harvati. in press. Voles (Rodentia, Mammalia) as a proxy to date the site Kyparissia 4 (Megalopolis Basin, Greece). *Paleoanthropology Book Series*. Tübingen University press.

von Koenigswald, W.. 1973. Veränderungen in der Kleinsäugerfauna von Mitteleuropa zwischen Cromer und Eem (Pleistozän). *Eiszeitalter und Gegenwart* 23/24: 159-167.

von Koenigswald, W.. 1980. Schmelzstruktur und Morphologie in dem molaren der Arvicoliden (Rodentia). *Abhandlungen Senckenbergische Naturforschende Gesellschaft* 59, no. 1: 1-94.

von Koenigswald, W. and Th. van Kolfschoten. 1996. The *Mimomys-Arvicola* boundary and the enamel thickness quotient (SDQ) *of Arvicola* as stratigraphic markers in the Middle Pleistocene. *The early Middle Pleistocene in Europe*, C. Turner (ed.): 211-26. Rotterdam: Balkema.

In Dusseldorp, G.L., Chu, W., Bakels, C.C. and M.A. Soressi (eds) 2025. Intent on the Paleolithic. Papers in honour of Prof. Dr. Wil Roebroeks. *Analecta Praehistorica Leidensia* 55. Leiden: Sidestone Press, pp. 147-158.

Exploring landscape burning as the earliest active use of fire

Femke H. Reidsma, Amanda G. Henry

Abstract

Fire is a key component of human technology, but when and how we first developed the ability to use and create fire remains heavily debated. Some argue that control of fire developed early and quickly became an essential tool for cooking food, based on morphological changes that first appeared in *Homo erectus*. Others emphasize the lack of widespread and consistent traces of fire in the early archaeological record, and argue that control of fire came relatively late, post 500 ka. In looking for archaeological evidence for fire use, most studies focus on the co-occurence of fire traces and other signs of human activity (e.g., stone tools, cut-marked bones). This assumes that our ancestors first focused on the idea of the hearth, either bringing the fire back to the activity area, or performing activities near a spatially-constrained fire. However, recent research has highlighted the potential importance of intentional burning of landscapes as a tool that could similarly provide nutritional and energetic benefits to the earliest hominin fire users. If landscape burning was the first type of intentional fire use, then we should not expect the co-occurrence of fire traces and other signs of hominin activity. Inspired in part by Wil Roebroeks' emphasis on off-site activities and longstanding interest in fire, here we provide a suggestion for a new approach to explore the archaeological record for evidence of intentional landscape burning. Future analyses using this approach may help us better understand the deep history of fire use, and resolve the potential mismatch between the early appearance of morphological traits associated with the consumption of cooked food and the late co-occurrence of fire traces and hominin activity in the archaeological record.

Keywords: landscape modification; anthropogenic fire; evolution; off-site fire

Introduction

Fire use has long been recognized as a defining trait of our species. It provides light, warmth, and the power to transform food and tools (McCauley *et al.* 2020). While it may be less present in the everyday life of many present-day people, though not in all parts of the world, our species is completely reliant on fire, particularly on the benefits of cooked foods (Wrangham and Conklin-Brittain 2003). Given its importance in our evolutionary history, fire has been the subject of multiple studies, with many researchers focused on identifying traces of ancient fire use and understanding how and when our ancestors first learned to use and control it.

Several researchers have proposed various consecutive stages in hominin interaction with fire (e.g., Burton 2009; Sandgathe 2017; Pruetz and Herzog 2017; Parker *et al.* 2016). Pruetz and colleagues (Pruetz and Herzog 2017; Pruetz and LaDuke 2010) have suggested that hominins first had to be able to conceptualize

Femke H. Reidsma
Faculty of Archaeology, Leiden University, Leiden, The Netherlands
femke.reidsma@gmail.com

Amanda G. Henry
Faculty of Archaeology, Leiden University, Leiden, The Netherlands
a.g.henry@arch.leidenuniv.nl

fire, that is, to understand the behaviour of fire and recognize the benefits of it. This was followed by the ability to maintain a fire, in which hominins could contain, feed, and extinguish a fire. In some studies this type of interaction with fire is considered "use" or "opportunistic use" of fire (Sandgathe 2017; Shimelmitz *et al.* 2014), and suggests that hominins were able to acquire some of the benefits of fire, but were not fundamentally reliant on it and / or were unable to create it. Even without the ability to create fire, hominins at this point could have had some degree of control of the fire, for example by choosing where the fire takes place, and choosing fuel types or heating conditions to influence fire properties such as heat, smoke, or light. The next step was gaining the ability to produce a fire. Confusingly, some also call this "control" of fire, focusing on having the knowledge and technology to create fire when desired (Sandgathe 2017). Others further distinguish between the ability to control and produce fire on the one hand, and "habitual" and "obligate" use on the other hand. The former terms focus on the technological and conceptual ability to manage and create fire, while "habitual" emphasizes the behaviour of regularly creating and using fire and fire products (Roebroeks and Villa 2011; Shimelmitz *et al.* 2014). "Obligate" refers to the biological adaptation to and reliance on fire (e.g., Wrangham and Conklin-Brittain 2003).

Our understanding of how past fire use developed is based on two main lines of evidence: inferences drawn from morphological changes to hominin skeletal material attributed to the ingestion of cooked foods, and archaeological data, informed by ethnographic studies. Skeletal changes, including a decrease in tooth and gut size with a concomitant increase in brain size, occurred early in the genus *Homo,* around 2 Mya. Based on energetic studies, these changes have been attributed to a reliance on cooked foods, which provide more net energy to the consumer (Carmody and Wrangham 2009b; 2009a; Wrangham 2017). This evidence has been interpreted to mean that hominins gained control of fire quite early and quickly became obligate fire users. This interpretation is often sharply contrasted with the archaeological evidence, which suggests a slower and potentially less linear process. Tracing the origins of fire use in the deep past relies on the presence and identification of different heated materials. These fire proxies include charred organic matter, ash, heated bones, heated lithics and stones, and heated sediment. Under unique conditions fire products such as tar or cooked food remains can also preserve and offer insight into specific types of fire use (Henry, Brooks *et al.* , and Piperno 2011; Mazza *et al.* 2006; Kabukcu *et al.* 2023; Niekus *et al.* 2019). While evidence for the use of fire becomes increasingly frequent and widespread during the Middle Pleistocene (from around 400 ka onwards), in the Lower Pleistocene record traces for fire are far more scarce and often contentious (Hlubik *et al.* 2019; James 1989; Roebroeks *et al.* 2021; Roebroeks and Villa 2011; Scott and Hosfield 2021). While the scarcity of the earlier record may be partially attributable to preservation bias, the overall mismatch between the morphological and archaeological traces may be due in part to the kinds of fire proxies that we accept as evidence for human use of fire, their visibility in the archaeological record, and the kinds of behaviours that are implied by this evidence.

As summarized by Hlubik and colleagues (2019; Table 1 and references therein), evidence for fire between 1.5 and 0.8 Ma is limited to a handful of sites, some of which have circular discolored sediments, and others with paleomagnetic and thermoluminescent conditions that are consistent with heating. Some have evidence for association between heated sediments and thermally-altered bone or thermally-altered artifacts. Notably, the implicit criteria of whether a fire proxy is truly evidence of human controlled fire is that it either 1) has the same shape and distribution of a hearth (a spatially-constrained fire), or 2) shows co-occurrence of fire and human activity.

The approach of accepting only "hearths" as evidence for early hominin use of fire was developed in part to avoid confusing intentional fire use with evidence of wildfires. Naturally-occurring fires are very common, and so some fire traces would be expected across landscapes. These wildfires could sweep across areas with hominin activities, leaving broad, diffuse patterns of fire traces at the same location as stone tools and bones. Therefore, spatially-constrained fire traces are seen as "better" evidence for intentional use of fire. It is important to note, however, that this approach is also used because scatters of stone tools and bones are more easily identified in the archaeological record, while fire traces often require specialized and targeted analyses to identify (e.g., Mallol and Henry 2017). The bones and stones are identified first, and then researchers look for fire at that location. However, recent work, particularly that done by Wil Roebroeks and colleagues (e.g., Scherjon *et al.* 2015; Pop *et al.* 2016; Roebroeks and Villa 2011; Roebroeks *et al.* 2021; 1992), has demonstrated that a focus on on-site activities limits our understanding of the past, and that only by incorporating the potential of off-site activities can we gain a broader and deeper understanding of human behaviour across their landscapes.

Here we argue that the focus on on-site, spatially-constrained fire traces has biased our understanding of the earliest hominin interaction with fire. We seek to emphasize the possibility that intentional landscape burning was an early stage in hominin fire use that provided many of the benefits commonly attributed only to spatially-constrained fire. In the sections below, we argue how studies of animals and the ethnographic record indicate that landscape burning provides significant nutritional and energetic benefits. Landscape burning has been considered as a method of controlling the environment (Scherjon *et al.* 2015), but evidence for this behaviour has only really been explored in the more recent archaeological record (Roebroeks *et al.* 2021; Thompson *et al.* 2021). The idea that landscape burning was the first active use of fire has been hinted at by others, particularly by Chazan (2017) and Parker and colleagues (2016), but the implications for how we should approach the earliest archaeological record, and specifically which traces we should be looking for, have not yet been addressed. This paper will present a possible strategy to test it in the archaeological record.

A note on terminology

In this paper we draw a distinction between wildfires – naturally occurring fires that burn across large areas – and landscape burning – intentional placement of fire by humans in order to burn parts of a landscape. Furthermore, we acknowledge that anthropogenic fires, whether created or only curated, can occur both on-site, that is, at the same location as concentrations of other archaeological materials, and off-site, away from other evidence of human activity. While on-site fires usually have a small, spatially-constrained footprint, off-site fires can both be similarly delimited, such as in the case of single-use hearths or ephemeral cooking fires (e.g., Holdaway *et al.* 2017), or widespread. We refer to both on-site and off-site delimited fires as "spatially-constrained fire", and widespread off-site anthropogenic fires as "landscape burning".

Modern evidence for the importance of landscape burning

Animal use of landscape fires

Wildfire has a very deep history, with evidence from plants suggesting that natural burning of landscapes was common from at least the Late Cretaceous period (Bond and Scott 2010; Hill *et al.* 2016; Keeley *et al.* 2011). Angiosperms in particular have adapted to tolerate and even benefit from fire regimes, for example by developing heat-induced germination of seeds that allows the quick recolonization of freshly-burned and therefore more nutrient-rich landscapes with less competition. Animals, too, adapted to frequent fires. Given their mobility, it is unsurprising that many animal taxa move away from or avoid fires. Interestingly, many taxa also display a variety of behaviours that range from calm acceptance of approaching fires (Pruetz and Herzog 2017), to active use of burned landscapes (Pausas and Parr 2018) and even direct manipulation of fires (Bonta *et al.* 2017). Animals that engage more actively with fire and burned landscapes reap nutritional and energetic benefits from these behaviours. Landscape fires flush insects, lizards and small mammals out of hiding, and many birds of prey and other animals in fire-prone habitats regularly forage along the edges of wildfires, while others feast on the killed insects after the fire has passed (Bonta *et al.* 2017 and citations therein). Some of these "fire-foraging" raptors have been observed to actively pick up burning sticks to spread fires beyond natural boundaries such as streams (Bonta *et al.* 2017). This behaviour has been interpreted as the intentional creation of a high return-rate habitat (e.g., one where food is more easily found and captured). Other animals may avoid active fires, but make heavy use of recently burned landscapes, again because of the foraging opportunities they provide. Herbivores preferentially consume the so-called "green flush", which is the high-protein and low-fiber new growth that occurs after fire (Wilsey 1996). Additionally, ash is rich in minerals, and some herbivores return to burned areas to acquire these vital salts (Eby and Ritchie 2016). Primates in fire-prone habitats, including vervet monkeys and chimpanzees, have been observed to preferentially forage in recently burned landscapes (Herzog *et al.* 2014; Pruetz and Herzog 2017). Several primate species have been observed to seek out foods in burned areas, including burned fruits.

Overall, the majority of observations about animal interactions with fires indicate that they use these burned landscapes for their nutritional or energetic benefits. These include increased nutrient density (green flush or salt availability), and decreased search time for prey (flushed or killed prey). Some have suggested that burned areas also reduce other energetic costs by making travel easier (Pruetz and Herzog 2017). For herbivores at least, other hypotheses for the use of these burned landscapes such as avoidance of parasites or preferences for warmer microclimates, are not supported (Eby and Ritchie 2016). Given that so many animal taxa have developed a reliance on burned landscapes for their nutritional benefits (Pausas and Parr 2018), it is not unsurprising that humans have also recognized and made considerable use of these advantages.

Human use of landscape fires

The energetic and nutritional benefits of spatially-constrained fires for modern humans have been extensively studied (e.g., McCauley *et al.* 2020). One of the main benefits comes from the transformation of foods via cooking. Cooked foods, including meat, tubers and fat-rich nuts, are easier to digest (Boback *et al.* 2007; Carmody *et al.* 2011; Carmody and Wrangham 2009b; Groopman, Carmody *et al.* 2015), and therefore provide more calories for other vital tasks, and potentially contribute to the ability to grow a larger brain (Aiello and Wheeler 1995; Carmody and Wrangham 2009b). Fire also produces warmth, making it easier to maintain a core body temperature and thus making marginal environments more habitable (Hosfield 2016), and light, which can extend the length of time available for necessary tasks (Attwell *et al.* 2015). While these benefits may be tempered by costs related to fuel collection (Henry *et al.* 2018), it is still clear that spatially-constrained fires provide considerable advantages.

In contrast, the value of landscape burning for modern humans has only occasionally been considered. Two reviews of ethnographic literature highlight how landscape or off-site fires are also primarily used for their energetic and nutritional benefits. Mills (1986) focused explicitly on what she called "prescribed burning", that is, the intentional creation of broad landscape fires. Across multiple different hunter-gatherer groups, the most common use of landscape burning was as a direct hunting strategy, in the form of drives or surrounds. The second most common use was to foster the growth of preferred food items such as grass seeds, vines, roots, bulbs, nuts and berries. The preference of animals for burned landscapes was also widely recognized, as the third most common use of landscape fire was to entice more game into the area by burning off unwanted plants and triggering a green flush. The top three uses of landscape fires are therefore directly related to the acquisition of food items. Numbers four and five also have nutritional / energetic benefits, but less direct. Landscape burning was occasionally used to get rid of obstructing or otherwise useless plants in order to make travel easier and preferred foods more visible. Finally, they were occasionally documented as a means to deter pests, parasites, and hazardous animals from preferred foraging areas, or to drive prey species into another, preferred, area.

In their more recent summary, Scherjon and colleagues (2015) increased the scope of the analyses to include both more ethnographic records, but also a wider variety of off-site fire uses such as torches and other spatially-constrained fire, not just landscape burning. The driving of game is identified as the most common use of off-site fires. Direct killing of animals and stimulation of particular flora are also relatively important, and in total, direct procurement of food accounts for the vast majority of off-site fire use. Future planning in the form of landscape burning to promote preferred prey animals is also relatively common, but less so than some social uses of fire, such as a mode of communication.

Interaction with and use of landscape fires therefore has multiple clear benefits for fire-making modern humans. Parker and colleagues (2016: 55) have elegantly argued that at least some of these benefits would apply even to early hominins, provided they could conceptualize the use of fire and potentially move it or even maintain it. They particularly emphasize that the benefits arising from the use of burned landscapes

> *"included improved efficiency in travel, resource detection and acquisition, and a reduction of food processing costs."*

Furthermore, they argue that

> *"Fire control in the form of simple landscape burning and purposeful cooking followed as a consequence of targeting the foraging benefits of fire."*

The above points all suggest that landscape burning is of potentially the same importance as on-site or spatially-constrained off-site fires, primarily by providing nutritional and energetic advantages. Thus far, however, we have not adapted our approach to the study of early fire use in order to effectively identify this behaviour in the Lower Pleistocene record.

How to identify landscape burning in the deep past

What to look for: fire chemistry and proxies

Fire is a chemical reaction that requires an external heat source, an organic fuel, and sufficient oxygen to be ignited and continue to burn (Emmons and Atreya 1982). The heat that is produced during combustion affects the physical and chemical properties of the materials it encounters. The outcome of this encounter depends on the characteristics of the fire (e.g., temperature, oxygen availability, exposure time), the properties of the material, and the conditions in which the material is heated (e.g., direct vs. indirect) (Abdolahzadeh *et al.* 2023; Aldeias et al. 2016; Braadbaart and Poole 2008; Reidsma *et al.* 2016; van Hoesel *et al.* 2019). These processes are universal and knowledge of them can be used to predict what fire proxies certain types of fire use are likely to leave behind. However, due to the nature of the chemical reaction, different types and uses of fire, both natural and anthropogenic, can leave very similar traces. Wildfires and landscape burning can be expected to have very similar proxy signals, as both involve the burning of vegetation in large areas and spread through the landscape in similar ways. The main proxy materials that remain after a landscape fire is extinguished are charred plant material and ash, and sometimes patches of heated soil, with variation in frequency and distribution depending on factors like vegetation type, available fuel biomass, fire intensity, and exposure time (e.g., Hudspith *et al.* 2018; Stahlschmidt *et al.* 2023; Finkelstein *et al.* 2005; Goldberg, Miller *et al.* 2017).

Natural fire events and their impact on past landscapes are identified through variation in the frequency of direct proxies like charred plant material and/or fire-related biomarkers such as polycyclic aromatic hydrocarbons (PAHs) preserved in sediment archives (e.g., Abu Hamad, Jasper *et al.* 2012; Nikulina *et al.* 2022; Synnott *et al.* 2021). The presence of natural fires can also be traced indirectly through the impact they have on the landscape. Wildfires open up the landscape by changing the type of vegetation that can thrive in an area, and are a known driver for the spread of savannah grassland (e.g., Nikulina *et al.* 2022; Karp *et al.* 2021; Karp *et al.* 2018; Foster *et al.* 2018). In addition, intentional clearing of vegetation by fire can enhance erosion and affect sedimentation processes and rate (e.g., Synnott *et al.* 2021; Ravi *et al.* 2012; Thompson *et al.* 2021). These landscape changes can be detected through a combination of palaeoenvironmental, geomorphological, and geochronological techniques, with the relative abundance of plant taxa in palynological data indicating different vegetation types (e.g., open vs. closed) being particularly informative (Nikulina *et al.* 2022; Thompson *et al.* 2021).

While very similar in behaviour and resulting proxy signal, there are some minor differences between wildfires and landscape burning. First, landscape burning likely occurs at a smaller scale than wildfires. This means that proxy data is needed that reflects the local or regional scale (Scherjon *et al.* 2015; Nikulina *et al.* 2022; Daniau *et al.* 2010). Second, wildfires are climate driven and often seasonal (e.g., Finkelstein *et al.* 2005; Baker 2022; Brittingham *et al.* 2019), while landscape burning can be done irrespective of those factors (Scherjon *et al.* 2015). However, it should be noted that in the absence of the ability to produce fire hominins would still depend on naturally available fires to conduct their landscape burning. Finally, landscape burning will have a strong association with the presence of archaeology in the area (e.g., occurring in the same geological / archaeological levels, and in roughly the same regions), whereas natural fire will likely be more strongly correlated with changes in palaeoenvironmental proxies (e.g., Brittingham *et al.* 2019; Nikulina *et al.* 2022).

Finding the evidence

Anthropogenic impact on landscapes through burning by hunter-gatherers has been suggested for a number of archaeological sites, most of which date to the late Pleistocene / early Holocene (see Nikulina *et al.* 2022; Roebroeks *et al.* 2021; Thompson *et al.* 2021). So far the oldest example of landscape burning comes from the Eemian lake-basin site of Neumark-Nord 2 (Germany) where it has been suggested that Neanderthals modified the landscape through burning (Roebroeks *et al.* 2021). This interpretation is based on the strong correlation between the micro-charcoal signal, palaeoenvironmental proxies that indicate a change to more open vegetation, and intensive hominin activity. This signal is contrasted by palaeoenvironmental data from comparable sites with little to no archaeology that closely match the Eemian succession of pollen assemblage zones commonly known for the wider region (Roebroeks *et al.* 2021). All cases where landscape burning is suggested, and particularly the Neumark-Nord example, rely on high-resolution multi-proxy palaeoenvironmental datasets and correlation with archaeology. However, this type of high-resolution data is often not available due to factors like preservation, research approach, and chronological constraints. Building a case for landscape burning for sites with less optimal conditions will be challenging, especially for sites of (much) greater antiquity.

We propose to look for evidence of landscape burning in the Lower Pleistocene record by supplementing the existing approach outlined above with biomarker geochemistry, particularly by identifying changes in the (isotopic) signal of leaf wax lipids and the frequency of PAH molecules. Both proxies are routinely used to detect changes in past climate, vegetation type, and natural fire regimes, and are known to preserve over geological timescales (e.g., Neupane et al. 2020; Karp *et al.* 2021). In addition, they can be extracted from the same sediment sample, making them easy to compare. Compound specific isotopic analysis of lipid biomarkers can be used to reconstruct past temperatures and precipitation levels, while identification of specific *n*-alkanes can be used to infer vegetation type and openness. These lipid biomarkers can provide a local and/or regional palaeoenvironmental signal (e.g., Tanner *et al.* 2007; Toney *et al.* 2020; Blyth *et al.* 2007). PAHs are molecules that are formed during the heating of organic materials and that will have a low (lPAH) or high (hPAH) molecular weight depending on their formation temperature. The ratio between these two groups of PAHs can be used to differentiate between on-site fire use (i.e., majority hPAHs) and fires in the landscape (i.e., majority lPAHs), based on existing studies that show that lPAHs will travel further than hPAHs (for details see Brittingham *et al.* 2019). PAHs can be used to trace fire in the landscape at both a local and regional scale (Nikulina *et al.* 2022).

As with other fire proxies, the PAH signal for landscape burning can be expected to be similar to that of wildfires. Identification of landscape burning using geochemical techniques therefore also requires a signal of local change against a regional baseline, and a strong correlation with evidence for hominin activity. A baseline signal can be obtained in three different ways: 1) by analysing a regional palaeoenvironmental record such as a lake core (cf. Thompson *et al.* 2021), 2) by analysing sites in similar contexts with less/no archaeology (cf. Roebroeks *et al.* 2021), or 3) by analysing a stratigraphic section that encompasses a period of hominin presence, as well as the period before and after. The chosen baseline will depend on the options available for the region and time period. Regardless, good chronological constraints are essential. The first step towards identifying landscape burning using geochemical techniques is to detect a change in the local PAH signal and to verify if the ratio between lPAHs and hPAHs reflects the signal of a natural fire, rather than on-site fire use. Subsequently, the PAH and biomarker palaeoenvironmental signals (i.e., climate and vegetation type) are compared with the archaeology to determine if the PAHs and vegetation type correlate more strongly with the presence of hominin activity in the landscape than they do with the climate. A strong correlation with layers with archaeological evidence will support landscape burning as a likely explanation for the observed pattern. Integrating the geochemical evidence with other fire and palaeoenvironmental proxies, when available, will strengthen the case. Furthermore, PAHs and lipid biomarkers can also serve as additional independent proxies for the identification of landscape burning in more recent contexts. The addition of PAHs would be particularly valuable as the only direct fire proxy included in existing landscape burning studies is micro-charcoal frequency. To our knowledge PAHs and lipid biomarkers have not been used to address questions related to landscape burning in the archaeological record. The suggested approach will require testing on an archaeological case with existing evidence for landscape burning, but in our opinion provides a promising methodology to trace anthropogenic landscape modification by fire further back in time.

Discussion

While we have no archaeological evidence yet to support landscape burning as the first hominin active use of fire, the ethnographic record and evidence from animal ecology strongly suggest this as a possible step. Animals and humans use landscape fires overwhelmingly for the increased energetic and nutritional benefits they provide, suggesting that these aspects, and not warmth or light, were the driving forces behind the human relationship with fire. Given that landscape burning could be a source of higher-quality food items (pre-cooked animal foods; potentially also fruits and vegetable material), the use of landscape burning might then help explain the mismatch between the rather patchy archaeological record for spatially-constrained, on-site fires and the morphological changes that imply an early reliance on cooked foods.

The potential value and use of landscape burning has thus far not been considered in archaeological narratives about the early development of fire use. Many of the steps described so far seem to focus on spatially-constrained use. Current narratives for the initial development of fire as a tool propose that "maintenance" or "harvesting" of fire occurs sometime after "conceptualization". We see room between these two steps for a first interaction that involves intentional setting of landscape fires in order to maximize energetic return from a habitat. This represents a smaller cognitive step than that from recognizing the benefits of wildfire to using spatially-constrained fire for cooking. The former requires only one step – intentionally moving the fire to obtain increased benefits in a new location – while the latter involves two – moving and containing the fire and moving the food items to a central area before consumption. Repeated spreading of wildfires could quickly allow a hominin to recognize which kinds of fuels were most likely to ignite from a thrown or carried ember, increasing their knowledge of fire properties and providing some degree of control over the fire. This knowledge of fire management could have then translated to the ability to curate it. In such a scenario, hominins could have maintained fire in a single location as a source of starting material for later landscape burning, and not for direct on-site use. Hominin activities could take place nearby simply as a means to ensure the fire was regularly fed. The archaeological signal would appear similar to those from "hearths" as we know them, but the use would have been more like a stone tool cache – a place to keep a resource for later use. It is also possible that the ability to start a fire may have developed in order to create new landscape fires in seasons where wildfires were unavailable.

Shifting the focus of early fire research from on-site use to landscape burning requires a methodology that is in line with the proxy evidence that is characteristic for this type of fire. The approach we describe (tracking the variation in frequency and nature of PAHs and lipid biomarkers) supplements existing strategies for the identification of landscape burning with geochemical techniques commonly used in the study of pre-Pleistocene wildfires. It is therefore well suited for the identification and analysis of proxies of considerable age. The re-conceptualization of the earliest active use of fire also has implications for the types of on-site traces that we can expect at Lower Pleistocene find localities. In the absence of fire curation and/or production, on-site fire evidence will likely be restricted to heated materials that are foraged from the burned landscape and brought back to a central location. These may include heated animal remains and fragments of toasted fruits and nuts. The presence of patches of heated sediment might then be interpreted as an initial form of fire curation, provided natural causes like tree-stump fires can be ruled out. Frequent passive and / or active use of natural fires may have increased the risk of accidental heating of archaeological materials, either by natural or anthropogenically propagated fire, contributing to the archaeological visibility of hominin interaction with fire.

While landscape burning as the first hominin active use of fire remains a hypothesis at this stage, the archaeological approach that we propose here could provide new insight into currently underexplored off-site behaviours throughout human history. Present day and recent historical hunter-gatherer groups regularly use off-site fires for multiple reasons (Scherjon *et al.* 2015); by documenting this kind of fire use in the past we gain a closer understanding of these otherwise invisible behaviours, during any time period.

Acknowledgements

We thank the organizers of this special issue, Marie Soressi, Corrie Bakels, Wei Chu, and Gerrit Dusseldorp, for providing us the opportunity to celebrate Wil Roebroeks' contribution to the field of archaeology. Wil's emphasis on off-site activities and fire uses beyond hearths inspired us to consider ancient use of landscape fires. Beyond these ideas, however, Wil has inspired and supported us both in ways that we cannot begin to count.

References

Abdolahzadeh, A., G. M. Leader, Li Li, D. I. Olszewski and Th. G. Schurr. 2023. Heat Exposed Lithics: An Experimental Approach to Quantifying Potlids by Temperature. *Journal of Archaeological Science: Reports* 48: 103894. DOI:10.1016/j.jasrep.2023.103894.

Abu Hamad, A. M. B., A. Jasper and D. Uhl. 2012. The Record of Triassic Charcoal and Other Evidence for Palaeo-Wildfires: Signal for Atmospheric Oxygen Levels, Taphonomic Biases or Lack of Fuel? *International Journal of Coal Geology* 96-97: 60-71. DOI:10.1016/j.coal.2012.03.006.

Aiello, L. C., and P. Wheeler. 1995. The Expensive-Tissue Hypothesis. *Current Anthropology* 36: 199-221.

Aldeias, V., H. L. Dibble, D. Sandgathe, P. Goldberg and Sh. J. P. McPherron. 2016. How Heat Alters Underlying Deposits and Implications for Archaeological Fire Features: A Controlled Experiment. *Journal of Archaeological Science* 67: 64-79. DOI:10.1016/j.jas.2016.01.016.

Attwell, L. K. Kovarovic and J. Kendal. 2015. Fire in the Plio-Pleistocene: The Functions of Hominin Fire Use, and the Mechanistic, Developmental and Evolutionary Consequences. *Journal of Anthropological Sciences* 93: 1-20. DOI:10.4436/JASS.93006.

Baker, S. J.. 2022. Fossil Evidence That Increased Wildfire Activity Occurs in Tandem with Periods of Global Warming in Earth's Past. *Earth-Science Reviews* 224: 103871. DOI:10.1016/j.earscirev.2021.103871.

Blyth, A.. , A. Asrat, A. Baker, P. Gulliver, M. J. Leng and D. Genty. 2007. A New Approach to Detecting Vegetation and Land-Use Change Using High-Resolution Lipid Biomarker Records in Stalagmites. *Quaternary Research* 68: 314-24. DOI:10.1016/j.yqres.2007.08.002.

Boback, Scott M., Christian L. Cox, Brian D. Ott, Rachel Carmody, Richard W. Wrangham, and Stephen M. Secor. 2007. Cooking and Grinding Reduces the Cost of Meat Digestion. *Comparative Biochemistry and Physiology, Part A* 148: 651-656.

Bond, W. J., and A. C. Scott. 2010. Fire and the Spread of Flowering Plants in the Cretaceous. *New Phytologist* 188: 1137-50. DOI:10.1111/j.1469-8137.2010.03418.x.

Bonta, M., R. Gosford, D. Eussen, N. Ferguson, E. Loveless and M. Witwer. 2017. Intentional Fire-Spreading by 'Firehawk' Raptors in Northern Australia. *Journal of Ethnobiology* 37: 700-718. DOI:10.2993/0278-0771-37.4.700.

Braadbaart, F, and I. Poole. 2008. Morphological, Chemical and Physical Changes during Charcoalification of Wood and Its Relevance to Archaeological Contexts*Journal of Archaeological Science* 35: 2434-2445. DOI:10.1016/j.jas.2008.03.016.

Brittingham, A., M. T. Hren, G. Hartman, K. N. Wilkinson, C. Mallol, B. Gasparyan and D. S. Adler. 2019. Geochemical Evidence for the Control of Fire by Middle Palaeolithic Hominins. *Scientific Reports* 9: 1-7. DOI:10.1038/s41598-019-51433-0.

Burton, F. D.. 2009. *Fire: The Spark That Ignited Human Evolution*. University of New Mexico Press.

Carmody, R. N., G. S. Weintraub and R. W. Wrangham. 2011. Energetic Consequences of Thermal and Nonthermal Food Processing. *PNAS* 108: 19199-19203. DOI:10.1073/pnas.1112128108.

Carmody, R. N. and R. W. Wrangham. 2009a. Cooking and the Human Commitment to a High-Quality Diet." *Cold Spring Harbor Symposia on Quantitative Biology* 74: 427-434.

Carmody, R. N. and R. W. Wrangham. 2009b. "The Energetic Significance of Cooking. *Journal of Human Evolution* 57: 379-91. DOI:10.1016/j.jhevol.2009.02.011.

Chazan, M.. 2017. Toward a Long Prehistory of Fire. *Current Anthropology* 58: S351-S359. DOI:10.1086/691988.

Daniau, A.-L., F. d'Errico and M. F. Sánchez Goñi. 2010. Testing the Hypothesis of Fire Use for Ecosystem Management by Neanderthal and Upper Palaeolithic Modern Human Populations. *PloS One* 5: e9157. DOI:10.1371/journal.pone.0009157.

Eby, S. and M. E. Ritchie. 2016. Alternative hypotheses for mammalian herbivore preference of burned areas in a savannah ecosystem. *African Journal of Ecology* 54: 471-478. DOI:10.1111/aje.12306.

Emmons, H. W. and A. Atreya. 1982. The Science of Wood Combustion. *Proceedings of the Indian Academy of Sciences Section C: Engineering Sciences* 5: 259-268. DOI:10.1007/BF02904581.

Finkelstein, D. B., L. M. Pratt, T. M. Curtin and S. C. Brassell. 2005. Wildfires and Seasonal Aridity Recorded in Late Cretaceous Strata from South-Eastern Arizona, USA. *Sedimentology* 52: 587-599. DOI:10.1111/j.1365-3091.2005.00712.x.

Foster, C. N., P. S. Barton, C. I. MacGregor, J. A. Catford, W. Blanchard and D. B. Lindenmayer. 2018. Effects of Fire Regime on Plant Species Richness and Composition Differ among Forest, Woodland and Heath Vegetation. *Applied Vegetation Science* 21: 132-143. DOI:10.1111/avsc.12345.

Goldberg, P., C. E. Miller and S. M. Mentzer. 2017. Recognizing Fire in the Paleolithic Archaeological Record. *Current Anthropology* 58: S175-S190. DOI:10.1086/692729.

Groopman, E. E., R. N. Carmody and R. W. Wrangham. 2015. Cooking Increases Net Energy Gain from a Lipid-Rich Food. *American Journal of Physical Anthropology* 156: 11-18. DOI:10.1002/ajpa.22622.

Henry, A. G., A. S. Brooks and D. R. Piperno. 2011. Microfossils in Calculus Demonstrate Consumption of Plants and Cooked Foods in Neanderthal Diets (Shanidar III, Iraq; Spy I and II, Belgium). *PNAS* 108: 486-91. DOI:10.1073/pnas.1016868108/-/DCSupplemental.www.pnas.org/cgi/doi/10.1073/pnas.1016868108.

Henry, A. G., Th. Büdel and P.-L. Bazin. 2018. Towards an Understanding of the Costs of Fire. *Quaternary International* 493: 96-105. DOI:10.1016/j.quaint.2018.06.037.

Herzog, N. M., C. H. Parker, E. R. Keefe, J. Coxworth, A. Barrett and K. Hawkes. 2014. Fire and Home Range Expansion. A Behavioral Response to Burning among Savanna Dwelling Vervet Monkeys (Chlorocebus Aethiops). *American Journal of Physical Anthropology* 154: 554-560. DOI:10.1002/ajpa.22550.

Hill, Robert S., Gregory J. Jordan, Robert S. Hill, and Gregory J. Jordan. 2016. "Deep History of Wildfire in Australia." *Australian Journal of Botany* 64: 557-563. DOI:10.1071/BT16169.

Hlubik, S., R. Cutts, D. R. Braun, F. Berna, C. S. Feibel and J. W.K. Harris. 2019. Hominin Fire Use in the Okote Member at Koobi Fora, Kenya: New Evidence for the Old Debate. *Journal of Human Evolution* 133: 214-229. DOI:10.1016/J.JHEVOL.2019.01.010.

Hoesel, A. van, F. H. Reidsma, B. J.H. van Os, L. Megens and F. Braadbaart. 2019. Combusted Bone: Physical and Chemical Changes of Bone during Laboratory Simulated Heating under Oxidising Conditions and Their Relevance for the Study of Ancient Fire Use. *Journal of Archaeological Science: Reports* 28: 102033. DOI:10.1016/j.jasrep.2019.102033.

Holdaway, S. J., B. Davies and P. C. Fanning. 2017. Aboriginal Use of Fire in a Landscape Context: Investigating Presence and Absence of Heat-Retainer Hearths in Western New South Wales, Australia. *Current Anthropology* 58: S230-S242. DOI:10.1086/691436.

Hosfield, R.. 2016. Walking in a Winter Wonderland? Strategies for Early and Middle Pleistocene Survival in Midlatitude Europe. *Current Anthropology* 57: 653-682. DOI:10.1086/688579.

Hudspith, V. A., R. M. Hadden, A. I. Bartlett and C. M. Belcher. 2018. Does Fuel Type Influence the Amount of Charcoal Produced in Wildfires? Implications for the Fossil Record. *Palaeontology* 61: 159-171. DOI:10.1111/pala.12341.

James, S. R.. 1989. Hominid Use of Fire in the Lower and Middle Pleistocene: A Review of the Evidence. *Current Anthropology* 30: 1-11. DOI:10.1086/203705.

Kabukcu, C., C. Hunt, E. van Hill, E. Pomeroy, T. Reynolds, G. Barker and E. Asouti. 2023. Cooking in Caves: Palaeolithic Carbonised Plant Food Remains from Franchthi and Shanidar. *Antiquity* 97: 12-28. DOI:10.15184/aqy.2022.143.

Karp, A. T., A. K. Behrensmeyer and K. H. Freeman. 2018. Grassland Fire Ecology Has Roots in the Late Miocene. *Proceedings of the National Academy of Sciences* 115: 12130-12135. DOI:10.1073/pnas.1809758115.

Karp, A.T., K. T. Uno, P. J. Polissar and K. Freeman. 2021. Late Miocene C4 Grassland Fire Feedbacks on the Indian Subcontinent. *Paleoceanography and Paleoclimatology* 36: e2020PA004106. DOI:10.1029/2020PA004106.

Keeley, J. E., J. G. Pausas, P. W. Rundel, W. J. Bond and R. A. Bradstock. 2011. Fire as an Evolutionary Pressure Shaping Plant Traits. *Trends in Plant Science* 16: 406-411. DOI:10.1016/j.tplants.2011.04.002.

Mallol, C. and A. Henry. 2017. Ethnoarchaeology of Paleolithic Fire: Methodological Considerations. *Current Anthropology* 58: S217-S229. DOI:10.1086/691422.

Mazza, P. P. Anthony, F. Martini, B. Sala, M. Magi, M. Perla Colombini, G. Giachi, F. Landucci, C. Lemorini, F. Modugno and E. Ribechini. 2006. A New Palaeolithic Discovery: Tar-Hafted Stone Tools in a European Mid-Pleistocene Bone-Bearing Bed. *Journal of Archaeological Science* 33: 1310-1318. DOI:10.1016/j.jas.2006.01.006.

McCauley, B., M. Collard and D. Sandgathe. 2020. A Cross-Cultural Survey of On-Site Fire Use by Recent Hunter-Gatherers: Implications for Research on Palaeolithic Pyrotechnology. *Journal of Paleolithic Archaeology* 3: 566-584. DOI:10.1007/s41982-020-00052-7.

Mills, B.. 1986. Prescribed Burning and Hunter-Gatherer Subsistence Systems. *Haliksa'i UNM Contrib Anthropol* 5: 1-26.

Neupane, P. C., M. R. Gani, N. D. Gani and Y. Huang. 2020. Neogene Vegetation Shift in the Nepalese Siwalik, Himalayas: A Compound-Specific Isotopic Study of Lipid Biomarkers. *The Depositional Record* 6: 192-202. DOI:10.1002/dep2.91.

Niekus, M. J.L.Th, P. R.B. Kozowyk, G. H.J. Langejans, D. Ngan-Tillard, H. Van Keulen, J. Van Der Plicht, K. M. Cohen *et al.* 2019. Middle Paleolithic Complex Technology and a Neandertal Tar-Backed Tool from the Dutch North Sea. *PNAS* 116: 22081-22087. DOI:10.1073/PNAS.1907828116.

Nikulina, A., K. MacDonald, F. Scherjon, E. A. Pearce, M. Davoli, J.-C. Svenning, E. Vella *et al.* 2022. Tracking Hunter-Gatherer Impact on Vegetation in Last Interglacial and Holocene Europe: Proxies and Challenges. *Journal of Archaeological Method and Theory* 29: 989-1033. DOI:10.1007/s10816-021-09546-2.

Parker, C. H., E. R. Keefe, N. M. Herzog, J. F. O'connell and K. Hawkes. 2016. The Pyrophilic Primate Hypothesis. *Evolutionary Anthropology: Issues, News, and Reviews* 25: 54-63. DOI:10.1002/evan.21475.

Pausas, J. G. and C. L. Parr. 2018. Towards an Understanding of the Evolutionary Role of Fire in Animals." *Evolutionary Ecology* 32: 113-125. https:.//doi.org/10.1007/s10682-018-9927-6.

Pop, E., W. Kuijper, E. van Hees, G. Smith, A. García-Moreno, L. Kindler, S. Gaudzinski-Windheuser and W. Roebroeks. 2016. Fires at Neumark-Nord 2, Germany: An Analysis of Fire Proxies from a Last Interglacial Middle Palaeolithic Basin Site. *Journal of Field Archaeology* 41: 603-617. DOI:10.1080/00934690.2016.1208518.

Pruetz, J. D. and N. M. Herzog. 2017. Savanna Chimpanzees at Fongoli, Senegal, Navigate a Fire Landscape. *Current Anthropology* 58 (S16): S337-50. DOI:10.1086/692112.

Pruetz, J. D. and T. C. LaDuke. 2010. Brief Communication: Reaction to Fire by Savanna Chimpanzees (Pan Troglodytes Verus) at Fongoli, Senegal: Conceptualization of 'Fire Behavior' and the Case for a Chimpanzee Model. *American Journal of Physical Anthropology* 141: 646-650. DOI:10.1002/ajpa.21245.

Ravi, S., M. C. Baddock, T. M. Zobeck and J. Hartman. 2012. Field Evidence for Differences in Post-Fire Aeolian Transport Related to Vegetation Type in Semi-Arid Grasslands. *Aeolian Research*, Aeolian processes: Biophysical drivers and biogeochemical implications, 7: 3-10. DOI:10.1016/j.aeolia.2011.12.002.

Reidsma, F. H., A. van Hoesel, B. J. H. van Os, L. Megens and F. Braadbaart. 2016. Charred Bone: Physical and Chemical Changes during Laboratory Simulated Heating under Reducing Conditions and Its Relevance for the Study of Fire Use in Archaeology. *Journal of Archaeological Science: Reports* 10: 282-292. DOI:10.1016/j.jasrep.2016.10.001.

Roebroeks, W., D. De Loecker, P. Hennekens and M. Van Ieperen. 1992. ' A Veil of Stones': On the Interpretation of an Early Middle Palaeolithic Low Density Scatter at Maastricht-Belvédère (The Netherlands). *Analecta Praehistorica Leidensia* 25: 1-16.

Roebroeks, W., K. MacDonald and F. Scherjon. 2021. Establishing patterns of early fire use in human evolution. The Beef behind All Possible Pasts. In: S. Gaudzinski-Windheuser and O. Jöris (eds.), *The Tandem Festschrift in Honour of Elaine Turner and Martin Street*: 7-16. Heidelberg: Propylaeum (Monographien des Römisch-Germanischen Zentralmuseums Band 157). DOI:10.11588/propylaeum.950

Roebroeks, W., K. MacDonald, F. Scherjon, C. Bakels, L. Kindler, A. Nikulina, E. Pop, and Sabine Gaudzinski-Windheuser. 2021. Landscape Modification by Last Interglacial Neanderthals. *Science Advances* 7: 1-15. DOI:10.1126/sciadv.abj5567.

Roebroeks, W. and P. Villa. 2011. On the Earliest Evidence for Habitual Use of Fire in Europe. *PNAS* 108: 5209-5214. DOI:10.1073/pnas.1018116108.

Sandgathe, D. M.. 2017. Identifying and Describing Pattern and Process in the Evolution of Hominin Use of Fire. *Current Anthropology* 58: S360-S370. DOI:10.1086/691459.

Scherjon, F., C. Bakels, K. MacDonald and W. Roebroeks. 2015. Burning the Land: An Ethnographic Study of Off-Site Fire Use by Current and Historically Documented Foragers and Implications of the Interpretation of Past Fire Practices in the Landscape. *Current Anthropology* 56: 299-326. DOI:10.1086/681561.

Scott, R. V. and R. Hosfield. 2021. Fire in the Round: A Holistic Approach to the Lower Palaeolithic Record. *Journal of Archaeological Science: Reports* 37: 102938. DOI:10.1016/j.jasrep.2021.102938.

Shimelmitz, R., S. L. Kuhn, A. J. Jelinek, A. Ronen, A. E. Clark and M. Weinstein-Evron. 2014. 'Fire at Will': The Emergence of Habitual Fire Use 350,000 Years Ago. *Journal of Human Evolution* 77: 196-203. DOI:10.1016/j.jhevol.2014.07.005.

Stahlschmidt, M. C., S. M. Mentzer, S. Heinrich, A. Cooper, M. N. Grote, P. J. McNeill, J. C.-B. Wilder, and T. E. Steele. 2023. Impact of a Recent Wildfire on Tortoises at Cape Point, South Africa, and Implications for the Interpretation of Heated Bones in the Archaeological Record. *Archaeological and Anthropological Sciences* 15: 126. DOI:10.1007/s12520-023-01806-4.

Synnott, D. P., L. Schwark, K. Dewing, P. Kent Pedersen and H. Sanei. 2021. Evidence for Widespread Wildfires and Their Environmental Impact in the Late Cretaceous Canadian Arctic. *Global and Planetary Change* 203: 103515. DOI:10.1016/j.gloplacha.2021.103515.

Tanner, B. R., M. E. Uhle, J. T. Kelley and C. I. Mora. 2007. C3/C4 Variations in Salt-Marsh Sediments: An Application of Compound Specific Isotopic Analysis of Lipid Biomarkers to Late Holocene Paleoenvironmental Research. *Organic Geochemistry* 38: 474-84. DOI:10.1016/j.orggeochem.2006.06.009.

Thompson, J. C., D. K. Wright, S. J. Ivory, J.-H. Choi, S. Nightingale, A. Mackay, F. Schilt *et al.* 2021. Early Human Impacts and Ecosystem Reorganization in Southern-Central Africa. *Science Advances* 7: eabf9776. DOI:10.1126/sciadv.abf9776.

Toney, J. L., A. García-Alix, G. Jiménez-Moreno, R. S. Anderson, H. Moossen and O. Seki. 2020. New Insights into Holocene Hydrology and Temperature from Lipid Biomarkers in Western Mediterranean Alpine Wetlands. *Quaternary Science Reviews* 240: 106395. DOI:10.1016/j.quascirev.2020.106395.

Wilsey, B. J. 1996. Variation in Use of Green Flushes Following Burns among African Ungulate Species: The Importance of Body Size. *African Journal of Ecology* 34: 32-38. DOI:10.1111/j.1365-2028.1996.tb00591.x.

Wrangham, R.. 2017. Control of Fire in the Paleolithic: Evaluating the Cooking Hypothesis. *Current Anthropology* 58: S303-S313. DOI:10.1086/692113.

Wrangham, R. and N.L. Conklin-Brittain. 2003. 'Cooking as a Biological Trait.' *Comparative Biochemistry and Physiology Part A: Molecular and Integrative Physiology* 136: 35-46. DOI:10.1016/S1095-6433(03)00020-5.

In Dusseldorp, G.L., Chu, W., Bakels, C.C. and M.A. Soressi (eds) 2025. Intent on the Paleolithic. Papers in honour of Prof. Dr. Wil Roebroeks. *Analecta Praehistorica Leidensia* 55. Leiden: Sidestone Press, pp. 159-170.

"Classical" geography, Carl Sauer and the evolution of human fire use

Jan Kolen

Abstract

This paper discusses a forgotten line of geographical research that developed in the mid-20th century, aiming at the reconstruction of the evolution of human fire use. This geographical tradition had striking similarities with the current debate on early human fire use in archaeology, both in terms of scope and results. The main topics dealt with are: the "discovery" of fire by early hominins in East Africa; the origins of fire as an on-site focus of activities and social interaction in the Palaeolithic; the use of fire as a landscaping technology in (prehistoric) hunters and gatherers; and the importance of the study of fire use for understanding the time depth of the "Anthropocene". Particular attention will be paid to the work of the American-German geographer Carl Ortwin Sauer (1889-1975).

Keywords: Sauer; geography; fire ecology; human evolution; sustainability

Introduction - From Wil Roebroeks to Carl Sauer

Wil has regularly initiated groundbreaking developments and insights in the archaeological research of our earliest ancestors. In the 1990s he caused a stir with his "short chronology" of the earliest hominin colonization of Europe (Roebroeks and van Kolfschoten 1994). A systematic and critical screening of the earliest archaeological sites and their dates showed that colonization only took place "late", around 500,000 years ago, and not more than a million years ago, as previously thought.

In 2011, Wil published an influential article together with Paola Villa, in which he also applied the critical screening of archaeological data and the "shortening" of chronology to the use of fire by the earliest inhabitants of Europe (Roebroeks and Villa 2011). In Europe, the habitual use of fire would not have really occurred until about 400,000 BP, which suggests that the use of fire may not have been of decisive importance for the colonization of the northern latitudes, with their temperate (humid) and cold climates and significant differences between the seasons. Since the publication of this important paper, an interest in the earliest human use of fire has not left Wil. With a growing team of students and colleagues, he has addressed a range of interrelated issues, such as the use of fire as a landscaping technology (Scherjon *et al.* 2015) and the earliest evidence thereof (Roebroeks *et al.* 2021), and the social mechanisms that may have been responsible for the cultural and geographical spread of innovations in the use of fire (MacDonald *et al.* 2021).

The evolution of human fire use is an issue that has a long tradition of research and discussion in archaeology and evolutionary anthropology. However, it is apparently unknown in archaeological circles that this issue has been raised

Jan Kolen
Faculty of Archaeology, Leiden University, The Netherlands
j.c.a.kolen@arch.leidenuniv.nl

Figure 1. Carl Ortwin Sauer in Berkeley, 1920.

before, and is also intensively discussed, in the field of geography. This applies in particular to the work of the influential geographer Carl Sauer, who between ca. 1935 and 1965 devoted a number of studies wholly or largely to the (deep) history of anthropogenic fire and its connection with the landscapes, habitats and biogeographical regions inhabited and modified by humans. The most important of these (Sauer 1936, 1938, 1947, 1958, 1961, 1962) are also included in the now classic overview of Sauer's work edited by John Leighly: *Land and Life. A selection from the writings of Carl Ortwin Sauer* (Leighly [ed.], 1963). Of the articles in this volume, the evolution of fire use is most prominently discussed in Sauer's *Fire and Early Man* (1961).

Carl Ortwin Sauer (1889-1975; Fig. 1) is widely regarded as one of the most important geographers of the twentieth century and as the founder of cultural geography: the study of the relationships between cultures and cultural practices and landscapes and geographical space (Williams *et al.* 2014). Sauer grew up in the United States but was a child of German parents. This background allowed him to build bridges between North American and European intellectual traditions. As a novice geographer, he quickly became impressed by the work of the German geographers Alfred Hettner and Otto Schluter respectively. His earliest geographical studies focused on the morphology of the landscape as an interim outcome of the ongoing interaction between people and their environment, between culture and nature (Sauer 1925). In that context, he paid special attention to the influencing of plant communities by humans. Because of his strong interest in the relationships between culture and landscape, he sought collaboration with the influential anthropologists of his time, who often emerged from the same German intellectual environment in North America, such as Franz Boas and Alfred L. Kroeber. Sauer zoomed in on the way in which human land use and the cultural landscape on the one hand, and specific cultural phenomena and the "mosaic" of cultures on the other, continually interact (Sauer 1925, in Leighly [ed.] 1963).

Sauer has always had a special interest in archaeology. Initially this interest was limited mainly to Latin America, but in his later career it broadened to more general themes on a global scale, such as his research into the origins and dispersals of agriculture (Sauer 1952). In the 1950s and 1960s, he also became increasingly interested in human evolution and the new discoveries in that field in Africa (particularly the

Olduvai Gorge) and China (Zhoukoudian). He became convinced that *Homo* was originally a "riparian species" (Sauer 1962), which developed partly thanks to the "discovery" and use of fire (Sauer 1961). Between about 1935 and 1965, Sauer published a series of articles in which he specifically addressed the evolution of human fire use and its influence on the natural landscape, the regions inhabited by hunters and gatherers and – ultimately – the face of the entire planet (e.g. Thomas *et al.* [eds] 1956). Throughout these studies, he emphasized the importance of fire as one of the most crucial potential "tools" in the history of humankind – and therefore important for an understanding of human environmental and cultural evolution. He accepted a similar role for stone and stone tool technologies, although the importance of these, according to him, has been exaggerated by archaeologists thanks to its material persistence and therefore visibility in the archaeological record. It is clear on many occasions that he would prefer to speak, in a sense, of a "Wooden Age" instead of a Stone Age when referring to the earliest stages of human cultural evolution:

> "[A]*n interpretation based on stone items alone is misleading; for some, or possibly many groups, may have directed their attention with plant and animal materials rather than with stone. It has been repeatedly suggested that cultures based mainly on wood and fiber may have been earlier than those based on stone. There are tribes still living, with a fairly rich culture, that work almost exclusively with plant and animal materials; the Bororo of Brazil are one well-known example". (1947, in Leighly [Ed.] 1963: 161).*

In this contribution I will reflect on discussions about human fire use in "classical" cultural geography by focusing on Sauer's innovative studies on the subject. This is not only interesting from a science-historical perspective, but also because Sauer's work on this point is surprisingly current and we as archaeologists can perhaps draw inspiration from his "forgotten" geographical hypotheses. I discuss Sauer's ideas about the topic in three sections: the beginning and earliest development of fire use, the earliest use of fire as a landscaping technology – also referred to as "human fire ecologies", and the social significance of fire. I consistently start with a brief overview of the innovative research done by Wil and his team and then go into more detail about Sauer's studies, where I want to let him speak as much as possible and therefore make frequent use of quotes from his publications. Although this is unusual in a scientific paper, it does more justice to the originality of Sauer's work. Quotations are cited from the aforementioned collection of Sauer's key publications in Leighly ([ed.] 1963), but with reference to the original year of publication. At the end of this paper I return from Sauer to Roebroeks, by briefly discussing their respective interest in "Man's Role in Changing the Face of the Earth" (Thomas *et al.* eds] 1956) and the future of the Earth as a "Liveable Planet" – the focus of the interdisciplinary research program at Leiden University in which Wil has been participating with his own "fire projects" since 2017.

Harvesting roasted meat: the beginnings of human fire use

Roebroeks and Villa (2011) propose a "short chronology" of habitual fire use in Europe, basing themselves on a thorough re-evaluation of archaeological data from over 100 Lower and Middle Paleolithic sites. They conclude that the earliest reliable indications for habitual fire use, e.g. in the form of deliberately constructed fireplaces, accumulations of burned materials (flint, bone, wood) in caves, and rare evidence for the production of plastic adhesives for stone tools, all date from after c. 400,000 BP. This conclusion goes against the "common sense" belief that the controlled use of fire, at least for warming the body and processing raw food, was a prerequisite for successfully colonizing and permanently inhabiting northern regions such as Europe. If Roebroeks and Villa are correct, the pioneers of Europe must have had other strategies to protect themselves against the cold climate and seasonal changes of the northern latitudes. Their conclusion further questions claims that early hominins in Africa used – and perhaps even produced – fire on a frequent basis much earlier, before they "colonized" Europe (e.g. Gowlett *et al.* 1981; Clark and Harris 1985; Wrangham 2009).

The question therefore is: what preceded the relatively late habitual use of fire? When and how did early hominids first start using fire, and how long did it take for occasional and opportunistic use to give way to frequent and controlled use? These questions interested Sauer. Like Roebroeks and Villa, Carl Sauer, in a 1961 paper on *Fire and Early Man*, critically reflected on early claims for fire use by early hominins in Africa and on the idea of an

Australopithecus Prometheus. In fact, he opted for a short chronology as well:

> *"The old view that the use of fire was developed very early, following upon that of tools and speech, lately has been sharply challenged by the archeologic record. It is objected that the early sites, such as those in Olduvai Gorge, show no evidence of the employment of fire. The earliest generally admitted site of fire-using man is that of Choukoutien inhabited by Peking Man* [...]. *Thereafter* [...] *followed a long interval without use of fire" (1961, in Leighly [ed.] 1963, 293).*

Nevertheless, Sauer considered East Africa as the region where fire was somehow "discovered". The environmental preferences of early hominins in this region, according to him, covered combinations of wetlands and areas with active volcanism. He did not believe much of the then current hypothesis that *Australopithecus* and *Homo habilis* were adapted to the vast African savannah (1961, in Leighly [ed.] 1963: 303). Instead, he became convinced that *Homo* was originally a "riparian species" – adapted to land adjacent to the sea, rivers and lakes. He considered the seashore as early *Homo's* "optimum habitat":

> *"To primitive collector and modern naturalist alike, the borders of the sea are richly rewarding. Between high and low tide a wide assemblage of life forms useful to man is to be had for the taking. These differ from place to place according to the bottom, whether it is sand, silt, or rock, according to the qualities and motions of the water, and by the extent of daily exposure above the water. Marine mammals and fish are stranded on occasion. In warm oceans sea turtles visit sandy beaches to deposit their eggs. From back beach to sea cliff a different and varied fauna and flora yield edible shoots, fruits, eggs, and nestlings. The shallow sea holds yet another assemblage" (1962, in Leighly [ed.] 1963: 309).*

Sauer added that these "affordances" were not only characteristic of coastal landscapes, but also of the river plains and lake shores, which were often situated at a relatively short distance from the coast or were directly connected to it. In the distribution and location of the earliest archaeological sites in Africa he saw confirmation of this hypothesis, although he did not overlook the favourable preservation conditions of local sedimentary environments (294). According to Sauer, the fact that the use of coastal, fluvial and lacustrine habitats was combined with a preference for volcanic landscapes was especially evident from the new discoveries in, among others, the Olduvai Gorge. The positive effect of volcanism on the productivity of wetland habitats, and therefore on the use options of early hominins, was significant:

> *"A volcanic outbreak may wipe out life over an area, but also open it to a fresh stocking by a pioneer vegetation* [...]. *The new vegetation in turn attracts the entry and increase of a new fauna profiting by the abundance of seed, leaf, and shoot. Insects, lizards, birds, small mammals, and even ruminants may thus benefit.* [...] *Such temporary fire gardens and feedlots may be inferred to have added to the foraging success of the early human inhabitants who were at hand" (1961, in Leighly [Ed.] 1963: 291).*

Areas with active volcanism still had another advantage. They supplied food – and especially meat – that was already naturally prepared for consumption:

> *"Volcanic outbreaks kill by heat and by smothering.* [...] *After the fire the burned tract provides carcasses to be picked up and dug up, perhaps tubers and roots that were roasted in the ground. Such opportunity for mass harvest provided also the earliest experience of the taste of roast flesh and skin and of cooked plant starches and proteins. The omnivorous human digestive tract* [...] *are basic to the direction (and success) of hominid evolution" (1961, in Leighly [ed.] 1963: 291-292).*

Sauer argued that early hominins remained faithful to their environmental preferences for hundreds of thousands of years, even when they left Africa and eventually entered southern Europe via West Asia. This seemed again evident from the distribution of palaeolithic sites in these newly colonized areas, but according to Sauer it was also logical for another reason. If early hominins wanted to retain fire, it was not enough to rely on the frequency of lightning strikes in areas without volcanic activity (298-299). However, Sauer did hypothesize that early hominins in Asia and the Mediterranean, and at the same time also in Africa, gradually became more successful in actively tapping fire from wildfires:

> *"The step from watching the effects of fire to using fire was bound to be taken. The awareness of fire was strong; attention became inquiry and this led to experiment. It may have taken time, perhaps a long time, before someone picked up a brand, carried it, and set a fire" (1961, in Leighly [ed.] 1963, 295).*

Sauer inferred that this last step was "forced" by the migration of *Homo erectus* into the cold regions of Southeast Asia, which marked a temporary breakthrough in the evolution of fire use. He and his contemporaries considered Zhoukoudian, as mentioned earlier, the earliest site with convincing indications for habitual fire use, "a pioneer of housewarming" (294), although we now know that fire use at this site is disputed (Goldberg *et al.* 2001; but see Gao *et al.* 2017). Yet, according to Sauer, a setback followed: "a long interval without use of fire, its continuing and regular employment only beginning with the Neanderthal and Fontechevade people at or prior to the onset of the Wurm Glaciation" (1961, in Leighly [ed.] 1963: 293-294). It was therefore only *after* the colonization of cold Europe, according to Sauer, that humankind possessed the full package of domesticated fire, including cooking.

Sauer's conclusions correspond remarkably well with those of Roebroeks and Villa 50 years later. He is convinced of a short chronology for human fire use and sees the habitual use of fire only appearing late in human history, even though he assumes that the northern latitudes would have been difficult to colonize and inhabit without regular use of fire (contra Roebroeks and Villa). He developed a series of hypotheses that together suggest a three-phase development of fire use. Over half a century before the publication of Parker *et al*'s (2016) popular "pyrophilic primate hypothesis", Sauer proposes what we might call a "natural barbecue hypothesis" for the earliest hominins in Africa. Essential to the habitat choice of these hominins was a high probability of encountering natural fires, so that the foraging strategy could be partly based on harvesting roasted meat. In the second phase, the step was taken towards actively tapping natural fires in order to transport the fire, keep it if possible and use it according to one's own needs. Sauer is not clear about where this transition might have taken place, but he seems to suggest that this must have been a post-Out-of-Africa event. Only after the colonization of Europe did the complete package become permanent. From then on, Neanderthals and "moderns" were able not only to produce fire where and when they saw fit, but also to use it for a wide range of activities, from roasting meat and cooking food to warming the body, extending the day and facilitating group formation and communication.

It is obviously problematic to demonstrate the first two phases in the archaeological record. But it is likely that they marked the first beginning of human use of fire, although perhaps not strictly in the order proposed by Sauer. The tapping and targeted use of transported fire occurs not only in human populations, but also in (some) other animal populations. Since time immemorial, Australian Aborigines have told stories about birds of prey that take smouldering branches from forest fires to start fires elsewhere and hunt small mammals and reptiles (Bonta *et al.* 2017). Transporting, keeping and exchanging fire was still a common strategy among many hunters, gatherers and small-scale farmers in the recent (ethnographically documented) past. The Pikunii, a branch of the Blackfeet Nation, are just one out of many examples: they carried fire with them from camp to camp along historical migration routes (https://www.firelab.org/project/carrying-fire-pikunii- way). Sauer's third phase is well attested in the archaeological record, not only with the significant increase in deliberately constructed hearths and accumulations of burned materials in caves (Roebroeks and Villa 2011), but also by the recent discovery of Andrew Sorensen (Sorensen *et al.* 2018) that Middle Paleolithic bifaces, in combination with pyrite, may have been used in Europe to produce fire.

Clearing the land: fire as a landscaping tool

For more than half a century, it has been common knowledge that prehistoric hunter-gatherers in Europe used fire not only in their dwellings and within the boundaries of their campsites, but also for a variety of uses *beyond* their campsites and even on an extensive landscape scale. Late Mesolithic communities in particular made use of fire as a response to the deciduous forest becoming increasingly dense and therefore less productive, although most researchers will now accept that this was practiced throughout the earlier stages of the Mesolithic as well (cf. Bos and Urz 2003; Innes *et al.* 2013; Scherjon *et al.* 2015). Palaeobotanical research, combined with the systematic analysis of other environmental proxies, such as (micro) charcoal data, and ethno-historical and experimental observations, strongly suggest that early Holocene woodlands were subject to periodic

Figure 2. Anthropogenic burning in Hadza country. Photo James F. O'Connell, University of Utah.

human disturbance of this kind. By thinning the forest matrix locally by using fire, hunters and gatherers manipulated the vegetation and its succession, in this way improving conditions for edible plant foods and game.

These insights raise the question when humans exactly started to influence land cover in "off-site" situations using fire (cf. Fig. 2), either intentionally as a landscaping tool or by accident (Daniau *et al.* 2010, Kaplan *et al.* 2016).

The issue regarding the time depth of human fire ecologies has been the subject of several Leiden studies in the past decade (e.g. Scherjon *et al.* 2015; Roebroeks and Bakels 2015; Roebroeks *et al.* 2021; Nikulina et al. 2022). Scherjon *et al.* (2015) presented a comparative study of fire use in off-site contexts among historically documented and recent groups, which can serve as a frame of reference for fire use by prehistoric hunters and gatherers, including early hominins. They show that recent hunters and gatherers used fire off-site for a broad range of purposes, varying from hunting and gathering, communication and travel to ritual and fun. Leiden researchers also conduct empirical research into the very earliest "candidates" for such forms of fire use in Europe. The results of a study for the Eemian interglacial site Neumark-Nord are spectacular in this respect, as fire-using Neanderthals probably already exerted a significant influence there on the wider environment 130-115,000 years ago (Roebroeks and Bakels 2015).

The influence of human use of fire on vegetation and the landscape was also a common thread through Carl Sauer's work, regardless of the period or area in which he worked. Sauer's interest in the use of fire at the landscape scale, particularly in large-scale human deforestation by fire, initially focused on the tropical forests of the Americas, although in his research he also paid attention to the temperate forest areas of North America and other biomes, such as the expansion and persistence of grasslands:

> *"There is abundant evidence that the vegetation of the East [of North America] did not consist of monotonous primeval forests, but that there was an abundance of open country that was not*

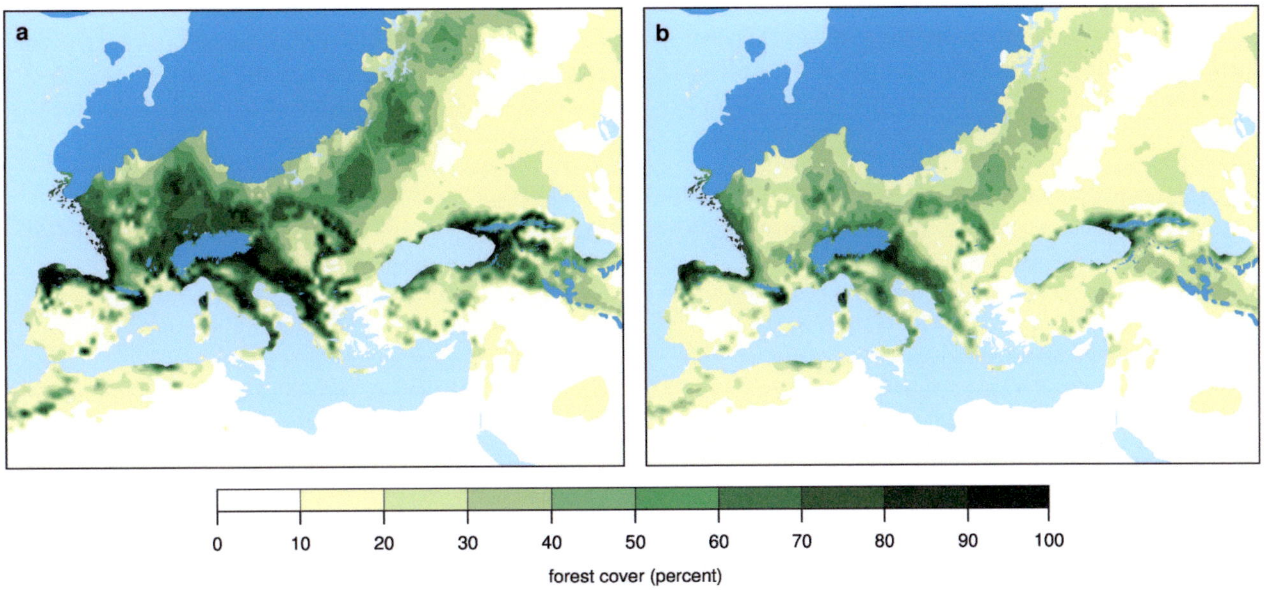

Figure 3. An example of an anthropogenic fire impact model: Kaplan *et al.* 2016. A= Expected (modelled) forest cover in Europe during the Last Glacial Maximum without human reduction caused by fire use; B= Expected (modelled) forest cover in Europe during the Last Glacial Maximum with human reduction caused by fire use (Kaplan *et al.* 2016, Fig. 4).

climatically determined, the glades, cow pens, and prairies being in the main deliberate or casual clearings. Fire, accidental and deliberate, has undoubtedly greatly enlarged the open spaces through this long period" (1930, in Leighly [ed.] 1963: 47).

Sauer located the origins of fire use as an off-site strategy, or even as deliberate landscaping technology, early in human evolution, although his conclusions at this point remain hypothetical:

"The glimpse into a remote past afforded by the Peking finds is sufficient for us to ascribe to man an ancient role in the modification of vegetation. He had habitual campsites; he wore paths out from them that became bordered by trailside weeds that took advantage of the added sunlight and tolerated trampling and other disturbance. [...] If fires were set primordially, at first to facilitate collecting and then as a hunting device, a most potent aid to vegetative modification was assured" (1947, in Leighly [ed.] 1963: 159-160).

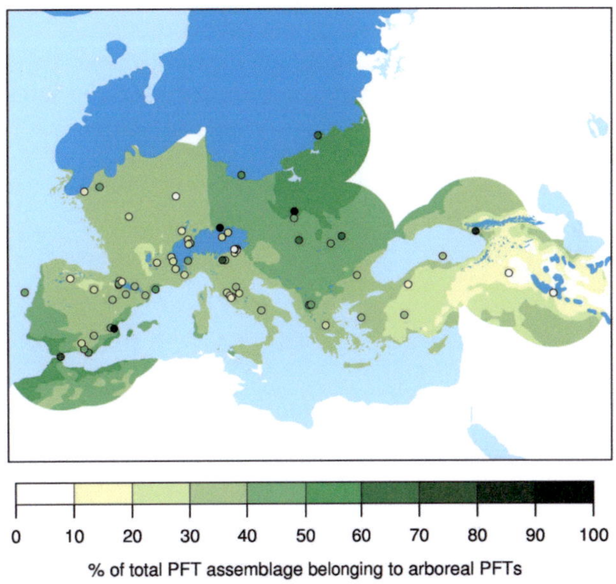

Figure 4. Empirically reconstructed (estimated) tree cover at the Last Glacial Maximum: the tree cover surface is based on the 3D interpolation of PFT scores from individual fossil pollen sites (circles). (Kaplan *et al.* 2016, Fig. 1).

It seems that Sauer has been a bit too uncritical and optimistic with his "long chronology" of off-site fire use, but with the recent studies for Neumark-Nord his hypothesis remains in view. Unintentionally, however, Sauer has another message in store for archaeology:

> *"If the activity of man in any of these directions was sporadic, no permanent alteration of plant complexes might result; but if activity was maintained in the same direction, cumulative and possibly permanent effects were registered in the association and perhaps the evolution of plants. A major question, therefore, is whether areas, when once they had been occupied, remained under continuous human occupation. A subsidiary question is whether they were subjected to increasing exploitation* [...]*" (1947, in Leighly [ed.] 1963: 160).*

With this comment, Sauer makes it clear that archaeologists do not always register the impact of off-site fire use if they focus exclusively on the inductive research of proxies at the level of sites and their immediate environment. The extent to which human-induced fire ecologies remain visible in the biogeographical and archaeological record depends on the intensity, frequency and duration of off-site fire use on a landscape scale – parameters that are in turn co-dependent on variables such as population density, mobility and climate. Therefore, archaeologists can only advance this research theme if they link the results of inductive research to deductive models (cf. Kaplan *et al.* 2016; see Fig. 3-4).

Hearth and home: fire and the evolution of social life

There is one aspect of the evolution of human fire use that has (yet) received little attention in the work of Roebroeks *et al.* and that is the earliest *social* implications of fire control. A recent article by MacDonald *et al.* (2021) on Middle Pleistocene fire use as a signal of widespread cultural diffusion in human evolution, touches on this theme the most. They emphasize that the archaeological signal of fire use becomes clearly visible around 400,000 BP, not only in western Eurasia, but also in Africa, and in "different subpopulations of the wider hominin metapopulation" (p. 1). This pattern is followed somewhat later by another innovation, the Levallois technology. It seems that from 400,000 BP onwards, social networks were in place to share and spread these innovations relatively quickly over large areas. However, although this study addresses the social conditions for the diffusion of human fire use, it leaves unaddressed the social *implications* of the evolution of fire use.

Goudsblom, a Dutch sociologist who in the early 1980s decided to write "cool book about fire", explained that the domestication of fire also immediately leads us to the roots and heart of social life: the family, the group, its interpersonal relations, and its rituals and celebrations (Goudsblom 1992; 1985). Fire control is a human ability that is not simply innate, but is learned and passed on in groups. Important tasks are of course producing, maintaining and possibly transporting fire, but these are not purely technical skills; they also require some specialization and social division of labour (Goudsblom 1985: 15). Moreover, Goudsblom argues, "the fire seems to have been the place where the development of human communication and language received particularly strong impulses. The members of the group returning from their daily exploration and supply trips met each other and their children by the fire, where they tried to communicate their experiences through sounds and gestures." (Goudsblom 1985: 16).

This hypothesis about the evolution of language has a long tradition in archaeology as well (see, among others, White and Brown 1973), but was only developed into a more coherent, evolutionary-anthropological theory by Robin Dunbar (1996). Gossip has a similar effect in groups of people as grooming in groups of primates: servicing relationships and maintaining interpersonal ties and strategic alliances. According to Dunbar, language evolved from this and the locus of gossip par excellence, the hearth, must have played a prominent role (see also Wrangham 2009; Dunbar and Gowlett 2014). Archaeologists are implicitly aware that the hearth of prehistoric hunters and gatherers must have attracted (strong) stories, memories and information relating to situations "beyond the horizon". Dunbar's studies gave this assumption a more solid basis, which was also translated by other innovators in the field, notably Clive Gamble, into socio-spatial patterns that the archaeologist must actually be able to "see" in the archaeological record and excavation plans (Gamble 1999: 68-76; see also Binford 1983).

It is surprising that Sauer has dared to draw some interesting conclusions about these social implications of fire use quite early, although geography as a discipline does not generally operate at this local scale of human space and social behaviour. Inspired by the sensational discovery of the "Peking Man site" at Zhoukoudian (China) in the 1920s, he devoted a large part of an article in 1947 to "Inferences from the culture of Peking man" (1947, in Leighly [ed.] 1963: 158-160), in which he concludes the following about the use of fire by this *Homo erectus*:

> *"Fire was basic to their culture.* [...] *The number of hearths suggests that they may have lived in family groups, each accustomed to collect its own place (hearth = home)* [...], *which met at the family fire, forming together a sort of primitive community that stayed put as long as seasonal changes in food supply permitted" (1947, in Leighly [ed.] 1963: 158-159).*

Later Sauer continued this line chronologically (Sauer 1956). Neanderthals already had the complete fire repertoire, he suggested, with the hearth symbolizing a "home" and family, and facilitating communication and exchange:

> *"Speech, tools and fire are the tripod of culture and have been so, we think, from the beginning. About the hearth, the home and workshop are centred.* [...] *About the fireplace, social life took form, and the exchange of ideas was fostered" (Sauer 1956: 54).*

It is one of the few spots in Sauer's reflections on the evolution of fire use where he seems to miss the point. In fact, we do not find the archetypal pattern of the central hearth with surrounding seating and corresponding organization of activities convincingly anywhere in the campsites of early hominins – or even in those of Neanderthals. Actually, quite the opposite is the case. Neanderthal hearths were usually small, tucked away in low, narrow recesses or along the walls of caves and abris, and in many cases probably provided space for only a single individual – to warm, eat and rest to sleep. Examples of this have been found at Hauteroche, in the Abric Romani rock shelter, Vilas Ruivas (in the open air), Canalettes and many other Neanderthal sites (Mellars 1996; Kolen 1999). Where we find the remains of Neanderthal living structures, such as at Nice-Lazaret, Arcy-sur-Cure and Molodova, which do not appear to have been real huts or homes, fires were not so much placed in the center, but along the periphery of the structures (Kolen 1999). The "social-life-around-the-hearth model" thus appears to have been a rarity or coincidence – not an evolutionarily significant phenomenon – before about 30,000 BP. But we can hardly blame Sauer in 1947 for "missing" that anomaly on our anthropological model of expectations. This model is still applied (uncritically!) by many archaeological colleagues to the use of fire and social life of Neanderthals and their predecessors.

Epilogue: waves of interest and the "Liveable Planet" perspective

Like no other, Sauer emphasized the crucial and productive role of fire use in human evolution, in the long process of human niche construction and in the development of biomes that we now generally qualify as "natural" and ecologically valuable. However, he also saw the downside of human intervention in the planetary system. In 1955, along with William Thomas Jr., Marston Bates and Lewis Mumford, he was the driving force behind the organization of an influential conference on "Man's Role in Changing the Face of the Earth" in Princeton, New Jersey, funded by the Wenner Gren Foundation (Thomas *et al.* [eds.] 1956). Even before the Second World War, Sauer already published some "alarming" articles, such as an extensive piece on the "Theme of Plant and Animal Destruction in Economic History" (1938; see also Sauer 1930):

> *"Our Zeitgeist is congenial to debate and to the planning of our future by resolution. Perhaps such a mood marks the beginning of a great era, but even though we be far better social engineers than I think we are, it is still most important to keep track of the present as to its position on the long graph of history, to see where we stand on the trend curves of social history.* [...] [T]*here is a dominant geographic theme which deals with the growing mastery of man over his environment. Antiphonal to this is the revenge of an outraged nature on man. It is possible to sketch the dynamics of human history in terms of this antithesis" (1938, in Leighly [ed.] 1963: 145).*

By conducting systematic research into the long history of human intervention in ecosystems, archaeologists, together with biologists, anthropologists and earth scientists, can reconstruct the historical background and causes of current environmental issues.
This provides knowledge of the time depth and developmental history of the Anthropocene, however that epoch is defined, which may in turn help us design solutions for the future. After all, environmental issues often involve long-term processes with delayed and irreversible consequences, which go unnoticed within

the time window of individual generations, while the disciplines mentioned have developed the temporal frames to be able to trace and monitor them over thousands of years. Sauer already mentioned some examples of this:

> *"Perhaps as far back as Neolithic time, the first ominous discordance develops.* [...] *Specifically, ancient overgrazing by herding peoples is blamed for the bareness of much of the great interior of the Old World. The damage developed perhaps three to four thousand years ago. Lapse of time has brought no repair of this destruction. The dry lands of the center of the Old World are permanently and sadly diminished in their utility." (1938, in Leighly [ed.] 1963: 146)*

According to Sauer, disciplines like archaeology and geography should also place "red flags" if the world population is in danger of repeating itself, or if the use of resources deviates significantly from the previous history of use due to the accelerated pace and increasing scale. In this respect too, Sauer's reflections fit well with the interests of Wil, who has worked – along similar lines – for the Liveable Planet Program of Leiden University in recent years.

Sauer's warning observations quoted above could be a good conclusion to this contribution. However, I prefer a more optimistic conclusion about the archaeological study of human fire use. This also fits better with Wil's work, which – as much has become clear from his publications on this subject – is intrinsically motivated to map and better understand the cultural evolution and behaviour of early hominins. As he has previously done for the "short chronology" as a framework for understanding the colonization and earliest habitation of Europe (Roebroeks 1994), Wil has given the starting signal for renewed archaeological research into the roots of fire use more than a decade ago (Roebroeks and Villa 2011). Meanwhile, his research – with an entire team – is still on the front line of this theme. From a science-historical perspective, Wil already initiated the "fourth wave of interest" in this field. The archaeological interest in fire shows a series of waves, each having its own interdisciplinary focus, theoretical source of inspiration, "hero projects" and key sites.

The third wave (counting backwards), in the 1980s, developed in the study of the earliest hominins in Africa and was motivated by (supposed) evidence of habitual use of fire at several early sites, such as Chesowanja (Gowlett *et al.* 1981; Clark and Harris 1985). These were interpreted as an indication of a long chronology of human fire use, which would have started in Africa around 1.5 million years ago. Much later, this long chronology formed the basis for Wrangham's hypothesis that the habitual use of fire was a catalyst for brain development and more complex social interaction (Wrangham 2009).

The second wave developed in the 1950s and 1960s under the influence of new developments in paleoecology and some unique discoveries, such as the early Holocene site Star Carr in England (Clark 1954), where organic materials, and thus indirectly also environmental indicators and proxies for fire use, are exceptionally well preserved (Innes *et al.* 2013). These gave rise in the following decades to a focus on the influence of human fire ecologies on the forested habitats of Mesolithic hunters and gatherers in Europe and the use of fire in early Neolithic farming practices, such as slash-and-burn based horticultures and early forms of shifting cultivation.

But the first wave was Carl Sauer's. Sauer was interested in the use of fire in land use in all periods, worldwide, and on top of this developed a specific curiosity for the role of fire use in both the evolution of humankind and in the modification of our planet's surface. He was inspired by the archaeological discoveries of his time, by key sites such as Zhoukoudian and Olduvai Gorge, but also by new insights from cultural ecology, biogeography and ethnography. That first wave is still important for researchers of the fourth wave, as its "forgotten" research may contain unexpectedly useful hypotheses that can be tested with results from new research. In any case, it is certain that Wil, together with his colleagues, will continue the fourth wave thanks to new discoveries and empirical research. We are curious...

References

Binford, L.R.. 1983. *In Pursuit of the Past.* London: Thames and Hudson.

Bonta, M., R. Gosford, D. Eussen, N. Ferguson, E. Loveless and M. Witwer. 2017. Intentional Fire-Spreading by "Firehawk" Raptors in Northern Australia. *Journal of Ethnobiology* 37: 700-718.

Bos, J.A.A. and R. Urz. 2003. Late Glacial and early Holocene environment in the middle Lahn river valley (Hessen, central-west Germany) and the local impact of early Mesolithic people – pollen and microfossil evidence. *Vegetation History and Archaeobotany* 12: 19-36.

Carrying the Fire the Pikunii Way. https.//www.firelab.org/project/carrying-fire-pikunii-way.

Clark, J.G.D.. 1954. *Excavations at Star Carr: An Early Mesolithic Site at Seamer Near Scarborough, Yorkshire.* Cambridge: Cambridge University Press.

Clark, J.D. J.W.K. Harris. 1985. Fire and its Roles in Early Hominid Lifeways. *African Archaeological Review* 3: 3-17.

Daniau, A.-L., F. d'Errico and M. F. S. Goñi. 2010. Testing the Hypothesis of Fire Use for Ecosystem Management by Neanderthal and Upper Palaeolithic Modern Human Populations. *PLoS ONE* 5: e9157.

Dunbar, R.. 1996. *Grooming, Gossip and the Evolution of Language.* Cambridge: MA Harvard University Press.

Dunbar, R.I.M. and J.A.J. Gowlett. 2014. Fireside Chat – The Impact of Fire on Hominin Socioecology. In: R.I.M. Dunbar, C. Gamble and J.A.J. Gowlett (eds.), *Lucy to Language. The benchmark papers*: 277-296. Oxford: Oxford University Press.

Gamble, C., 1999. *The Palaeolithic Societies of Europe.* Cambridge: Cambridge University Press.

Gao, X., S. Zhang, Y. Zhang and F. Chen. 2017. Evidence of Hominin Use and Maintenance of Fire at Zhoukoudian. *Current Anthropology* 58: 267-277.

Goldberg, P., S. Weiner, O. Bar-Yosef, Q. Xu and J. Liu. 2001. Site formation processes at Zhoukoudian, China. *Journal of Human Evolution*, 41: 483-530.

Goudsblom, J. 1985. Vuur en beschaving. De domesticatie van vuur als een beschavingsproces. (2)

De functies van vuurbeheersing in pre-agrarische samenlevingen, *De Gids* 148: 3-21.

Goudsblom, J.. 1992. The Civilizing Process and the Domestication of Fire. *Journal of World History* 3: 1-12.

Gowlett, J.A.J., J. W. K. Harris, D. Walton and B. A. A. Wood. 1981. Early archaeological sites, hominid remains and traces of fire from Chesowanja, Kenya. *Nature* 294: 125-129.

Innes, J.B., J.J. Blackford and P. A. Rowley-Conwy. 2013. Late Mesolithic and early Neolithic forest disturbance: a high resolution palaeoecological test of human impact hypotheses. *Quaternary Science Reviews* 77: 80-100.

Kaplan J.O., M. Pfeiffer, J.C.A. Kolen and B.A.S. Davis. 2016. Large Scale Anthropogenic Reduction of Forest Cover in Last Glacial Maximum Europe. *PLoS ONE* 11: e0166726. DOI:10.1371/journal.pone.0166726.

Kolen, J.. 1999. Hominids without homes. On the nature of Middle Palaeolithic settlement in Europe. In: C. Gamble and W. Roebroeks (eds.), *The Middle Palaeolithic Occupation of Europe*: 139-175. Leiden: Leiden University Press.

Leighly, J. (ed.). 1963. *Land and Life. A selection from the writings of Carl Ortwin Sauer*. Berkeley and Los Angeles: University of California Press.

MacDonald, K., F. Scherjon, E. van Veen and W. Roebroeks. 2021. Middle Pleistocene fire use: The first signal of widespread cultural diffusion in human evolution. *PNAS* 118: e2101108118.

Mellars, P.. 1996. *The Neanderthal Legacy: An Archaeological Perspective from Western Europe.* Princeton: Princeton University Press.

Nikulina, A., K. MacDonald and F. Scherjon *et al.*. 2022. Tracking Hunter-Gatherer Impact on Vegetation in Last Interglacial and Holocene Europe: Proxies and Challenges. *Journal of Archaeological Method and Theory* 29: 989-1033.

Parker, C.H., E.R. Keefe, N.M. Herzog, J.F. O'Connell and K. Hawkes. 2016. The pyrophilic primate hypothesis, *Evolutionary Anthropology* 25: 54-63.

Roebroeks, W. and T. van Kolfschoten. 1994. The earliest occupation of Europe: a short chronology, *Antiquity* 68: 489-503.

Roebroeks, W. and P. Villa. 2011. On the earliest evidence for habitual use of fire in Europe. *PNAS.* 108: 5209-5214.

Roebroeks, W. and C. C. Bakels. 2015 'Forest Furniture' or 'Forest Managers'? On Neanderthal presence in Last Interglacial environments. In: F. Coward, R. Hosfield, M. Pope and F. Wenban-Smith (eds.), *Settlement, Society and Cognition in Human Evolution: Landscapes in Mind*: 174-188. Cambridge: Cambridge University Press.

Roebroeks, W., K. MacDonald, F. Scherjon, C. Bakels, L. Kindler, A. Nikulina, E. Pop and S. Gaudzinski-Windheuser. 2021. Landscape modification by Last Interglacial Neanderthals. *Science Advances* 17: abj5567.

Sauer, C.O.. 1925. The Morphology of Landscape. *University of California Publications in Geography* 2: 19-54.

Sauer, C.O.. 1930. Historical Geography and the Western Frontier. In: J.F. Willard and C.B. Goodykoontz (eds.), *The Trans-Mississippi West: Papers Read at a Conference Held at the University of Colorado June 18-June 21, 1929 (Boulder, 1930)* 277-289. Boulder: University of Colorado.

Sauer, C.O.. 1936. American Agricultural Origins: A Consideration of Nature and Culture. In: R.H. Lowie (ed.), *Essays in Anthropology Presented to A.L. Kroeber in Celebration of his Sixtieth Birthday,* June 11: 279-297. Berkeley.

Sauer, C.O.. 1938. Theme of Plant and Animal Destruction in Economic History, *Journal of Farm Economics* 20: 765-775.

Sauer, C.O.. 1947. Early Relations of Man to Plants. *Geographical Review* 37: 1-25.

Sauer, C.O.. 1952. *Agricultural Origins and Dispersals.* New York: American Geographical Society.

Sauer, C.O.. 1956. The Agency of Man on the Earth. In: W.L. Thomas, C.O. Sauer, M. Bates and L. Mumford (eds.), *Man's Role in Changing the Face of the Earth – Volume 1*: 49-69. Chicago and London: The University of Chicago Press.

Sauer, C.O. 1958. Man in the Ecology of Tropical America. *Proceedings of the Ninth Pacific Science Congress* 20: 105-110.

Sauer, C.O.. 1961. Fire and Early Man. *Paideuma – Mitteilungen zur Kulturkunde* 7: 399-407.

Sauer, C.O.. 1962. Seashore – Primitive Home of Man? *Proceedings of the American Philosophical Society* 106: 41-47.

Scherjon, F., C. Bakels, K. MacDonald and W. Roebroeks. 2015. Burning the land: An ethnographic study of off-site fire use by current and historically documented foragers and implications for the interpretation of past fire practices in the landscape. *Current Anthropology* 56: 299-326.

Sorensen, A.C., E. Claud and M. Soressi. 2018. Neandertal fire-making technology inferred from microwear analysis. *Scientific Reports* 8: 10065.

Thomas, W.L., C.O. Sauer, M. Bates and L. Mumford (eds.). 1956. *Man's Role in Changing the Face of the Earth – Volume 1*. Chicago and London: The University of Chicago Press.

Williams, M., D. Lowenthal and W.M. Denevan. 2014. *To Pass On a Good Earth: The Life and Work of Carl O. Sauer*. Charlottesville: University of Virginia Press.

White, E. and D.M. Brown. 1973 *The first men.* Time-Life Books, Series Emergence of man.

Wrangham, R.. 2009. *Catching Fire: How Cooking Made Us Human*. New York City: Basic Books.

In Dusseldorp, G.L., Chu, W., Bakels, C.C. and M.A. Soressi (eds) 2025. Intent on the Paleolithic. Papers in honour of Prof. Dr. Wil Roebroeks. *Analecta Praehistorica Leidensia* 55. Leiden: Sidestone Press, pp. 171-188.

Protection against fire usage-associated toxic compounds in Neanderthal and Denisovan hominins, compared to ancient and extant modern humans

Jac M.M.J.G. Aarts, Gerrit M. Alink

Abstract

Two studies are reviewed, exploring the potential of Neanderthal and Denisovan genome analysis regarding smoke detoxification capacities to clarify the chronology of fire use. At 35 loci within 19 genes it was determined which gene variant occurred: the low- or high-risk variant, classified based on epidemiological studies in contemporary humans reporting the relation between tobacco smoking and/or exposure to genotoxic food heating products, relevant gene variants, and the risk of male, female or foetal reproduction-toxic effects, or based on mechanistic studies on protection against these toxicants. Ancient hominins predominantly carried the protective, low-risk allele (25/35 loci), indicating that they were likely efficient detoxifiers, but 23/25 protective alleles were ancestral. Protection against fire use-related toxicants was apparently piggybacking on evolutionarily much older genetic traits already present in great apes, possibly involved in detoxifying plant toxins or hazardous skin photo-oxidation products. Counterintuitively, modern humans tend to have more and newly derived high-risk alleles at higher frequencies, already 45 kya in a hunter-gatherer from Ust'-Ishim, long before the agricultural transition.

The second study followed up a contradicting publication suggesting that modern humans were less vulnerable to smoke, based on the responsiveness of the aryl hydrocarbon (Ah) receptor (AHR), a key regulator of important detoxification genes such as cytochrome P4501A1 (CYP1A1), found being 150-1000 times lower in modern humans compared to Neanderthals, but when expressed in heterologous rat cells using expression constructs with altered codon usage, lacking the untranslated regions (UTRs). We repeated this comparative study in human HeLa cells using the complete, unaltered AHR1 mRNA sequence including the 5'- and 3'-UTRs, and observed, in this setup, no significant difference between the effective concentration 50% (EC_{50}) for CYP1A1 mRNA induction by 2,3,7,8-tetrachlorodibenzo-p-dioxin (TCDD). The known incompatibility between the modern human AHR and a non-human cellular background is probably an important factor in explaining this huge discrepancy.

Keywords: fire usage; smoke toxicity; evolution of detoxification; polycyclic aromatic hydrocarbons (PAHs); Neanderthals; hominins; aryl hydrocarbon (Ah) receptor (AHR); 2,3,7,8-tetrachlorodibenzo-p-dioxin (TCDD); mRNA untranslated regions (UTRs); codon usage

Jac M.M.J.G. Aarts
Human Origins Group, Faculty of Archaeology, Leiden University, Leiden, The Netherlands
jm.aarts@ziggo.nl

Gerrit M. Alink
Human Origins Group, Faculty of Archaeology, Leiden University, Leiden, The Netherlands
gerritalink3@gmail.com

Introduction

Background of the research project

Mastering the production and the systematic use of fire was a major step in the technological evolution of mankind (Wrangham 2009). It conveyed a range of benefits such as protection against cold weather conditions, increasing the nutritional value of food by cooking or roasting, deterrence of predators, increased social interaction and extended light to perform tasks during darkness (Gowlett 2006; Carmody and Wrangham 2009; Wrangham 2009; Gowlett and Wrangham 2014). It also came with costs, mainly the adverse health effects of exposure to toxic combustion products from smoke as well as to genotoxic compounds formed in heated food items (Aarts *et al.* 2016), but also the workload of fuel collection (Henry *et al.* 2018).

Estimates of when in history the regular use of fire started vary widely: from around 1.8 million years ago (Mya) at the earliest (Wrangham 2009), until as late as 30 to 40 thousand years ago (kya), with the ascent of anatomically modern humans (Sandgathe *et al.* 2011). In western Europe the traces of fire usage are very scarce before 350 kya, with a sharp increase from that time period on, suggesting that the intentional production of fire had not developed earlier (Roebroeks and Villa 2011). Because of the great uncertainty regarding the chronology of routine fire use, the research team at Leiden University in the Netherlands, led by Professor Wil Roebroeks, sought for supplementary approaches, including promising techniques from other disciplines.

A wide scientific scope

The group from Leiden had already scrutinized the archaeological records for the available evidence indicating the production of fire at will (Roebroeks and Villa 2011; Sorensen, Roebroeks, and van Gijn 2014; Sorensen 2017; Sorensen, Claud, and Soressi 2018), including indirect evidence such as ignition enhancers facilitating fire making (Heyes *et al.* 2016). Given the lack of conclusive evidence from the archaeological record, any "archaeology-free" independent approach to shed light on this issue would be very welcome. Looking for such a supplementary approach, Wil Roebroeks envisioned the potential of genetic analysis in general, and in particular when focussed on the capacities to deal with toxic combustion and food-heating products. These toxic compounds are carried along with smoke and other gaseous exhausts from burning organic fuel (for example wood or other plant materials) (Bruce *et al.* 2000; Mishra *et al.* 2013) and, as a result, will end up in the food. They are also formed *in situ* by the chemical conversion of heat-labile constituents in cooked or roasted food items (Skog, Johansson, and Jägerstad 1998). Exposure to such fire usage-related toxic compounds could therefore take place through inhalation of combustion products and by ingestion of contaminated food items.

Being aware of the potential of such a genetic approach and the availability of reliable genomic information from a number of ancient hominin individuals, Wil Roebroeks launched an initiative to enable such investigations, teaming up with the late Katharine MacDonald, Fulco Scherjon, and Alison C. Smith, all affiliated to the Faculty of Archaeology at Leiden University, and with Gerrit M. Alink, Harm Nijveen, Henk J. Franssen, and Jac M.M.J.G. Aarts from Wageningen University, thereby adding expertise in molecular biology (Jac Aarts/Henk Franssen), bioinformatics (Harm Nijveen) and toxicology (Gerrit Alink/Jac Aarts) to the archaeological research group. Financial support was obtained by Wil Roebroeks acquiring two prestigious awards, a Spinoza Grant from the Dutch Research Council and the Professor Prize, program 2013, from the Royal Netherlands Academy of Arts and Sciences.

Overview of the research project and summary of the results

The research team first focused on the capacities to convert toxic compounds related to fire usage into less harmful or non-harmful metabolites and to counteract their harmful effects. This involves, for example, detoxification by certain enzymes, called biotransformation enzymes, catalysing this conversion (Lodovici *et al.* 2004), inhibition of the bioactivation of toxicant precursors (Xue and Warshawsky 2005), alleviating the oxidative stress caused (Sobinoff *et al.* 2012), repair of induced DNA damage (Ji *et al.* 2010), and proper regulation of the process by transcription factors and other regulatory proteins (Köhle and Bock 2007). The genes in the Neanderthal and Denisovan genome encoding the proteins involved in protection against toxic combustion and food-heating products were studied, and the gene variants observed were compared to the variants found in the present-day human population to determine their protectiveness. This study of 35 genetic variants in 19 different detoxification-related genes showed, surprisingly, that ancient hominins were essentially relatively efficient detoxifiers as compared to present-day humans. These results were published in PLoS ONE (Aarts *et al.* 2016). Shortly after this article was accepted, a related publication was issued (Hubbard *et al.* 2016), presenting a functional analysis of the Neanderthal aryl hydrocarbon (Ah) receptor (AhR), a ligand-activated transcription factor and key regulator of important detoxification genes. This study, based on only one detoxification-linked gene, came to the completely opposite conclusion, that Neanderthals, as compared to modern humans, were less capable of handling toxic smoke due to an overactive AHR, and suggested that a much less responsive AHR could have represented an evolutionary advantage for modern humans. However, their experimental approach used non-human (rodent) cells and an incomplete and modified AhR mRNA sequence to compare the Neanderthal and modern human AhR, although it has been reported that the human Ah receptor behaves aberrantly in mouse cells due to its functional incompatibility with the non-human co-factors in these cells (Moriguchi *et al.* 2003). Therefore our research team repeated the analyses of Hubbard *et al.* (2016) in human cells, using the complete and authentic Ah receptor mRNA sequence derived from the Altai-Neanderthal genome (Prüfer *et al.* 2014), with a totally different result. With this setup no significant functional difference between the Neanderthal and the modern human AHR was observed. This revised functional study of the Neanderthal AhR was published in Molecular Biology and Evolution (Aarts *et al.* 2021), the same journal that had also published the study by Hubbard *et al.* (2016).

Detoxification-related gene variants in Neanderthal and Denisovan genomes

Working hypothesis

Our working hypothesis was that the ability to produce and effectively use fire would obviously lead to frequent exposure to toxic compounds, both through the inhalation of smoke, and through the ingestion of genotoxic food-heating products. Regarding their known negative health effects, frequent exposure would result, according to population genetics, in positive selection pressure on new gene variants conferring more efficient protection against the toxic compounds associated with the use of fire. Individuals carrying these protective gene variants on one or both chromosomes (respectively, heterozygous or homozygous for the associated genetic trait) would therefore have increased reproductive success and become more frequent in the population.

The epidemiology of tobacco smoking applied to ancient hominin genomes

Numerous epidemiological studies on tobacco smoking have proven the negative health effects of smoke due to the presence of polycyclic aromatic hydrocarbons (PAHs) (Freeman and Cattell 2014), such as benzo[*a*]pyrene, and other compounds with dioxin-like toxicity, and have provided deep insight into which genetic variants in relevant detoxification-related genes are more or less protective against these negative health effects (Lammer *et al.* 2004; Shi *et al.* 2008; Gu *et al.* 2010; Qin *et al.* 2012). The same applies to the well-studied PhIP (2-amino-1-methyl-6-phenylimidazo[4,5-b]pyridine) (Brittebo *et al.* 1994) and other genotoxic heterocyclic amines which are formed, besides PAHs, during the heating of food (Lan *et al.* 2004; Eichholzer *et al.* 2012). These detrimental effects are mainly the enhancement of malignancies in the lungs and other organs (DeMarini 2004; Lodovici *et al.* 2004), and decreased male (Gu *et al.* 2010) as well as female fertility (Neal, Foster, and Younglai 2008) and damage to the developing foetus (Mishra, Retherford, and Smith 2013). Since ancient hominins may not often have reached a sufficiently old age to develop cancers (Caspari and Lee 2004), and cancers tend to occur after an individual has already reproduced, we focused on epidemiological studies that reported negative reproduction effects of smoking or toxicants from heated food, which would certainly result in genetic selection pressure within Neanderthal and Denisovan populations, and in particular on those studies that present statistical or mechanistic evidence for increased or decreased protection by certain extant human gene variants against the negative health effects of these fire usage-related toxicants (for example: Lammer *et al.* 2004; Ji *et al.* 2008; Shi *et al.* 2008; Gu *et al.* 2010; Ji *et al.* 2010; Qin *et al.* 2012).

Results and Discussion

Protective detoxification-related gene variants are predominant in ancient hominin genomes and are mostly ancestral

To test our working hypothesis, we analysed one high-coverage (Altai) (Prüfer *et al.* 2014) and six low-coverage Neanderthal genomes (Green *et al.* 2010), and one high-coverage Denisovan genome (Meyer *et al.* 2012). The genetic variant was determined occurring in the available genomic sequences at 35 polymorphic loci in 19 detoxification-associated genes that had been reported in the literature to affect the odds of reproduction-toxic effects, such as impaired fertility or foetal defects, when smoking tobacco or in relation to genotoxic food-heating products.

At 25 out of these 35 loci, the more protective (low-risk) allele was consistently observed in both the Neanderthal and the Denisovan genomes (Table 1 and 2), whereas at only six loci both species carried the less protective (high-risk) variant (not shown; see Aarts *et al.*, 2016: Table 3). At four loci both hominin species were contradictory as to which of the two risk variant alleles was present (Table 1 and 2). All loci of both ancient hominin species studied here were certified being homozygous, except for one *GSTP1* locus in Vi33.26 (Table 2), and the GSTM1 and GSTT1 loci; the latter are possibly homozygous (WT/WT) as well, but also WT/null still confers a lower-risk as compared to null/null (Crosbie *et al.* 2009; Nørskov *et al.* 2011). Their populations were generally small and isolated (Prüfer *et al.* 2014), resulting in a high degree of inbreeding and homozygosity.

At only two out of the abovementioned 25 loci the protective allele is derived (Table 2); the remaining 23 protective variants turned out to be the ancestral allele.

Therefore, genetic analysis of 35 loci within 19 relevant genes of the Neanderthal and Denisovan genome clearly showed that there is a pronounced predominance of the more protective variants, which confer a relatively low risk of reproduction-toxic effects when exposed to fire usage-associated toxicants. This implies that these ancient hominins were probably very efficient detoxifiers. In agreement with our working hypothesis, this is to be expected if they were frequently using fire. Interestingly, we observed, in contradiction to our hypothesis, that for 23 out of the 25 loci the protective variant appeared to be the ancestral allele also found in chimpanzees and gorillas (Table 1 and 2). Therefore, the predominance of these protective variants in the population apparently did not occur due to selection pressure on derived new alleles resulting from fire usage, but was evolutionarily inherited. In the observed situation, the protective alleles appear to have already existed before the split between the hominin and the great ape lineages, and the paucity of newly derived alleles of these detoxification-related genes suggests that there was no need for adaptation towards increased efficiency in these ancient hominins.

The predominance of protective alleles in the ancient genomes cannot be accounted for by coincidence

The assumption that the observed predominance of protective alleles would present a pure coincidence (denial of any bias) would imply that the low- and high-risk allele each have an equal (50%) chance to be observed. The chance to find 48 (23 homozygous loci plus at least one GSTM1 and GSTT1 protective allele per high-coverage genome), or an even more biased (higher) number of protective alleles at 35 loci, can be calculated (Berman, n.d.) based on the binomial probability distribution to amount to 0.2%, which demonstrates that coincidence, and therefore the absence of bias, can already be dismissed as highly unlikely from a statistical point of view. Moreover, the 35 loci studied are all located in genes involved in protection against specific toxicants, and are therefore functionally related. For that reason, they will obviously evolve in concert in the same physiological direction, for example when under selection pressure for becoming more efficient; in other words, they are supposed not to evolve randomly, and certainly not in opposite physiological directions. Hence, it is not only for statistical reasons that it is highly unlikely that the observed predominance of 25 out of 35 loci exclusively carrying the low-risk allele would be a coincidence, even though this conclusion is based on high-coverage genomic information from only two individuals supplemented with partial low-coverage information from six other individuals (see also Aarts *et al.*, 2016: 9, and Supporting Information, S1 Text, section B). Instead, this clean-cut bias will most likely have a physiological reason.

Protection against fire use-associated toxicants in ancient hominin appears to ride along on much older detoxification capacities, most likely directed to plant toxins and possibly to hazardous photo-oxidation products

That the detoxification capacities conferred by the gene variants studied here mainly evolved before the hominin and great ape lineages diverged is an interesting new observation, indicating that they were apparently needed long before because of other reasons than resistance against toxicants associated with fire use. A plausible hypothesis would be that these capacities originally evolved for protection against plant toxins that were prominently present in the diet of the ancestors of great apes and ancient hominins (Bjeldanes *et al.* 1991; Lee-Hilz *et al.* 2006;

Treutter 2006; Korkina *et al.* 2008; Head *et al.* 2011). Another reason may be the formation in the skin of hazardous compounds, such as 6-formylindolo[3,2-b] carbazole (FICZ), upon exposure to ultraviolet (UV) light (Wei *et al.* 1998; Wincent *et al.* 2009). Such photo-oxidation products of the amino acid tryptophan display dioxin-like toxicity (Rannug *et al.* 1987; Katiyar *et al.* 2000), may cause oxidative DNA damage (Park, Shigenaga, and Ames 1996; McCormick and Thomason 2014) and, on the other hand, are potent endogenous Ah receptor-mediated regulators of important physiological processes such as circadian rhythms (Mukai and Tischkau 2007; Wincent *et al.* 2012), and therefore have to be kept well-balanced. Detoxification of smoke and food-heating toxicants would then just have been piggybacking on these older capabilities that were apparently already sufficiently effective given the level of selection pressure. Adaptation to further increased efficiency apparently did not bring about increased evolutionary fitness.

Less protective, high-risk alleles emerged and increased in frequency along the modern human lineage

In the ancient hominin genomes, the low-risk variant was exclusively observed at 25 out of 35 loci. However, in the present-day human population the less protective, high-risk allele is always found besides the low-risk allele, and often at substantial frequencies (Aarts *et al.* 2016: S1 Text, Section I), which strongly suggests that the number and frequency of the high-risk alleles increased in modern humans as compared to ancient hominins. The number of loci in the two high-coverage ancient genomes studied (from the Altai-Neanderthal and Denisovan, respectively) carrying exclusively the low-risk allele (referred to as "low-risk loci"), and the number of loci carrying one or two high-risk alleles (referred to as "high-risk loci") were counted and compared to the distribution of the number of low-/high-risk loci in a global population of 2,504 modern humans studied by The 1000 Genomes Project Consortium (2015), which covered 32 of the 35 gene loci investigated here. Both ancient hominin species turned out to have an extremely low number of high-risk loci as compared to present-day humans. Only 2% of the 1000 Genomes population showed an equally low or lower number of high-risk loci than the Altai-Neanderthal (with 10 high-risk loci out of 32), and the Denisovan (with 8 high-risk loci out of 32) was even falling at the extreme end of the distribution, as only 0.24% of the 1000 Genomes population had also 8 high-risk loci, and no individual was observed with less than 8. Therefore, we conclude that the substantially higher number of loci holding less protective, high-risk alleles observed in the present-day human population is a genuine difference between ancient hominin and modern human genomes, and not a matter of coincidence. The fact that these high-risk alleles are mostly derived (i.e. evolutionarily derived from, and thus different from the ancestral allele), and therefore newly appeared at some point along the human lineage, is consistent with an increase in number of these high-risk alleles. Moreover, the high frequencies which some of these high-risk alleles have attained, up to the 80% range (Aarts 2016: S1 Text, Section I), suggests that not only the number, but also the allele frequencies of these high-risk alleles have increased along the modern human lineage.

To address the question as to when the derived high-risk alleles arose and when the overall gain in frequency of the high-risk alleles started, we analysed the genotype at the 35 loci under study (Table 1 and 2) in a range of ancient anatomically modern humans (AMHs): from Upper-Palaeolithic hunter-gatherers found at Ust'-Ishim and Mal'ta (MA-1) in Siberia, respectively around 45 and 24 kya old (Fu *et al.* 2014; Raghavan *et al.* 2014), via one from North America (Anzick-1) dating to around 12.6 kya (Rasmussen *et al.* 2014), to Holocene examples including a Neolithic (NE1; 7.2 kya) and a Bronze Age (BR2; 3.2 kya) individual from Hungary (Gamba *et al.* 2014), an approximately 4,000-year old Palaeo-eskimo (Saqqaq) (Rasmussen *et al.* 2010), and a more recent 100-year old Aboriginal Australian (Rasmussen *et al.* 2011).

Within the group of 21 loci in Table 1 for which could be certified that they are homozygous for the ancestral low-risk allele, a derived high-risk allele emerged at 8 of these loci in the oldest AMH studied, the individual found at Ust'-Ishim. Furthermore, in Ust'-Ishim 15 out of the 32 loci covered by the 1000 Genomes Project held one or two high-risk alleles, against 10 of 32 loci in the Altai-Neanderthal and 8 of 32 in the Denisovan. We did not observe a clear gradual increase over time based on the younger AMH genomes analysed (Table 1, 2; Table 3 of Aarts *et al.*, 2016), but the Ust'-Ishim data are consistent with an early start of the increase in number and allele frequency of the less protective, high-risk variants in the modern human lineage, already before 45 kya.

Gene	Function[1)]	Extant Humans				Ancient Hominins	
		Polymorphism					
		RefSNP Number	HGVS Name[2)]	Low[3)]	High[3)]	Nea	Den
AHR1	Regulation	rs2066853	p.Arg554Lys	A	G	A	A
		rs2282885	c.66-3946A>G	A	G	A	A
AHRR1	Regulation	rs2292596	p.Pro189Ala	C	G	C	C
CYP1A1	Detox. 1	rs4646903	c.*1189T>C	T	C	T	T
		rs1048943	p.Ile462Val	A	G	A	A
CYP1B1	Detox. 1	rs1056836	p.Leu432Val	G	C	G	G
EPHX1	Detox. 1	rs1051740	p.Tyr113His	T	C	T	T
EPHX2	Detox. 1	rs1042064	c.*93T>C	C	T	C	C
GSTA4	Detox. 2	rs316133	c.415-48C>G	G	C	C	G
		rs3756980	c.139+176T>C	T	C	C	T
GSTM1	Detox. 2	-	Null[4)]	WT	Null	WT	WT
GSTP1	Detox. 2	rs1138272	p.Ala114Val	C	T	C	C
		rs762803	c.232+13C>A	C	A	C	C
GSTT1	Detox. 2	-	Null[4)]	WT	Null	WT	WT
NAT1	Detox. 2	rs1057126	c.*215A>T	T	A	T	A
		rs4986782	p.Arg187Gln	G	A	G	G
		rs5030839	p.Arg187Ter	C	T	C	C
		rs56379106	p.Arg64Trp	C	T	C	C
		rs56318881	p.Arg33Ter	C	T	C	C
		rs56172717	p.Asp251Val	A	T	A	A
NAT2	Detox. 2	rs1801280	p.Ile114Thr	T	C	T	T
		rs1799930	p.Arg197Gln	G	A	G	G
CAT1	Anti-ox.	rs1001179	c.-330C>T	C	T	C	C
SOD2	Anti-ox.	rs4880	p.Ala16Val	C	T	C	C
		rs5746136	c.*441G>A	G	A	G	G
XPA1	Repair	rs1800975	c.-4A>G	G	A	G	G

Table 1. Polymorphisms for which the low-risk gene variant observed in Neanderthal and/or Denisovan is the ancestral allele (Aarts *et al.* 2016). Column headers: Nea = Neanderthal; Den = Denisovan; Chimp = Chimpanzee; Gor = Gorilla; Ust'-Ishim/Mal'ta (MA-1)/Anzick-1 = Siberian/Siberian/North-American pre-Holocene hunter-gatherer; NE1/BR2 = Neolithic/Bronze Age Hungarian individual; Saqqaq = Palaeo-eskimo; Aus = Aboriginal Australian. Cell shading: light grey = low-risk variant; dark-grey = high-risk variant. **1)** Detox 1 = Detoxification phase 1; Detox 2 = Detoxification phase 2; Anti-ox = Oxidative stress management; Regulation = Regulation of detoxification gene expression; Repair = Repair of DNA damage. **2)** SNP nomenclature as recommended by the Human Genome Variation Society (HGVS) (den Dunnen *et al.* 2016). **3)** Allele associated with a relatively low-risk, respectively, high-risk of adverse reproduction effects based on epidemiological or biochemical studies (details in Aarts *et al.*, 2016, S1 Text of the Supporting Information). **4)** Null = deletion of the entire gene sequence. High risk phenotype only

Great Apes		Modern Humans						
Chimp	**Gor**	**Ust'-Ishim (45 kya)**	**MA-1 (24 kya)**	**Anzick-1 (12 kya)**	**NE1 (7.2 kya)**	**Saqqaq (3.9 kya)**	**BR2 (3.2 kya)**	**Aus (0.1 kya)**
A	A	A G	A	G	G	G	G	–
A	A	A	A	A	A G	A G	A G	A
C	C	C G	C G	C	G	C G	C G	C G
T	T	T	–	T C	T	T	T C	C
A	A	A	A	A G	A	A	A G	A
G	G	G	C	C	C	G C	C	C
T	T	T	C	T	T C	T C	T	T
C	C	C T	C	T	T	C	T	T
G	G	G	–	G	G	G	G	–
T	T	T	T	T	T	T	T	T
WT	WT	WT	WT	Null	WT	Null	WT	Null
C	C	C	C	C	C	C	C	C
C	C	C A	C	C A	A	C	C	C
WT	WT	WT	WT	WT	Null?[5]	Null	WT	Null
T Diff.[6]	Del.[6]	T	T	A	T A	T	T	–
G	G	G	G	G	G	G	G	–
C	C	C	C	C	C	C T	C	–
C	C	C	C	C	C	C	C	C
C	C	C	C	C	C	C	C	C
A	A	A	A	A	A	A	A	–
T	T	T	T	T	T C	T	T	T
G	G	G	G	G A	G	G	A	–
C	C	C T	T	C	C	C	C	C
C	C	C T	T	C T	C	C	T	C T
G	G	G A	A	G	G	G	G	G
G	G	G A	G	G A	G A	G	G A	G A

when homozygous for null. WT = wild-type allele. One or two WT alleles cannot be discriminated based on sequencing data only. WT/null also confers at least intermediate protection and a lower risk as compared to null/null (Crosbie *et al.* 2009; Nørskov *et al.* 2011). **5)** Most likely Null for GSTT1, since only reads observed in intron regions, not in exon regions.

6) This A/T polymorphism is the 3'-terminal base of a tandem repeat of four AAT elements that is found full length in modern humans, Neanderthals, and the Denisovan, but is variable in length in chimpanzees (0 – 4 AAT elements, following position 14,328,777 in assembly CSAC 2.1.4/panTro4). Only the full length 4 x AAT variant encompasses the position corresponding to human polymorphism rs1057126. Diff. indicates that some of the PanMap chimpanzees are clearly or possibly (indicated by ? in Aarts *et al.*, 2016, S2 Table) heterozygous for variants encompassing 0 – 4 AAT elements. Del. indicates that gorilla misses all 4 AAT elements (= 0 x AAT).

Reproduced from Aarts *et al.* (2016).

Gene	Function[1]	Extant Humans				Ancient Hominins	
		Polymorphism[2]		Low[3]	High[3]	Nea	Den
		RefSNP Number	HGVS Name[2]				
EPHX1	Detox 1	rs2234922	p.His139Arg	A	G	A	A
GSTP1	Detox 2	rs1695	p.Ile105Val	A	G	A[4] G	A
ERCC1	Repair	rs3212986	c.*197G>T	G	T	G	G

Conclusions

In conclusion, the ancient hominin genomes studied possessed predominantly protective variants at polymorphic gene loci of contemporary humans found to be involved in protection against smoke and food-heating toxicants. This indicates that these ancient hominins were apparently efficient detoxifiers. The protective gene variants were mainly ancestral, showing that, contrary to our working hypothesis, there was no need for adaptation since the divergence of the chimpanzee and human lineage 7-8 million year ago (Langergraber *et al.* 2012), even though fire-making and usage must have become common practice at some point during the evolution of modern humans. Apparently, the protection against toxicants related to fire use could piggyback on evolutionarily older capacities that possibly evolved to deal with plant toxins and/or hazardous photo-oxidation products formed in the skin upon exposure to ultraviolet (UV) light.

The genetic patterns observed dismiss the supposed positive selection pressure by fire usage, but it is possible that positive selection played a role in maintaining the ancestral protective gene variants against the forces of genetic drift.

Surprisingly, as compared to ancient hominins, present-day humans tend to possess a higher number of less protective, high-risk alleles at a higher frequency, suggesting that the capacities conferring protection against fire usage-related toxicants are decreasing instead of increasing along the modern human lineage, and this probably started already before 45 kya, long before the dietary changes and expansions associated with the introduction of agriculture. This may indicate that the total toxic burden did not increase significantly upon the onset of fire use, possibly because cooking decreased exposure to dietary toxins and/or reduction of UV-induced hazardous compounds occurred due to concomitant cultural adaptations such as clothing and housing. Furthermore, Neolithic population expansions could have played a role, since rapid expansions (indicating relatively prosperous conditions) tend to enhance the proportion of rare and deleterious variants, because the purifying selection on less favourable variants is reduced (Henn *et al.* 2016).

The ancestral Ah receptor variant of ancient hominins has disappeared completely

The human *AHR1* gene encodes the aryl hydrocarbon (Ah) receptor (AHR), a key regulator of a range of major detoxification-related genes (Köhle and Bock 2007), such as *CYP1A1* (Corchero *et al.* 2001; Nukaya and Bradfield 2009), encoding cytochrome P450 subclass 1A, variant 1, an important biotransformation enzyme involved, among others, in both the bioactivation and detoxification of polycyclic aromatic hydrocarbons (PAHs) (Nebert *et al.* 2013; Divi *et al.* 2014) present in smoke (Freeman and Cattell 2014). The modern human AHR has a valine (Val) residue at position 381 in the ligand-binding domain of its amino acid chain, resulting in a lower binding affinity for prototype AHR agonists (Ema *et al.* 1994), a shifted ligand selectivity (Ciolino *et al.* 1999; Vrba *et al.* 2012), transcription activation agonist efficiency (Flaveny *et al.* 2009) and target gene range (Flaveny, Murray, and Perdew 2010) as compared to the exhaustively studied mouse homolog *Ahr*, which carries alanine (Ala) at homologous position 375 of its amino acid chain (Ema *et al.* 1994).

All ancient hominins studied here, if sequence information is available (for Neanderthals Altai / Vi33.16; Denisovan) (Prüfer *et al.* 2014 / Green *et al.* 2010; Meyer 2012), display the ancestral AHR variant with alanine at position 381, also found

Great Apes		Modern Humans						
Chimp	**Gor**	**Ust'-Ishim (45 kya)**	**MA-1 (24 kya)**	**Anzick-1 (12 kya)**	**NE1 (7.2 kya)**	**Saqqaq (3.9 kya)**	**BR2 (3.2 kya)**	**Aus (0.1 kya)**
G	G	A	–	A	A	A	G	A G
G	G	A G	A	A G	G	A	A	A
T	T	G	G	G	G T	G T	G	G

Table 2. Polymorphisms for which the low-risk gene variant observed in Neanderthal and/or Denisovan is a derived allele (Aarts *et al.* 2016). Column headers and cell shading as for Table 1. **1-3)** See footnotes of Table 1. **4)** Neanderthal individual Vi33.26 is possibly heterozygous low-/high-risk (A/G) based on a single, high-quality read of each type; the high-coverage Altai Neanderthal and possibly Vi33.25 (one high-quality G read) are homozygous for the high-risk G allele. Reproduced from Aarts *et al.* (2016).

at the homologous position in all chimpanzee genomes evaluated (The Chimpanzee Sequencing and Analysis Consortium 2005; Auton *et al.* 2012) and the gorilla reference genome (Scally *et al.* 2012). The Val381 AHR is a newly derived variant that apparently reached fixation in the modern human population, since the ancestral Ala381 has, up to now, never been observed in modern humans (Green *et al.* 2010).

Functional analysis of AHR variants in heterologous cells tends to produce artefacts

The "Ala381Val polymorphism" has become meaningful since we know about the A381 variant occurring in ancient hominins. It was excluded from the genetic analyses described above because there is no evaluation available of the risk profile of this polymorphism in a human cellular context. This is even more problematic, since it has been reported that a human Ah receptor (Val381) had still a lower transactivation capacity than the AHR from DBA/2, a mouse strain carrying a homologous natural Val375 mutant AHR (Moriguchi *et al.* 2003). This has been attributed to poor co-operation of the human AHR with the necessary co-factors provided by a mouse cellular background, which is not surprising given the evolutionary distance between mouse and human, with the most recent common ancestor dated at about 75 Mya (Members of the Mouse Genome Analysis Group *et al.* 2002). Therefore, it could be very misleading to evaluate the relative risk for reproduction toxicity conferred by the ancient hominin Ala381 AHR based on data from a non-human species, more specifically, to extrapolate the functional difference observed in a mouse's cellular background between the wild-type Ala375 and the mutant Val375 mouse AHRs to assess the difference between the ancient hominin Ala381 and the modern human Val381 AHR.

Functional comparison of the Neanderthal and modern human AHR in human cells

The HeLa human cervix epithelial adenocarcinoma cell line has been reported to express very low endogenous AHR levels and has been validated for testing AHR gene variants before (Koyano *et al.* 2005). Therefore, HeLa was the cell line of choice for comparison of the responsiveness of the Neanderthal and modern human AHR in a human cellular background. Because they are isolated from a human cervix tumour, HeLa cells are essentially immortal and can be continuously cultured in plastic flasks. Fig. 1 shows that the AHR cDNA sequences used in this study comprised the complete mRNA transcript sequence, including the 5'- and 3'-untranslated regions (UTRs), while maintaining the original codon usage to express the Altai-Neanderthal, respectively, the modern human AHR in HeLa cells by transient transfection with an expression construct under transcriptional control of a strong viral promoter/ enhancer. Also shown is the setup of a previous similar study by Hubbard *et al.* (2016) in BP8 rat hepatoma cells using Neanderthal and modern human AHR expression constructs encompassing only the amino acid sequence-encoding region of the AHR messenger RNA (mRNA) without the 5'- and 3'-UTRs, and with altered codon use aiming at optimised instead of natural AHR protein production levels in mammalian cells.

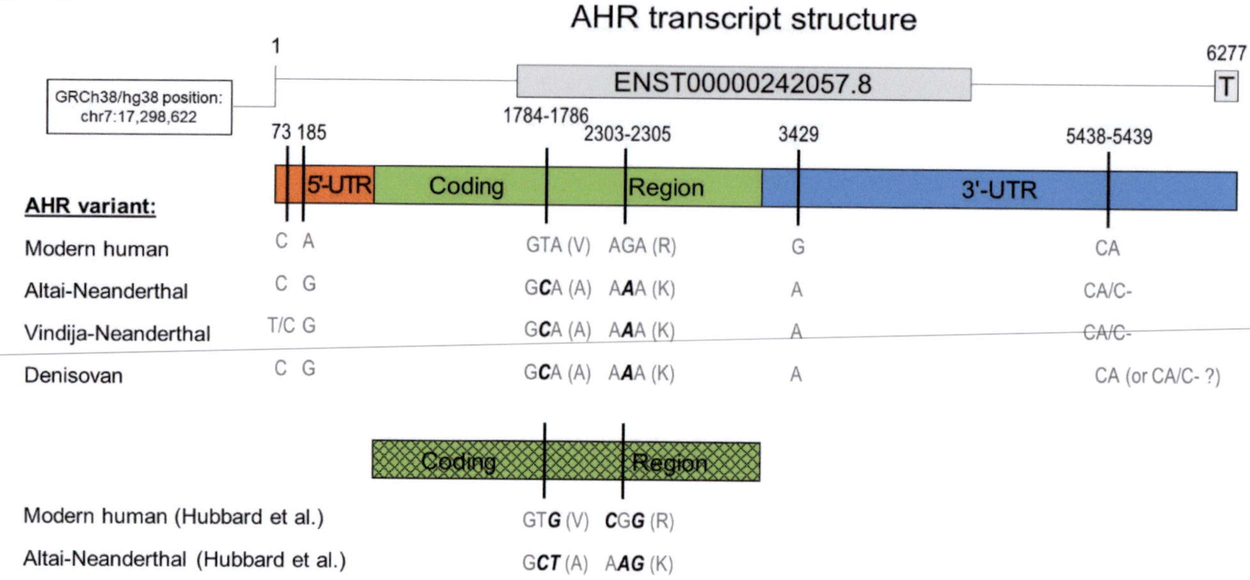

Figure 1. Map of the modern human and ancient hominin AHR mRNA sequence variants studied by Aarts *et al.* (2021) versus by Hubbard *et al.* (2016). The modern human AHR variant tested was based on manually expert-curated sequence information from the Ensembl and RefSeq databases (details in Materials and Methods of Aarts *et al.* (2021)). The bar gives a scaled representation of this 6277 bp sequence including the 5′-untranslated region (UTR), the AHR protein-coding region, and the 3′-UTR. The numbers above are the coordinates of the variant sites within the Ensembl AHR reference transcript ENST00000242057.8. Below it, the base, or codon plus encoded amino acid (V = valine; A = alanine; R = arginine; K = lysine) present in each AHR variant is indicated. Cross-hatching indicates optimization for mammalian codon use and minimal secondary mRNA structure; absence of hatching indicates that the original codon usage is maintained. Codon bases in bold italics differ naturally, respectively, have been changed in the synthetic sequences tested by Hubbard *et al.* (2016), as compared to the modern human reference. Apart from the indicated variable sites the reported ancient AHR transcript sequences are identical to the modern reference sequence (see Supplementary Data SI1, SI10 of Aarts *et al.* (2021)). Reproduced from Aarts *et al.* (2021).

Results and Discussion

No significant functional difference between Neanderthal and modern AHR

CYP1A1 mRNA expression in HeLa cells transfected with the Neanderthal or modern human AHR upon exposure to a concentration range of 2,3,7,8-tetrachlorodibenzo-*p*-dioxin (TCDD) was measured by real-time reverse transcription polymerase chain reaction (RT-PCR). Fig. 2 shows that the dose-response curves for the Neanderthal and the modern human AHR-transfected HeLa cells almost coincide. No significant difference between the corresponding effective concentration 50% (EC_{50}) values was observed: The EC_{50} value [95% confidence interval] in pM TCDD was 4097 [2837-5834] and 3450 [2443-4806] for HeLa cells transfected with the modern human and Neanderthal AHR expression construct, respectively (see Table 1 in Aarts *et al.*, 2016). A small but significant difference was found between the ancient and modern AHR regarding the extrapolated maximum of the dose-response curve reached at concentrations far above 10,000 pM, which are outside the physiologically relevant range.

The essentially similar responsiveness of the Neanderthal and modern human AHR observed in this study stands in sharp contrast to the tremendous difference between the Neanderthal and modern human AHR produced by the experimental approach chosen by Hubbard *et al.* (2016). The Neanderthal A381 AHR induced mRNA expression of its *CYP1A1* target gene at a 150-1,000 times lower AHR agonist concentration than the modern V381 AHR. Given the reported incompatibility between the modern human V381 receptor and an A375 mouse cellular background (Moriguchi *et al.* 2003), this discrepancy is likely due to the use of a heterologous "Ala-type" (rat BP8) cellular background, possibly

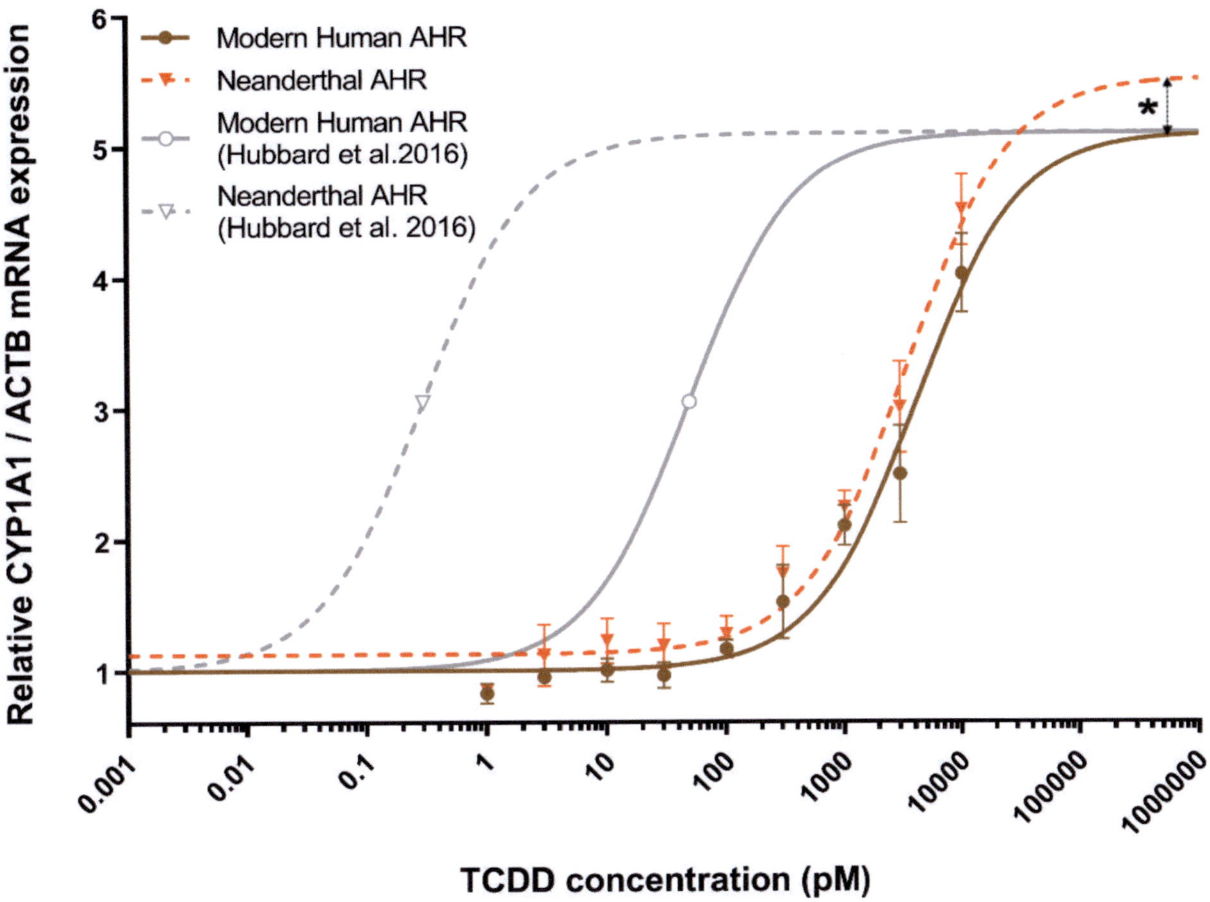

Figure 2. Dose-response relation observed for CYP1A1 mRNA induction by 2,3,7,8-tetrachlorodibenzo-*p*-dioxin (TCDD) in HeLa human cervix epithelial adenocarcinoma cells transfected with a modern human or Neanderthal AHR expression construct. Each curve is based on two data sets from different HeLa cell batches, and each data set comprised mostly 3, or at least 2 TCDD exposure replicates per data point. Cytochrome P450 1A1 (CYP1A1) and β-actin (ACTB) mRNA expression were measured in duplicate by quantitative reverse transcriptase polymerase chain reaction (qRT-PCR) and CYP1A1 mRNA expression was normalized to β-actin mRNA expression. The CYP1A1/ACTB ratios were subsequently normalized to the basal level for the modern human AHR, i.e. this value was set to 1 and all values expressed as a fold value relative to this response level, and plotted against the TCDD concentration in the culture medium during exposure. Error bars indicate the standard error of the mean (SEM). Dose-response curves were generated by fitting the one-site receptor-ligand binding equation to the experimental data, which produced best-fit estimates for the Effective Concentration 50% (EC_{50}), and for the basal and maximal values of TCDD-induced CYP1A1 mRNA expression (see Materials and Methods section and Table 1 of Aarts *et al.* (2021)). For comparison, full-range dose-response curves as reported by Hubbard *et al.* (2016) are shown in grey based on their observed EC_{50} values and a toxic equivalency factor (TEF) for TCDF = 0.1 (van den Berg *et al.* 2006) to account for the AHR agonist potency difference between TCDF used by them and TCDD (TEF = 1) used by Aarts *et al.* (2021). * = The maximal CYP1A1 induction levels attained by the modern human and Neanderthal AHR were statistically evaluated as different based on the calculated difference in corrected Akaike's Information Criterion (column ΔAICc in Table 1 of Aarts *et al.* (2021)). Reproduced from Aarts *et al.* (2021).

in combination with the expression of an AHR transcript with altered codon use and missing UTRs, and is also at odds with the maximal difference of a factor 10 in responsiveness reported in previous studies between the mouse and human "Val-type" and the mouse "Ala-type" receptors (Ema *et al.* 1994; Okey *et al.* 2005; Silkworth *et al.* 2005; Connor and Aylward 2006; Flaveny *et al.* 2009; Budinsky *et al.* 2010).

Limited Impact of the Ala381Val Substitution in the Modern Human Lineage

The results of our functional analysis of the ancient hominin AHR may seem at odds with previous comparisons of ligand, DNA binding and ensuing transcription activation potency of "Ala-" and "Val-type" AHRs (Roberts *et al.* 1990; Ema *et al.* 1994; Harper *et al.* 2002; Okey *et al.* 2005; Silkworth *et al.* 2005; van den Berg *et al.* 2006; Budinsky *et al.* 2010; Forgacs *et al.* 2013). However, transactivation of gene expression involves intricate interactions between functional domains within the AHR molecule and with certain functional domains of other proteins in the ultimately formed complex bound to the upstream regulatory region of a target gene (Whitelaw, Gustafsson, and Poellinger 1994; Ma *et al.* 1995; Baek *et al.* 2006; Billas and Moras 2013; De Vera *et al.* 2017, 1506; Seok *et al.* 2017, 5434). Previously, however, assays to measure ligand and DNA response element binding and transcription activation by the AHR, have used assay conditions that may have lacked important accessory proteins needed for natural AHR functioning, such as its dimerization partner, chaperones and co-factors (Pandini *et al.* 2007; Chiaro *et al.* 2008; Hubbard *et al.* 2015), or used non-human cell types or cellular extracts and other biochemical components of non-human origin (Bank *et al.* 1992; Ema *et al.* 1994; Flaveny *et al.* 2009), providing heterologous "Ala-type" functional conditions, which may essentially produce artefactual results for the human "Val-type" receptor. The utmost importance of a homologous cellular background observed in our study may explain these supposed discrepancies. Moreover, a limited impact on ligand binding of the Ala to Val mutation at position 381 in the AHR amino acid chain is corroborated by a number of molecular modelling studies of the AHR (Henry and Gasiewicz 2008; Bonati *et al.* 2017; Giani Tagliabue *et al.* 2019).

The V381 variant originated in the modern human and became fixed in the population (Green *et al.* 2010). This does not necessarily indicate functional importance, as implied by our results [Aarts *et al.*, 2021]. Alternative explanations for fixation are random genetic drift, facilitated by the small size of ancient hominin populations, or positive selection via genetic coupling to a nearby selective sweep region, found within 1 Mb distance (Green *et al.* 2010).

Final remarks

Both studies reviewed here (Aarts *et al.* 2016; 2021) are strongly at odds with a selective advantage of modern humans to deal with the toxic burden coming with fire use, neither due to their AHR variant, nor to any other aspect of their genetic make-up as investigated here. On the contrary, modern humans rather seem to have become less efficient detoxifiers.

Our study is a caveat against the use of heterologous cell systems and biochemical conditions for functional studies, in particular against future expression studies with ancient hominin gene products in non-human cells.

Our results do not exclude that differential responsiveness of the ancient and modern human AHR to other ligand categories, such as endogenous and diet-borne AHR agonist, may have provided some kind of evolutionary advantage. Future research should especially focus on this aspect and will hopefully be able to add more genetic depth as more high-quality ancient hominin genomes become available.

Methods

For methodological details the reader is kindly referred to the Materials and Methods sections of the original publications (Aarts *et al.* 2016; 2021).

Acknowledgements

The studies underlying this Festschrift contribution in honour of Professor Wil Roebroeks had as a goal to provide more insight into genetic adaptations to the use of fire as a tool to assess the start of fire use in hominin evolution. By introducing disciplines such as molecular biology and toxicology into archaeology an interesting new research line was initiated by Wil that opened unexpected and fascinating new avenues, such as the one described here.

Although they did not provide more insight into the chronology of fire use, the original goal of these works, and the data did surprisingly suggest an evolution of modern human detoxification capacities contradictory to our expectations, the studies presented here may have implications for understanding the ancient hominin niche, as well as modern human adaptations to their environment.

The first author is very grateful for being appointed by Wil at the Faculty of Archaeology to perform these studies in collaboration with others. On behalf of all members of the research team, he and the second author of this Festschrift contribution would like to express great gratitude to Wil for giving them the opportunity to participate in this exciting new research line.

This research project was financially supported by a Spinoza Grant (28-548) from the Dutch Research Council (Nederlandse Organisatie voor Wetenschappelijk Onderzoek; NWO) and the Professor Prize, program 2013, from the Royal Netherlands Academy of Arts and Sciences (Koninklijke Nederlandse Academie van Wetenschappen; KNAW).

References

Aarts, J. M. M. J. G., G. M. Alink, H. J. Franssen and W. Roebroeks. 2021. Evolution of Hominin Detoxification: Neanderthal and Modern Human Ah Receptor Respond Similarly to TCDD. *Molecular Biology and Evolution* 38: 1292-1305. DOI:10.1093/molbev/msaa287.

Aarts, J. M. M. J. G., G. M. Alink, F. Scherjon, K. MacDonald, A. C. Smith, H. Nijveen and W. Roebroeks. 2016. Fire Usage and Ancient Hominin Detoxification Genes: Protective Ancestral Variants Dominate While Additional Derived Risk Variants Appear in Modern Humans. *PLoS ONE* 11: e0161102. DOI:10.1371/journal.pone.0161102.

Auton, A., A. Fledel-Alon, S. Pfeifer, O. Venn, L. Ségurel, T. Street, E. M. Leffler *et al.*. 2012. A Fine-Scale Chimpanzee Genetic Map from Population Sequencing. *Science* 336: 193-198. DOI:10.1126/science.1216872.

Baek, S. H., K. A. Ohgi, C. A. Nelson, D. Welsbie, C. Chen, C. L. Sawyers, D. W. Rose and M. G. Rosenfeld. 2006. Ligand-Specific Allosteric Regulation of Coactivator Functions of Androgen Receptor in Prostate Cancer Cells. *PNAS* 103: 3100-3105. DOI:10.1073/pnas.0510842103.

Bank, P. A., E. F. Yao, C. L. Phelps, P. A. Harper and M. S. Denison. 1992. Species-Specific Binding of Transformed Ah Receptor to a Dioxin Responsive Transcriptional Enhancer. *European Journal of Pharmacology: Environmental Toxicology and Pharmacology* 228: 85-94. DOI:10.1016/0926-6917(92)90016-6.

Berg, M. van den, L. S. Birnbaum, M. Denison, M. De Vito, W. Farland, M. Feeley, H. Fiedler *et al.*. 2006. The 2005 World Health Organization Re-Evaluation of Human and Mammalian Toxic Equivalency Factors for Dioxins and Dioxin-like Compounds. *Toxicological Sciences* 93: 223-241. DOI:10.1093/toxsci/kfl055.

Berman, H.B. n.d. 'Binomial Probability Calculator'. [online] Available at: https://stattrek.com/online-calculator/binomial URL [Accessed Date: 6-9-2023].

Billas, I. and D. Moras. 2013. Controls of Nuclear Receptor Function in the Regulation of Transcription. *Journal of Molecular Biology* 425: 2317-2329. DOI:10.1016/j.jmb.2013.03.017.

Bjeldanes, L. F., J. Y. Kim, K. R. Grose, J. C. Bartholomew, and C. A. Bradfield. 1991. Aromatic Hydrocarbon Responsiveness-Receptor Agonists Generated from Indole-3-Carbinol in Vitro and in Vivo: Comparisons with 2,3,7,8-Tetrachlorodibenzo-p-Dioxin. *Proceedings of the National Academy of Sciences* 88: 9543-9547. DOI:10.1073/pnas.88.21.9543.

Bonati, L., D. Corrada, S. Giani Tagliabue and S. Motta. 2017. Molecular Modeling of the AhR Structure and Interactions Can Shed Light on Ligand-Dependent Activation and Transformation Mechanisms. *Current Opinion in Toxicology* 2: 42-49. DOI:10.1016/j.cotox.2017.01.011.

Brittebo, E. B., A. A. Karlsson, K. I. Skog, and I. M. Jagerstad. 1994. Transfer of the Food Mutagen PhIP to Foetuses and Newborn Mice Following Maternal Exposure. *Food & Chemical Toxicology* 32: 717-726.

Bruce, N., R. Perez-Padilla and R. Albalak. 2000. Indoor Air Pollution in Developing Countries: A Major Environmental and Public Health Challenge. *Bulletin of the World Health Organization* 78: 1078-1092.

Budinsky, Robert A., Edward L. LeCluyse, Stephen S. Ferguson, J. Craig Rowlands, and Ted Simon. 2010. 'Human and Rat Primary Hepatocyte CYP1A1 and 1A2 Induction with 2,3,7,8-Tetrachlorodibenzo-p-Dioxin, 2,3,7,8-Tetrachlorodibenzofuran, and 2,3,4,7,8-Pentachlorodibenzofuran'. *Toxicological Sciences* 118: 224-235. DOI:10.1093/toxsci/kfq238.

Carmody, R. N. and R. W. Wrangham. 2009. Energetic Significance of Cooking. *Journal Of Human Evolution* 57: 379-391. DOI:10.1016/j.jhevol.2009.02.011.

Caspari, R., and S.-H. Lee. 2004. Older Age Becomes Common Late in Human Evolution. *PNAS* 101: 10895-10900.

Chiaro, C. R. J. Luis Morales, K. S. Prabhu and G. H. Perdew. 2008. Leukotriene A4 Metabolites Are Endogenous Ligands for the Ah Receptor. *Biochemistry* 47: 8445-855. DOI:10.1021/bi800712f.

Ciolino, H. P., P. J. Daschner, and G. C. Yeh. 1999. Dietary Flavonols Quercetin and Kaempferol Are Ligands of the Aryl Hydrocarbon Receptor That Affect CYP1A1 Transcription Differentially. *Biochemical Journal* 340: 715-22.

Connor, K. T. and L. L. Aylward. 2006. Human Response to Dioxin: Aryl Hydrocarbon Receptor (AhR) Molecular Structure, Function, and Dose-Response Data for Enzyme Induction Indicate an Impaired Human AhR AU . *Journal of Toxicology and Environmental Health, Part B* 9: 147-171. DOI: 10.1080/15287390500196487.

Corchero, J., S. Pimprale, S. Kimura, and F. J. Gonzalez. 2001. Organization of the CYP1A Cluster on Human Chromosome 15: Implications for Gene Regulation. *Pharmacogenetics* 11: 1-6.

Crosbie, P. A. J., P. V. Barber, K. L. Harrison, A. R. Gibbs, R. M. Agius, G. P. Margison and Andrew C. Povey. 2009. GSTM1 Copy Number and Lung Cancer Risk. *Mutation Research/Fundamental and Molecular Mechanisms of Mutagenesis* 664: 1-5. DOI: 10.1016/j.mrfmmm.2009.01.006.

De Vera, I. M. S., J. Zheng, S. Novick, J. Shang, T. S. Hughes, R. Brust, P. Munoz-Tello *et al.* 2017. 'Synergistic Regulation of Coregulator/Nuclear Receptor Interaction by Ligand and DNA'. *Structure* 25: 1506-1518. DOI: 10.1016/j.str.2017.07.019.

DeMarini, D. M. 2004. Genotoxicity of Tobacco Smoke and Tobacco Smoke Condensate: A Review. *Mutation Research/Reviews In Mutation Research* 567: 447-474. DOI: 10.1016/j.mrrev.2004.02.001.

Divi, R. L., T. L. Einem Lindeman, M. E. Shockley, Ch. Keshava, A. Weston and M. C. Poirier. 2014. Correlation between CYP1A1 Transcript, Protein Level, Enzyme Activity and DNA Adduct Formation in Normal Human Mammary Epithelial Cell Strains Exposed to Benzo[a]Pyrene. *Mutagenesis* 29: 409-417. DOI: 10.1093/mutage/geu049.

Dunnen, J. T. den, R. Dalgleish, D. R. Maglott, R. K. Hart, M. S. Greenblatt, J. McGowan-Jordan, A.-F. Roux, T. Smith, S. E. Antonarakis and P. E. M. Taschner. 2016. HGVS Recommendations for the Description of Sequence Variants: 2016 Update. *Human Mutation* 37: 564-569. DOI: 10.1002/humu.22981.

Eichholzer, M., S. Rohrmann, A. Barbir, S. Hermann, B. Teucher, R. Kaaks and J. Linseisen. 2012. Polymorphisms in Heterocyclic Aromatic Amines Metabolism-Related Genes Are Associated with Colorectal Adenoma Risk. *International Journal of Molecular Epidemiology and Genetics* 3: 96-106.

Ema, M., N. Ohe, M. Suzuki, J. Mimura, K. Sogawa, S. Ikawa, and Y. Fujii-Kuriyama. 1994. Dioxin Binding Activities of Polymorphic Forms of Mouse and Human Arylhydrocarbon Receptors. *Journal of Biological Chemistry* 269: 27337-27343.

Flaveny, C. A., I. A. Murray, C. R. Chiaro and G. H. Perdew. 2009. Ligand Selectivity and Gene Regulation by the Human Aryl Hydrocarbon Receptor in Transgenic Mice. *Molecular Pharmacology* 75: 1412-1420. DOI: 10.1124/mol.109.054825.

Flaveny, C. A., I. A. Murray and G. H. Perdew. 2010. Gene Regulation by the Human and Mouse Aryl Hydrocarbon Receptor. *Toxicological Sciences* 114: 217-225. DOI: 10.1093/toxsci/kfp308.

Forgacs, A. L., E. Dere, M. M. Angrish and T. R. Zacharewski. 2013. Comparative Analysis of Temporal and Dose-Dependent TCDD-Elicited Gene Expression in Human, Mouse, and Rat Primary Hepatocytes. *Toxicological Sciences* 133: 54-66. DOI: 10.1093/toxsci/kft028.

Freeman, D. J. and F. C. R. Cattell. 2014. Woodburning as a Source of Atmospheric Polycyclic Aromatic Hydrocarbons. *Environmental Science & Technology* 24: 1581-1585. DOI: 10.1021/es00080a019.

Fu, Q., H. Li, P. Moorjani, F. Jay, S. M. Slepchenko, A. A. Bondarev, P. L. F. Johnson *et al.* 2014. Genome Sequence of a 45,000-Year-Old Modern Human from Western Siberia. *Nature* 514: 445-449. DOI: 10.1038/nature13810.

Gamba, C., E. R. Jones, M. D. Teasdale, R. L. McLaughlin, G. Gonzalez-Fortes, V. Mattiangeli, L. Domboroczki *et al.* 2014. Genome Flux and Stasis in a Five Millennium Transect of European Prehistory. *Nature Communications* 5: 5257. DOI: 10.1038/ncomms6257.

Giani Tagliabue, S., S. C. Faber, S. Motta, M. S. Denison and L. Bonati. 2019. Modeling the Binding of Diverse Ligands within the Ah Receptor Ligand Binding Domain. *Scientific Reports* 9: 1-14. DOI: 10.1038/s41598-019-47138-z.

Gowlett, J. A. J. 2006. The Early Settlement of Northern Europe: Fire History in the Context of Climate Change and the Social Brain. *Comptes Rendus Palevol* 5: 299-310. DOI: 10.1016/j.crpv.2005.10.008.

Gowlett, J. A. J. and R. W. Wrangham. 2014. Earliest Fire in Africa: Towards the Convergence of Archaeological Evidence and the Cooking Hypothesis. *Azania: Archaeological Research in Africa* 48: 5-30. DOI: 10.1080/0067270x.2012.756754.

Green, R. E., J. Krause, A. W. Briggs, T. Maricic, U. Stenzel, M. Kircher, N. Patterson *et al.* 2010. A Draft Sequence of the Neandertal Genome. *Science* 328: 710-722. DOI:10.1126/science.1188021.

Gu, A., G. Ji, P. Zhu, Y. Zhou, G. Fu, Y. Xia, L. Song, S. Wang and X. Wang. 2010. Nucleotide Excision Repair Polymorphisms, Polycyclic Aromatic Hydrocarbon Exposure, and Their Effects on Sperm Deoxyribonucleic Acid Damage and Male Factor Infertility. *Fertility and Sterility* 94: 2620-2625. DOI:10.1016/j.fertnstert.2010.04.059.

Harper, P. A., J. M. Y. Wong, M. S. M. Lam and A. B. Okey. 2002. Polymorphisms in the Human AH Receptor. *Chemico-Biological Interactions* 141: 161-187. DOI:10.1016/S0009-2797(02)00071-6.

Head, J. S., C. Boesch, L. Makaga and M. M. Robbins. 2011. Sympatric Chimpanzees (*Pan troglodytes troglodytes*) and Gorillas (*Gorilla gorilla gorilla*) in Loango National Park, Gabon: Dietary Composition, Seasonality, and Intersite Comparisons. *International Journal of Primatology* 32: 755-775. DOI:10.1007/s10764-011-9499-6.

Henn, B. M., L. R. Botigué, S. Peischl, I. Dupanloup, M. Lipatov, B. K. Maples, A. R. Martin *et al.* 2016. Distance from Sub-Saharan Africa Predicts Mutational Load in Diverse Human Genomes. *PNAS* 113: E440-E449. DOI:10.1073/pnas.1510805112.

Henry, A. G., T. Büdel and Pierre-Louis Bazin. 2018. Towards an Understanding of the Costs of Fire. *Quaternary International* 493: 96-105. DOI:10.1016/j.quaint.2018.06.037.

Henry, E. C. and T. A. Gasiewicz. 2008. Molecular Determinants of Species-Specific Agonist and Antagonist Activity of a Substituted Flavone towards the Aryl Hydrocarbon Receptor. *Archives of Biochemistry and Biophysics* 472: 77-88. DOI:10.1016/j.abb.2008.02.005.

Heyes, P. J., K. Anastasakis, W. de Jong, A. van Hoesel, W. Roebroeks and M. Soressi. 2016. Selection and Use of Manganese Dioxide by Neanderthals. *Scientific Reports* 6: 22159. DOI:10.1038/srep22159.

Hubbard, T. D., I. A. Murray, W. H. Bisson, T. S. Lahoti, K. Gowda, S. G. Amin, A. D. Patterson, and G. H. Perdew. 2015. Adaptation of the Human Aryl Hydrocarbon Receptor to Sense Microbiota-Derived Indoles. *Scientific Reports* 5: 12689. DOI:10.1038/srep12689.

Hubbard, T. D., I. A. Murray, W. H. Bisson, A. P. Sullivan, A. Sebastian, G. H. Perry, N. G. Jablonski and G. H. Perdew. 2016. Divergent Ah Receptor Ligand Selectivity during Hominin Evolution. *Molecular Biology and Evolution* 33: 2648-2658. DOI:10.1093/molbev/msw143.

Ji, G., A. Gu, Y. Xia, C. Lu, J. Liang, S. Wang, J. Ma, Y. Peng and X. Wang. 2008. ERCC1 and ERCC2 Polymorphisms and Risk of Idiopathic Azoospermia in a Chinese Population. *Reproductive Biomedicine Online* 17: 36-41. DOI:10.1016/S1472-6483(10)60290-8.

Ji, G., A. Gu, Y. Zhou, X. Shi, Y. Xia, Y. Long, L. Song, S. Wang and X. Wang. 2010.'Interactions between Exposure to Environmental Polycyclic Aromatic Hydrocarbons and DNA Repair Gene Polymorphisms on Bulky DNA Adducts in Human Sperm. *PLoS ONE* 5: e13145.

Katiyar, S. K., M. S. Matsui and Hasan Mukhtar. 2000.'Ultraviolet-B Exposure of Human Skin Induces Cytochromes P450 1A1 and 1B1. *Journal of Investigative Dermatology* 114: 328-333.

Köhle, C. and K. W. Bock. 2007. Coordinate Regulation of Phase I and II Xenobiotic Metabolisms by the Ah Receptor and Nrf2. *Biochemical Pharmacology* 73: 1853-1862. DOI:10.1016/j.bcp.2007.01.009.

Korkina, L. G., S. Pastore, C. De Luca and V. A. Kostyuk. 2008. Metabolism of Plant Polyphenols in the Skin: Beneficial Versus Deleterious Effects. *Current Drug Metabolism* 9: 710-729.

Koyano, S., Y. Saito, H. Fukushima-Uesaka, S. Ishida, S. Ozawa, N. Kamatani, H. Minami *et al.* 2005. Functional Analysis of Six Human Aryl Hydrocarbon Receptor Variants in a Japanese Population. *Drug Metabolism and Disposition* 33: 1254-1260. DOI:10.1124/dmd.105.004655.

Lammer, E. J., G. M. Shaw, D. M. Iovannisci, J. Van Waes and R. H. Finnell. 2004. Maternal Smoking and the Risk of Orofacial Clefts: Susceptibility With NAT1 and NAT2 Polymorphisms. *Epidemiology* 15: 150-156. DOI:10.1097/01.ede.0000112214.33432.cc.

Lan, C. M., T. H. Kao, and B. H. Chen. 2004. Effects of Heating Time and Antioxidants on the Formation of Heterocyclic Amines in Marinated Foods. *Journal of Chromatography B* 802: 27-37.

Langergraber, K. E., K. Prüfer, C. Rowney, C. Boesch, C. Crockford, K. Fawcett, E. Inoue *et al.* 2012. Generation Times in Wild Chimpanzees and Gorillas Suggest Earlier Divergence Times in Great Ape and Human Evolution. *PNAS* 109: 15716-21. DOI:10.1073/pnas.1211740109.

Lee-Hilz, Y. Y., A. M. Boerboom, A. H. Westphal, W. J. Berkel, J. M. Aarts, and I. M. Rietjens. 2006. Pro-

Oxidant Activity of Flavonoids Induces EpRE-Mediated Gene Expression. *Chemical Research in Toxicology* 19: 1499-1505.

Lodovici, M., C. Luceri, F. Guglielmi, Ch. Bacci, V. Akpan, M. L. Fonnesu, V. Boddi and P. Dolara. 2004. Benzo(a)Pyrene Diolepoxide (BPDE)-DNA Adduct Levels in Leukocytes of Smokers in Relation to Polymorphism of CYP1A1, GSTM1, GSTP1, GSTT1, and mEH. *Cancer Epidemiology Biomarkers & Prevention* 13: 1342-1348.

Ma, Q., L. Dong and J. P. Whitlock. 1995. Transcriptional Activation by the Mouse Ah Receptor: Interplay between Multiple Stimulatory and Inhibitory Functions. *Journal of Biological Chemistry* 270: 12697-12703. DOI:10.1074/jbc.270.21.12697.

McCormick, J. P. and T. Thomason. 2014. Near-Ultraviolet Photooxidation of Tryptophan. Proof of Formation of Superoxide Ion. *Journal of the American Chemical Society* 100: 312-313. DOI:10.1021/ja00469a068.

Members of the Mouse Genome Analysis Group, A. T. Chinwalla, L. L. Cook, K. D. Delehaunty, G. A. Fewell, L. A. Fulton, R. S. Fulton *et al.* 2002. Initial Sequencing and Comparative Analysis of the Mouse Genome. *Nature* 420: 520-562. DOI:10.1038/nature01262.

Meyer, M., M. Kircher, M.-T. Gansauge, H. Li, F. Racimo, S. Mallick, J. G. Schraiber *et al.* 2012. A High-Coverage Genome Sequence from an Archaic Denisovan Individual. *Science* 338: 222-226. DOI:10.1126/science.1224344.

Mishra, V., R. D. Retherford and K. R. Smith. 2013. Cooking Smoke and Tobacco Smoke as Risk Factors for Stillbirth. *International Journal of Environmental Health Research* 15: 397-410. DOI:10.1080/09603120500288913.

Moriguchi, T., H. Motohashi, T. Hosoya, O. Nakajima, S. Takahashi, S. Ohsako, Y. Aoki *et al.* 2003. Distinct Response to Dioxin in an Arylhydrocarbon Receptor (AHR)-Humanized Mouse. *PNAS* 100: 5652-5657.

Mukai, M. and S. A. Tischkau. 2007. Effects of Tryptophan Photoproducts in the Circadian Timing System: Searching for a Physiological Role for Aryl Hydrocarbon Receptor. *Toxicological Sciences* 95: 172-181. DOI:10.1093/toxsci/kfl126.

Neal, M., W. Foster and E. Younglai. 2008. The Detrimental Effect of Smoking on Female Fertility and IVF Success. *Current Women's Health Reviews* 4: 16-24. DOI:10.2174/157340408783572088.

Nebert, D. W., Z. Shi, M. Gálvez-Peralta, S. Uno and N. Dragin. 2013. Oral Benzo[a]Pyrene: Understanding Pharmacokinetics, Detoxication, and Consequences – *Cyp1* Knockout Mouse Lines as a Paradigm. *Molecular Pharmacology* 84: 304-313. DOI:10.1124/mol.113.086637.

Nørskov, M. S., R. Frikke-Schmidt, S. E. Bojesen, B. G. Nordestgaard, S. Loft and A. Tybjærg-Hansen. 2011. Copy Number Variation in Glutathione-S-Transferase T1 and M1 Predicts Incidence and 5-Year Survival from Prostate and Bladder Cancer, and Incidence of Corpus Uteri Cancer in the General Population. *The Pharmacogenomics Journal* 11: 292-299. DOI:10.1038/tpj.2010.38.

Nukaya, M. and C. A. Bradfield. 2009. Conserved Genomic Structure of the *Cyp1a1* and *Cyp1a2* Loci and Their Dioxin Responsive Elements Cluster. *Biochemical Pharmacology* 77: 654-659. DOI:10.1016/j.bcp.2008.10.026.

Okey, A. B., M. A. Franc, I. D. Moffat, N. Tijet, P. C. Boutros, M. Korkalainen, J. Tuomisto and R. Pohjanvirta. 2005. Toxicological Implications of Polymorphisms in Receptors for Xenobiotic Chemicals: The Case of the Aryl Hydrocarbon Receptor. *Toxicology and Applied Pharmacology* 207: 43-51. DOI:10.1016/j.taap.2004.12.028.

Pandini, A., M. S. Denison, Y. Song, A. A. Soshilov and L. Bonati. 2007. Structural and Functional Characterization of the Aryl Hydrocarbon Receptor Ligand Binding Domain by Homology Modeling and Mutational Analysis. *Biochemistry* 46: 696-708. DOI:10.1021/bi061460t.

Park, J. Y., M. K. Shigenaga and B. N. Ames. 1996. Induction of Cytochrome P4501A1 by 2,3,7,8-Tetrachlorodibenzo-p-Dioxin or Indolo(3,2-b)Carbazole Is Associated with Oxidative DNA Damage. *Proceedings of the National Academy of Sciences* 93: 2322-2327.

Prüfer, K., F. Racimo, N. Patterson, F. Jay, S. Sankararaman, S. Sawyer, A. Heinze *et al.* 2014. The Complete Genome Sequence of a Neanderthal from the Altai Mountains. *Nature* 505: 43-49.

Qin, Y., X. Han, Y. Peng, R. Shen, X. Guo, L. Cao, L. Song, J. Sha, Y. Xia and X. Wang. 2012. Genetic Variants in Epoxide Hydrolases Modify the Risk of Oligozoospermia and Asthenospermia in Han-Chinese Population. *Gene* 510: 171-174. DOI:10.1016/j.gene.2012.09.016.

Raghavan, M., P. Skoglund, K. E. Graf, M. Metspalu, A. Albrechtsen, I. Moltke, S.Rasmussen *et al.* 2014. Upper Palaeolithic Siberian Genome Reveals Dual

Ancestry of Native Americans. *Nature* 505: 87-91. DOI:10.1038/nature12736

Rannug, A., U. Rannug, H. S. Rosenkranz, L. Winqvist, R. Westerholm, E. Agurell and A. K. Grafstrom. 1987. Certain Photooxidized Derivatives of Tryptophan Bind with Very High Affinity to the Ah Receptor and Are Likely to Be Endogenous Signal Substances'. *The Journal of Biological Chemistry* 262: 15422-15427.

Rasmussen, M., S. L. Anzick, M. R. Waters, P. Skoglund, M. DeGiorgio, T. W. Stafford Jr, S. Rasmussen *et al.* 2014. The Genome of a Late Pleistocene Human from a Clovis Burial Site in Western Montana. *Nature* 506: 225-229. DOI:10.1038/nature13025 http://www.nature.com/nature/journal/v506/n7487/abs/nature13025.html#supplementary-information.

Rasmussen, M., X. Guo, Y. Wang, K. E. Lohmueller, S. Rasmussen, A. Albrechtsen, L. Skotte *et al.* 2011. An Aboriginal Australian Genome Reveals Separate Human Dispersals into Asia. *Science* 334: 94-98.

Rasmussen, M., Y. Li, S. Lindgreen, J. Skou Pedersen, A. Albrechtsen, I. Moltke, M. Metspalu *et al.* 2010. 'Ancient Human Genome Sequence of an Extinct Palaeo-Eskimo'. *Nature* 463: 757-62.

Roberts, E. A., K. C. Johnson, P. A. Harper and A. B. Okey. 1990.'Characterization of the Ah Receptor Mediating Aryl Hydrocarbon Hydroxylase Induction in the Human Liver Cell Line Hep G2. *Archives of Biochemistry and Biophysics* 276: 442-450. DOI:10.1016/0003-9861(90)90743-I.

Roebroeks, W. and Paola Villa. 2011. On the Earliest Evidence for Habitual Use of Fire in Europe. *PNAS* 108: 5209-5214. DOI:10.1073/pnas.1018116108.

Sandgathe, D. M., H. L. Dibble, P. Goldberg, S. P. McPherron, A. Turq, L. Niven and J. Hodgkins. 2011. Timing of the Appearance of Habitual Fire Use. *PNAS* 108: E298. DOI:10.1073/pnas.1106759108.

Scally, A., J. Y. Dutheil, L. W. Hillier, G. E. Jordan, I Goodhead, J. Herrero, A. Hobolth *et al.* 2012. Insights into Hominid Evolution from the Gorilla Genome Sequence. *Nature* 483: 169-175.

Seok, S.-H., W. Lee, L. Jiang, K. Molugu, A. Zheng, Y. Li, S. Park, C. A. Bradfield and Y. Xing. 2017. Structural Hierarchy Controlling Dimerization and Target DNA Recognition in the AHR Transcriptional Complex. *PNAS* 114: 5431-36. DOI:10.1073/pnas.1617035114.

Shi, Min, G. L. Wehby and J. C. Murray. 2008. Review on Genetic Variants and Maternal Smoking in the Etiology of Oral Clefts and Other Birth Defects. *Birth Defects Research Part C: Embryo Today: Reviews* 84: 16-29. DOI:10.1002/bdrc.20117.

Silkworth, J. B., A. Koganti, K. Illouz, A. Possolo, M. Zhao and S. B. Hamilton. 2005. Comparison of TCDD and PCB CYP1A Induction Sensitivities in Fresh Hepatocytes from Human Donors, Sprague-Dawley Rats, and Rhesus Monkeys and HepG2 Cells. *Toxicological Sciences* 87: 508-519. DOI:10.1093/toxsci/kfi261.

Skog, K. I., M. A. E. Johansson and M. I. Jägerstad. 1998. Carcinogenic Heterocyclic Amines in Model Systems and Cooked Foods: A Review on Formation, Occurrence and Intake. *Food and Chemical Toxicology* 36: 879-896. DOI:10.1016/S0278-6915(98)00061-1.

Sobinoff, A. P., V. Pye, B. Nixon, S. D. Roman,and E. A. McLaughlin. 2012. Jumping the Gun: Smoking Constituent BaP Causes Premature Primordial Follicle Activation and Impairs Oocyte Fusibility through Oxidative Stress. *Toxicology* 260: 70-80. DOI:10.1016/j.taap.2012.01.028.

Sorensen, A. C., E. Claud and M. Soressi. 2018. Neandertal Fire-Making Technology Inferred from Microwear Analysis. *Scientific Reports* 8: 10065. DOI:10.1038/s41598-018-28342-9.

Sorensen, A. C.. 2017. 'On the Relationship between Climate and Neandertal Fire Use during the Last Glacial in South-West France. *Quaternary International* 436 A: 114-128. DOI:10.1016/j.quaint.2016.10.003.

Sorensen, A., W. Roebroeks and A. van Gijn. 2014. Fire Production in the Deep Past? The Expedient Strike-a-Light Model. *Journal of Archaeological Science* 42: 476-86. DOI:10.1016/j.jas.2013.11.032.

The 1000 Genomes Project Consortium. 2015. A Global Reference for Human Genetic Variation. *Nature* 526: 68-74.

The Chimpanzee Sequencing and Analysis Consortium. 2005. Initial Sequence of the Chimpanzee Genome and Comparison with the Human Genome. *Nature* 437: 69-87.

Treutter, D.. 2006.'Significance of Flavonoids in Plant Resistance: A Review. *Environmental Chemistry Letters* 4: 147-157. DOI:10.1007/s10311-006-0068-8.

Vrba, J., V. Kren, J. Vacek, B. Papouskova, and J. Ulrichova. 2012. Quercetin, Quercetin Glycosides and Taxifolin Differ in Their Ability to Induce AhR Activation and CYP1A1 Expression in HepG2 Cells. *Phytotherapy Research* 26: 1746-1752. DOI:10.1002/ptr.4637.

Wei, Y.-D, H. Helleberg, U. Rannug and A. Rannug. 1998.'Rapid and Transient Induction of CYP1A1 Gene Expression in Human Cells by the Tryptophan Photoproduct 6-Formylindolo[3,2-b] Carbazole. *Chemico-Biological Interactions* 110: 39-55. DOI:10.1016/S0009-2797(97)00111-7.

Whitelaw, M. L., J. A. Gustafsson and L. Poellinger. 1994. Identification of Transactivation and Repression Functions of the Dioxin Receptor and Its Basic Helix-Loop-Helix/PAS Partner Factor Arnt: Inducible versus Constitutive Modes of Regulation. *Molecular and Cellular Biology* 14: 8343-8355.

Wincent, E., N. Amini, S. Luecke, H. Glatt, J. Bergman, C. Crescenzi, A. Rannug and U. Rannug. 2009. The Suggested Physiologic Aryl Hydrocarbon Receptor Activator and Cytochrome P4501 Substrate 6-Formylindolo[3,2-b]Carbazole Is Present in Humans. *Journal of Biological Chemistry* 284: 2690-2696. DOI:10.1074/jbc.M808321200.

Wincent, E., J. Bengtsson, A. Mohammadi Bardbori, T. Alsberg, S. Luecke, U. Rannug and A. Rannug. 2012. Inhibition of Cytochrome P4501-Dependent Clearance of the Endogenous Agonist FICZ as a Mechanism for Activation of the Aryl Hydrocarbon Receptor. *PNAS* 109: 4479-4484. DOI:10.1073/pnas.1118467109.

Wrangham, R.. 2009. *Catching Fire: How Cooking Made Us Human*. New York: Basic Books.

Xue, W. and D. Warshawsky. 2005.'Metabolic Activation of Polycyclic and Heterocyclic Aromatic Hydrocarbons and DNA Damage: A Review. *Toxicology and Applied Pharmacology* 206: 73-93.

In Dusseldorp, G.L., Chu, W., Bakels, C.C. and M.A. Soressi (eds) 2025. Intent on the Paleolithic. Papers in honour of Prof. Dr. Wil Roebroeks. *Analecta Praehistorica Leidensia* 55. Leiden: Sidestone Press, pp. 189-196.

Remarks on taxonomy, signal to noise, scaling, Palaeolithic life, lemmings and the joys of archaeology

Nicholas J. Conard

Abstract

Having known Wil Roebroeks for about forty years and having spent many hours discussing the fields of archaeology and palaeoanthropology, this paper draws on these discussions to consider some of the strengths and weaknesses of these fields. The paper argues against trying to force all researchers to use the same taxonomies. The discussion is critical of the many pseudoscientific numerical methods prevalent in archaeology and human evolution, while advocating the use of shifting scales of temporal, spatial, demographic and behavioural analyses. The paper advocates keeping archaeology fun and collegial but welcoming lively critical debate about the past and our current research agendas. Finally, researchers should slow down and focus on conducting research of lasting, high quality and destroying less of the limited archaeological record.

Keywords: taxonomy; pseudoscientific archaeology; scales of analyses; kindness and collegiality; preserving the archaeological record

Introduction

For about 20 years I collected data from archaeologists I knew and interacted with on digs, at meetings and elsewhere concerning the following three questions. 1) What is your favourite archaeological monograph? 2) What is your favourite archaeological article? and 3) Who is your favourite archaeologist? While I was particularly interested in learning from colleagues who were more experienced and knowledgeable than I was, I also asked younger colleagues and students in all sorts of settings and recorded what they said. Sometimes the topic came up on digs, after guest lectures, in academic settings or at parties. I noted the answers on legal pads, matchbooks, envelopes or whatever I had handy.

My answers were Squire and Davis' 1848 *Ancient Monuments of the Mississippi Valley: Comprising the Results of Extensive Original Surveys and Explorations*, which appeared in the Smithsonian Institute's Contributions to Knowledge Volume 1. I am from southern Ohio and without this volume, our record of the spectacular earthworks in the region would be much poorer. My favourite article is J. O. Brew's The Use and Abuse of Taxonomy from 1946 for reasons I will return to below. Flinders Petrie is my favorite archaeologist for many reasons, but perhaps most simply put, his fieldwork produced an enormous amount of extremely useful information, and he published in a timely manner.

Nicholas J. Conard,
Department of Early Prehistory and Quaternary Ecology, University of Tübingen

The many answers I received over the years to these questions have all been lost. A while back I gave my many handwritten notes on the topic to a student from Berkeley who said that he wanted to produce a database on the answers I collected. Meg Conkey had sent the student to dig with me in the Swabian caves. I could probably determine his name, but that is not important here. Unfortunately, he disappeared without a trace and without having created a database. My copious notes disappeared with him. I am sure I had Wil Roebroeks' answers to these questions along with those of Desmond Clark, Frank Hole, Patty Jo Wathson, Ofer Bar-Yosef, Gerhard Bosinski, Nikolai Praslov, Cathrine Perles, Meg Conkey, Olga Soffer, Cathrine Farizy, Francine David and many other colleagues and students.

The answers were sometimes funny, surprising and unusual. While some answers were light-hearted, most answers were thoughtful and often enlightening.

In this paper for Wil Roebroeks, I want to touch on a few trends in Palaeolithic archaeology, in palaeoanthropology and archaeology in general. I wish I could look up what Roebroeks' answers to these questions were, but they are sadly lost with those of many other archaeologists. Few of these trends and comments I mention are new, and some have cycled through numerous generations of archeologists in different iterations. I know from many conversations with Roebroeks that some of these matters are close to his heart and were subject to his wit and sharp tongue.

Standardization of Taxonomy

It seems like every generation struggles with the irregularities of archaeological taxonomy, the lack of reproducibility and the peculiarities that accompany many taxonomies. I know that Irving (Ben) Rouse, who earned his Ph.D. in the 1930s and was one of my Ph.D. supervisors, was concerned with such matters, and I suspect that these issues were far from new when he was a student. I was exposed to these debates already in junior high school and high school when I started digging with the Dayton Museum of Natural History, in Dayton, Ohio in 1976. There were – and still are – many odd aspects of the taxonomies used to try to create order in archaeology of Eastern North America. In the 1970s students still read works like *The Fort Ancient Aspect* by James B. Griffin (1943) and even W. C. McKern's (1939) Midwestern Taxonomic System. At the time I did not know what was odd about this effort to create order in the past, but I did have a sense that something was wrong.

When I went to Rochester, New York and studied with René Millon – the mapper of Teotihuacan – he laughed about what he considered the backward views of prehistory in the American heartland. René Millon was a Marxist, who had bigger fish to fry than how to codify the process of naming things in the past. He thought McKern's and Griffin's views and goals were ridiculous in this respect.

Sometime in Rochester, almost certainly in one of Millon's seminars, I read Brew's (1946) "The Use and Abuse of Taxonomy", which though long for an article, made quick work of these matters, and clearly demonstrated the futility of trying to create a universally valid and perfect system of taxonomy. For reasons Brew explains, taxonomies are only tools and nothing more and nothing less. I am comfortable with a relatively high degree of creative chaos. I also tend to agree with Brew that rather than reducing the number of taxonomies, it should be *the more the merrier* as long as the taxonomies help to answer relevant questions. The effort of forcing researchers to all use the same archaeological taxonomies does more harm than good and makes people stupider not smarter. Brew must have been both smart and brave to express his many arguments against standardized taxonomies in the 1940s. From my perspective his views are still important today.

Today, as in previous generations, one can easily find researchers who are trying to define the ideal way of classifying objects and cultural taxonomic units rather than trying to ask meaningful questions about the past. At present I know of attempts to find optimal archaeological taxonomies in multiple contexts. While I use taxonomies more or less every day, I will be surprised if anyone will ever develop the ideal taxonomy for the countless questions regarding material culture and the socio-economic behaviour of prehistoric people.

Signal to Noise and Pseudoscience

This is one of the topics Wil Roebroeks and I discussed numerous times, most memorably at nearly every SAA and European Society for the study of Human Evolution (ESHE) meeting we attended. Roebroeks showed disdain for what he called *possibilist archaeology*, myth making, using insufficient data, ill-suited statistics and poor logic. Possibilist archaeology is story telling about the past with little or no basis in fact, in a way in which archaeologists favour expedient or sensationalist explanations in the absence of convincing data and critical reflection.

Archaeology often produces inaccurate, inadequate and inappropriate data relative to the questions being asked, the hypotheses being tested and the statistics being employed (Cowgill 1977, 2015; Thomas 1978). Sometimes weak data based on sloppy excavations or poor sample sizes can be employed for something useful, but often weak data yield weak results regardless of how many Principal Component Analyses (PCAs) one applies to the data, how many two, three or more dimensional models are used to develop or test ideas. Roebroeks liked to see high signal to noise and was relentless in his condemnation of pseudoscience, problematic numerical methods, too many "significant figures", ridiculous null hypotheses and circular reasoning, all of which are common in archaeology and palaeoanthropology. Often the outliers, which were the most informative data points, were eliminated from studies to make the plots look better, to facilitate the number crunching and to optimize correlations.

Given the pace of the day-to-day rat race of academic archaeology and the frequent rush to publish seemingly important results, too few researchers find time to maintain high standards of quality control and to check seriously the logic of throwing away outliers or making bold claims with lousy data. While there is always room for food for thought and creative and speculative hypotheses, sloppy or unconvincing data and observations are often presented in archaeology with a maximum spin to try to attain the maximum impact, even if this impact is short-lived.

John Speth, who kindly commented on a draft of this paper, added the points paraphrased as follows, and with which I fully agree. 1) The overemphasis on testing while deemphasizing the need for evaluating and re-evaluating the underlying assumptions; and 2) saying one is testing a hypothesis when there is neither a null hypothesis, nor an alternative one. In such cases, the authors are really only trying to "prove" their pet idea, not trying to see if it stands up to a serious attempt to falsify it (Platt 1964). In terms of statistics and "significance," most studies focus on statistical significance but do not consider whether the significant results would have been culturally meaningful or relevant. Waguespack (2009) provides an example of this. She showed that arrows with stone points penetrated FBI target gels significantly deeper than arrows with a sharpened wooden tip. She also noted that according to the FBI, both types of arrows reach a lethal depth, so whether or not they had stone points really would not matter if killing game was the central issue.

Scaling and Palaeolithic Lifeways

Palaeolithic archaeologists and paleoanthropologists find themselves in a privileged position. While primatologists, linguists, neurologists, theologians and many other researchers often speculate about human evolution, only archaeologists and biological anthropologists regularly collect, study and interpret data that come from the deep past and provide a direct empirical record of hominin behaviour from the time the hominins lived. The wide variety of information to which we have access opens endless channels for enquiry.

Nonetheless, one often encounters researchers highly focused on a specific scale of analysis. This is well and good, but no scale has a fundamental priority over any other scale. I, for example, think it is possible to use archaeological data to reconstruct the actions of individuals, be they living, dead or extinct; or in other words to pursue palaeoethnology (Conard 1994; Conard and Adler 1997). Working on beautifully preserved sites in Holocene Ohio, Egypt and South Africa and across the Pleistocene Old World convinced me that such approaches are viable. The record provided by material cultural remains can complement the results gained from well preserved human skeletons that allow us to determine important aspects of the life histories of long-dead individuals (Trinkaus 1983; Nava *et al.* 2020).

There can be little doubt that long sequences of refitting lithics reflect the actions of individual knappers and that studies of lithic provenience can inform us about the movement of people and raw materials. Wil Roebroeks conducted pioneering work on both of these topics (Roebroeks 1988, Roebroeks *et al.* 1988). Tracks of footprints, butchered carcasses, complementary techno-functional studies of lithic, botanical and osseous artifacts can also bring us close to the lives of past peoples. Similarly, the actions related to producing and interacting with Palaeolithic art or making and playing Palaeolithic musical instruments bring us close to the lives, talents and beliefs of people who lived tens of thousands of years ago (Conard 2019). In some cases, whether at Schöningen, Wallertheim, Abric Romani or Pincevent, the spatial distribution between artifacts can reveal the actions and movements of individuals. Roebroeks used the spatial distributions of artifacts to address the importance of the tracks of individuals and groups across the landscape. To use Glynn Isaac's (1981a, 1981b) terms, at Maastricht-Belvédère and elsewhere, Roebroeks considered the Stone Age visiting cards these individuals left behind and how they created scatters and patches that inform researchers about past individuals and societies.

Elsewhere, I have discussed the importance of using a shifting scale of analysis and the benefits of being open to questions based on different temporal, spatial, demographic and behavioural scales (Conard 2001). Although all scales have their strengths and weaknesses, when researchers collect high resolution data, the same data can also be useful on larger scales. If, however, excavations are conducted with less rigor to save time or money and thereby fail to recover high quality data, the potential of a site will be irrevocably lost (Conard *et al.* 2023). Thus, outside the context of rescue excavations, it is usually best to excavate lower volumes of sediment using high resolution analytical methods. Current research on stratigraphic microfacies and the rapid advances in molecular archaeology (Courty 2001; Brown *et al.* 2021; Massilani *et al.* 2021) are only some examples of the many advances in the field that are rapidly coming into common use. Every indication suggests that the future will see ever more advances in field work and laboratory studies, so sampling all classes of materials and improving recovery should remain a high priority.

In a world of ever increasing specialization, pursuing research on a long diachronic scale and large spatial scales can, of course, be highly informative. Roebroeks' edited series on the European Lower Palaeolithic (Roebroeks and van Kolfschoten 1996), Middle Palaeolithic (Roebroeks and Gamble 1999) and Upper Palaeolithic (Roebroeks *et al.* 2000) represents a model of this kind of research. While we need specialists, I see little chance of specialists diminishing in number and importance. When it comes to generalists with broad knowledge and experience, I sometimes wonder whether they will someday be an endangered species of archaeologist. In my experience, the reward for using broad approaches and having broad interests is usually greater than the risk of becoming a poorly informed dilettante dabbling in too many areas. In this regard rare researchers like Roebroeks who have broad interests cross-cutting time and space, classes of material culture, environmental and geological knowledge will always be in great demand. I hope this volume includes a bibliography of Roebroeks' impressive scope of publication, because I am not going to provide exhaustive bibliographic references of his work here. A short look at Roebroek's diverse record of publications on multiple classes of artifacts, fauna, subsistence and settlement dynamics, symbolic behaviour, demography, pyrotechnology etc. will demonstrate this point.

Lemmings

Roebroeks often used environmental data to frame questions about the past, but he was never a vulgar environmental determinist. He knew his plants, mammals, micromammals and biostratigraphy. He, often working with Thijs van Kolfschoten, would examine patterns of geographic expansions and contractions of hominin ranges (Roebroeks *et al.* 1992; Dennell and Roebroeks 2005). He considered niches, ranges and demographics in the context of shifting Palaeolithic settlement dynamics. He knew his microfauna and his lemmings, but he never acted like a lemming regardless of whether he was critically assessing the early settlement of Europe or the importance of the Asian Palaeolithic record.

At times Roebroeks would express surprise at seeing so many people asking the same sorts of questions, using the same – at times problematic – methods and doing the same kinds of research. In the binary taxonomy of colleagues as those from whom one could learn much and colleagues from whom one could learn little, Roebroeks always fell solidly in the first camp. He is a researcher who thinks for himself and does not mind defending a controversial point of view or postulating unexpected ideas. He is also willing to change his perspective when new data or ideas come into focus and is much more often found on the innovative and creative side of debates in our field than holding on to established scientific positions. This is true for almost all issues from his discussions of technology and planning depth to his efforts to establish reliable standards for the early settlement of Eurasia. He shows no particular allegiance to either archaic hominins or modern humans in the often heated discussion of the evolution of behavioural modernity. Roebroeks is an intellectual leader in the field and plays a central role in questioning assumptions that often formed the foundations upon which our understanding of human evolution rested. Our field could use more such creative and independent thinkers.

Being an archaeologist

Knowing Roebroeks, I think he would try to make the best of any situation and could have done well in most fields. Still, he has certainly enjoyed being an archaeologist and through his actions conveyed

Figure 1. One of many festive archaeological evenings at Schloss Monrepos overlooking the Rhine Valley above Neuwied. Left to right, T. van Kolfschoten, W. Roebroeks and N. Conard. Photographer unknown, 1989.

the view that our work should be fun. Roebroeks' many achievements have earned him respect and a high degree of recognition in his profession and in neighbouring academic fields. Whatever he did, he has always seemed to me to love his work, and he went out of his way to be fair and generous to colleagues and students. Speaking for myself, seeing Wil Roebroeks always puts a smile on my face and underlines the wisdom of my decision to be an archaeologist (Fig. 1).

He gives credit where credit is due and motivates those around him to participate in our inspiring field or research without being naïve or being cut-throat or overly self-interested. In a period where archaeology is gradually becoming subject to the competitive rules of other areas of research, we will do well to enjoy the benefits of a field where every practitioner has unique data sets that are often non-reproducible, thereby making excessive competition and pettiness unnecessary. While animated critical debate on all matters is welcome and can and should at times be conducted with an edge and with wit, we archaeologists are in a unique position to do what we can to keep our field a friendly one where generosity and kindness can remain the norm. Wil Roebroeks seems to me to embody this approach to science, and I will long remember his scientific generosity and personal kindness. I am also sure that many others in our field share my experiences.

Finally, I wish I had not lost Wil's answers to the three questions posed at the top of this short sketch, and I think I will ask him again when I next see him. I am pretty sure that his answers will put another smile on my face.

Acknowledgements

I thank the editors of this volume for asking me to contribute a paper and for their patience.

Most importantly, I thank Wil Roebroeks for being one of the best of colleagues and friends one could hope for, a person of his word and a person of high character. Both smart and fun, and over the period since the 1980s we have touched upon many of topics discussed above and often laughing until our ribs ached. Concerning the topic of pseudoscience in archaeology, the line between laughing and crying was often thin.

I thank John Speth for taking a look at a rough draft of the paper and arguing for a bit of moderation. I followed his advice and eliminated some of the more pointed statements and all expletives, which Roebroeks will likely hold against me. In our exchange Speth provided his answers to the three questions posed at the top of the paper: Monograph: *Man the Hunter*; Article: Slobodkin, L.B., 1968. Toward a Predictive Theory of Evolution; Archaeologist: Richard Lee.

Speth added: "I love archaeology; always have. But I think what I love most about archaeology is being out in the boonies, getting stinking hot and dirty and physically exhausted, and then having a cold beer. That's what the joy of archaeology was about. I've always found most of the nuts and bolts of archaeology boring as hell. As a youngster I could never articulate what was missing. *Man the Hunter* and *New Perspectives in Archaeology* came out at about the same time. I loved *Man the Hunter*. *New Perspectives* put me to sleep better than any sleeping pill ever could. *Man the Hunter*, esp. the parts by Lee, were asking big questions about people, archaeology was hung up pushing miserable little sherds and flakes around the table and arguing about what to call them. I was always a fish out of water at Michigan because Binford's ghost was still alive and well in the hallways when I began as a grad student. But Binford's Archaeology as Anthropology wasn't Anthropology. It was watered down environmental determinism stripped of any role for culture. *Man the Hunter* had plenty of environment but it had plenty of Anthropology as well. That was where it was at for me… and still is".

References

Brew, J. O.. 1946. *The Archaeology of Alkali Ridge, Southeastern Utah*. Cambridge MA: Harvard University Press. (Papers of the Peabody Museum of American Archaeology and Ethnology 21).

Brown, S., T. Higham, V. Slon, S. Paabo, M. Meyer, K. Douka, F. Brock, D. Comeskey, N. Procopio, M. Shunkov, A. Derevianko and M. Buckley. 2016. Identification of a new hominin bone from Denisova Cave, Siberia using collagen fingerprinting and mitochondrial DNA analysis. *Scientific Reports* 6: 23559.

Conard, N. J.. 2001. The Future of Archaeology. In: Z. Kobylinski (ed.), *Quo vadis Archaeologia? Whither European Archaeology in the 21st Century?*: 106-117. Warsaw: Polish Academy of Sciences.

Conard, N. J.. 1994. On the Prospects for an Ethnography of Extinct Hominids. *Current Anthropology* 35: 281-282.

Conard, N. J. and D. S. Adler. 1997. Lithic Reduction and Hominid Behavior in the Middle Paleolithic of the Rheinland. *Journal of Anthropological Research* 53: 147-176.

Conard, N. J. 2019. Paleolithic art and design. In: B. Franzen (ed.), *40.000 a Museum of Curiostiy, 14. Fellbach Triennal*: 395-437. London: Koenig. (English and German)

Conard, N. J ., C. E. Miller, S. Riehl, and M. Zeidi. 2023. Testing the Palaeolithic Harvesting Hypothesis at Char-e Boof with Improved Botanical Recovery. In: T. Richter and H. Darabi (eds.), *The Epipalaeolithic and Neolithic in the Eastern Fertile Crescent. Revisiting the Hilly Flanks*: 350-370. Milton Park: Routledge.

Courty, M. A.. 2001. Microfacies Analysis Assisting Archaeological Stratigraphy. In: P. Goldberg, V.T., Holliday and C.R. Ferring (eds.), *Earth Sciences and Archaeology* 205-239. Boston, MA.: Springer.

Cowgill, G.L.. 1977. The trouble with significance tests and what we can do about it. *American Antiquity* 42: 350-368.

Cowgill, G.L.. 2015. Some things I hope you will find useful even if statistics isn't your thing. *Annual Review of Anthropology* 44: 1-14.

Dennell, R., W. Roebroeks. 2005. An Asian perspective on early human dispersal from Africa. *Nature*. 438: 1099-1104.

Griffin, J. B.. 1943. *The Fort Ancient Aspect: Its Cultural and Chronological Position in Mississippi Valley Archaeology*. Ann Arbor: University of Michigan Press. (Anthropological Papers 28).

Isaac, G. L.. 1981a. Stone Age visiting cards: approaches to the study of early land-use patterns. In I. Hodder, G. L. Isaac and N. Hammond (eds.), *Pattern of the past*: 131-155. Cambridge: Cambridge University Press.

Isaac, G. L.. 1981b. Archaeological tests of alternative models of early hominid behaviour: excavation and experiments. *Philosophical Transactions, Royal Society of London, series B*, 292: 177-188.

Lee, R. and I. DeVore (eds.). 1968. *Man the Hunter. The First Intensive Survey of a Single, Crucial Stage of Human Development – Man's Once Universal Hunting Way of Life*. Chicago: Aldine.

McKern, W.C..1939. The Midwestern Taxonomic Method as an Aid to Archaeological Cultural Study. *American Antiquity*, 4: 301-313.

Massilani, D., M. W. Morley, S. M. Mentzer, V. Aldeias, B. Vernot, C. Miller, M. Stahlschmidt, M. B. Kozlikin, M. V. Shunkov, A. P. Derevianko, N. J. Conard, S. Wurz, C. S. Henshilwood, J. Vasquez, E. Essel, S. Nagel, J. Richter, B. Nickel, R. G. Roberts, S. Pääbo, V. Slon, P. Goldberg and M. Meyer. 2021. Microstratigraphic preservation of ancient faunal and hominin DNA in Pleistocene cave sediments. *PNAS* 199: e2113666118.

Nava, A., F. Lugli, M. Romandini, F. Bandino, D. Evans, A.H. Helbling, G. Oxilia, S. Arrighi, E. Bortolini, D. Delpiano, R. Duches, C. Figus, A. Livraghi, G. Marciani, S. Silvestri, A. Cipriani, T. Giovanardi, R. Pini, C. Tuniz, F. Bernardini, I. Dori, A. Coppa, E. Cristiani, C. Dean, L. Bondioli, M. Peresani, W. Müller and S. Benazzi. 2020. Early life of Neanderthals. *PNAS* 117: 28719-28726.

Platt, J.R., 1964. Strong Inference. *Science* 146: 347-353.

Roebroeks, W. 1988. From Find Scatters to Early Hominid Behaviour: A Study of Middle Palaeolithic Riverside Settlements at Maastricht-Belvédère (The Netherlands). *Analecta Praehistorica Leidensia* 21.

Roebroeks, W., J. Kolen and E. Rensink. 1988. Planning depth, anticipation and the organization of middle palaeolithic technology. The "archaic natives" meet Eve's descendants. *Helinium* 28: 17-34.

Roebroeks, W. and T. van Kolfschoten (eds.). 1996. *The earliest occupation of Europe*. Leiden: Leiden University Press.

Roebroeks, W. and C. Gamble (eds.).1999. *The Middle Palaeolithic Occupation of Europe*. Leiden: European Science Foundation.

Roebroeks, W., M. Mussi, J. Svodoba and K. Fennema (eds.) 2000. *Hunters of the Golden Age: The mid upper palaeolithic of Eurasia 30,000-20,000 BP*. Leiden: European Science Foundation.

Roebroeks, W., N. J. Conard and T. van Kolfschoten. 1992. Dense Forests, Cold Steppes, and the Palaeolithic Settlement of Northern Europe. *Current Anthropology*. 33: 551-586.

Roebroeks, W., M. J. Sier, T. K. Nielsen, D. De Loecker, J. M. Parés, C. E. S. Arps and H. J. Mücher. 2012. Use of red ochre by early Neandertals. *PNAS*. 109: 1889-1894.

Roebroeks, W. and P. Villa. 2011. On the earliest evidence for habitual use of fire in Europe. *PNAS*. 108: 5209-5214.

Slobodkin, L.B.. 1968. Toward a Predictive Theory of Evolution. In: R.C. Lewontin (ed.), *Population Biology and Evolution*: 187-205. University of Syracuse Press, Syracuse.

Thomas, D.H.. 1978. The awful truth about statistics in archaeology. *American Antiquity* 43: 231-244.

Trinkaus, E.. 1983. *The Shanidar Neanderthals*. New York: Academic Press.

Waguespack, N.M., T. A. Surovell, A. Denoyer, A. Dallow, A. Savage, J. Hyneman and D. Tapster. 2009. Making a Point: Wood- Versus Stone-Tipped Projectiles. *Antiquity* 83: 786-800.

In Dusseldorp, G.L., Chu, W., Bakels, C.C. and M.A. Soressi (eds) 2025. Intent on the Paleolithic. Papers in honour of Prof. Dr. Wil Roebroeks. *Analecta Praehistorica Leidensia* 55. Leiden: Sidestone Press, pp. 197-212.

Hearths, huts, and hospitality: a dive into deep history

Clive Gamble

Abstract

The importance and chronology of hominin hearths has been much discussed, particularly by Wil Roebroeks. Similar attention has been paid to the appearance of artificial structures, especially circular huts. Less consideration has been paid to the significance of the earliest hearths inside huts. In Europe, fire has been controlled in hearths since the Late Middle Pleistocene. The oldest circular huts come much later, less than 35,000 years ago, while the evidence for hearths in these structures begins before the Last Glacial Maximum, increasing during the Lateglacial. In this contribution I present an explanation for why hearths in huts, the first recognizable domestic dwellings, are such a late feature in deep history. I propose that placing these containers inside one another indicates changes in the practice of hospitality; a universal law which underpins social and cultural life. To explore this proposition the spatial and social order of circular huts is examined with ethnoarchaeological studies from Siberia and North America. These reveal the segmentation of small circular spaces by gender, age, status, and the formalities of hospitality. This segmentation refers also to a cosmology that explains the natural order through circular dwellings in open environments. The paper concludes by considering more broadly the implications of the Lateglacial evidence for changes in human social life, when offering hospitality to strangers became a defining social act.

Keywords: Law of hospitality; mammoth bone houses; hearths, 4E cognition; Lateglacial Ukraine; Palaeolithic cosmology; containers

How can a hominin credited with language and 4**E** cognition; **E**xtended, **E**mbodied, **E**mbedded and **E**nactive (Newen *et al.* 2018) not build a house around a hearth? It seems such a cognitively simple idea, useful anywhere in the Palaeolithic Old World; a container concept that brings benefits of warmth and protection while creating a social focus to share well-roasted meat that, unknowingly, stimulates brain growth. Wil Roebroeks might not put the question quite like this but by collating and scrutinizing the evidence for fire (MacDonald *et al.* 2021; Roebroeks and Villa 2011a), technology (Roebroeks and van Kolfschoten 1994; Roebroeks, Kolen, and Rensink 1988; Gaudzinski-Windheuser *et al.* 2023) landscape use, and settlement (Roebroeks 1988; Roebroeks *et al.* 1992; Roebroeks 1996; Roebroeks and Bakels 2015), he has transformed the way we understand Palaeolithic data, establishing beyond reasonable doubt that absence of evidence is, in the case of hearths in huts, evidence of absence. What singles out Wil's approach is that when attending to the details he never forgets the broader evolutionary narratives they serve (Dennell and Roebroeks 2005; Roebroeks 2006). In this paper I want to further

Clive Gamble
Centre for the Archaeology of Human Origins (CAHO)
Avenue Campus, University of Southampton, SO17 1BF, UK

his approach by suggesting a solution to the hearths in huts issue. But first some background.

Crossing thresholds

The hominins I have in mind lived between 400,000 and 15,000 years ago during the Late Middle and Upper Pleistocene. For the majority of this period they were restricted to parts of the Old World, which elsewhere I have called *Terra* 2 (Gamble 2013), only breaking out in *Terra* 3 50,000 years ago. Europe is a well-researched part of the environmentally dynamic *Terra* 2 occupied by the first, but I am sure not the last, aDNA ancestor, Denisovan, and the anatomical species *H.heidelbergensis*, *H.neanderthalensis* and *H.sapiens*. These last three were all large-brained hominins (>1200cms^3) and during this time interval there is a marked up-tick in the energetically expensive evolutionary trend of encephalization (Aiello and Wheeler 1995; Roebroeks 2007). From these data (Gamble 2013, Table 4.2), it is predicted that individuals now associated in groups of sufficient size and complexity to render obsolete the primate means of affirming social bonds through finger-tip grooming. Language of some sort is the most probable development as the benefits from these richer social lives drove encephalization further (Aiello and Dunbar 1993). The vocal apparatus for speech is in place by 500,000 years ago and other developments such as dance and music-making can also be inferred (Gamble 2012; Morley 2013).

These lines of enquiry point to a threshold in hominin evolution that was crossed 400,000 years ago (Pope *et al.* 2018; Stiner 2002; MacDonald *et al.* 2021). Hominins were now capable of 4th and even 5th levels of intentionality, indicating a theory of mind capable of creating imaginary times, places and geographies as indicated by stone technology (Cole 2015), burials (Pettitt 2011), art and ornament (Finlayson 2019) and the appearance of persistent places in hominin landscapes (Shaw *et al.* 2016). The cognition that enabled this threshold to be crossed bears the hallmarks of **4E** (Menary 2010; Newen, De Bruin, and Gallagher 2018), which in deep history unites bodies, brains, materials and environments (Malafouris 2013; Gosden 2010; McNabb *et al.* 2004).

Today's fears over AI doing our thinking for us is the latest iteration of a **4E** based cognition that crossed a threshold in the Middle Pleistocene. As discussed by philosopher Ben Jeffares (2010, 517), the significance of hominin stone technology was that it structured not only their immediate cognitive environment but also the informational environment of successive generations. According to Lambros Malafouris (2013, 77), 'human thinking is, first and above all, thinking *through*, *with*, and *about* things, bodies and others'. Such cognition is embodied in our senses and extended beyond the envelope of our skin; embedded in the relationship between things and bodies; and enacted during activities as varied as walking down a track, spearing a rhinoceros or sitting round a hearth.

Fire and hearths

The deep history of fire use can now be traced in broad outline. John Gowlett (2010: Figure 17.4 and Box 17.1) describes the sequence of hominin fire-use as opportunistic, then conserved and lastly kindled. In broad chronological terms these transitions date, respectively, to 1.5Ma, 0.7Ma and 0.4Ma. In deep history, fire has received attention not just for its co-evolution with language (Dunbar 2014), but also for the transformative effects of cooking on hominin diets and digestion (Wrangham 2009; Roebroeks 2007), which had the unintended consequence of facilitating encephalization. To date the earliest evidence for cooking are the remains of freshwater fish in the 'phantom hearths' at Gesher Benot Ya'aqov, Israel, dated to 780ka (Zohar *et al.* 2022).

Roebroeks and Villa (2011a, Fig.2) identify a step increase in the evidence for kindled fire after 400ka during the Late Middle Pleistocene (MIS11-6, 427-126ka) (Rolland 2004; Roebroeks and Villa 2011a; MacDonald *et al.* 2021; Gowlett 2016). It takes the form of repeatedly used hearths, sometimes built with stones, and found within caves and open-air locales. The ability of these hominins to make fire has been questioned (Sandgathe *et al.* 2011) and answered (Roebroeks and Villa 2011b), while evidence from northern France (Hérisson *et al.* 2013) settles the issue. Hearths can be substantial, as at Qesem Cave, Israel (Karkanas *et al.* 2007), and multiple, as at Terra Amata, France (de Lumley 2009).

Persistent places

Following the lead of Mary Stiner (2002), the late Middle Pleistocene marks a significant threshold in hominin evolution (Pope *et. al.* 2018). Besides encephalization and hearths, a further feature is the appearance of persistent places in the landscape. Palaeolithic examples are La Cotte de St Brelade, Jersey (Shaw *et al.* 2016; Scott *et al.* 2023), Tabun, Hayonim (Kuhn *et al.* 2018; Stiner 2018) and Qesem

Cave (Barkai *et al.* 2018) in Israel. Persistent places in the Palaeolithic contain evidence from open, cave and abri locales for repeated use over long periods of time, sometimes spanning a glacial-interglacial cycle (Shaw *et al.* 2016, 1440). They are marked either by hearths, signifying controlled, kindled fire, as at Qesem, or by evidence for burning in the form of charcoal, burnt stone and bone that may have been less controlled, as at La Cotte. One feature of controlled fire and the persistent use of places is the obligatory management of the surrounding landscape for firewood, especially during colder summer climates with reduced woodland regeneration (Pryor *et al.* 2016). The threshold that hominins crossed at 400ka also transformed their relationship with the places where they lived. Feeding the fire was instrumental (Kuhn and Stiner 2019, 320).

Figure 1. (a) The museum at Kostenki (Anovsovka), Russia. (b) The walls of a circular *yaranga*, Kostenki 11-1, made with stacked mammoth jaws and long bones. Photo Clive Gamble

		structure type	m²	internal hearth	Reference
Before Last Glacial Maximum **20-30 ka**					
Ohalo hut 13	Israel	brush shelter	5		Nadel 2003
Ohalo hut 2	Israel	brush shelter	7		Nadel 2003
Ohalo hut 12	Israel	brush shelter	8.5		Nadel 2003
Ohalo hut 3	Israel	brush shelter	10		Nadel 2003
Ohalo hut 1 Floor II	Israel	brush shelter	11		Nadel 2003
Ohalo hut 1 floor III	Israel	brush shelter	13		Nadel 2003
Dolni Vestonice #1	Czech Republic	tent	13		Iakovleva 2015
Dolni Vestonice #2	Czech Republic	tent	13	yes	Iakovleva 2015
Pavlov #5	Czech Republic	tent	13	yes	Iakovleva 2015
Dolni Vestonice II #3	Czech Republic	tent	13		Iakovleva 2015
Milovice G	Czech Republic	tent	13	yes	Iakovleva 2015
Pushkari	Ukraine	tent	13	yes	Iakovleva 2015
Pushkari	Ukraine	tent	13	yes	Iakovleva 2015
Gagarino	Russia	tent	24	yes	Iakovleva 2015
Kostenki 4 Upper layer 2 dwellings	Russia	tent	28		Iakovleva 2015
Kostenki 2	Russia	*yaranga*	50	yes	Iakovleva 2015
Kostenki (Anovsovka 2) 11-1	Russia	*yaranga*	63	yes	Iakovleva 2015
Lateglacial **10-20 ka**					
Gontsy #2	Ukraine	*yaranga*	5		Iakovleva 2015
Gontsy #4	Ukraine	*yaranga*	10		Iakovleva 2015
Dobranichevka #1	Ukraine	*yaranga*	13		Iakovleva 2015
Dobranichevka #2	Ukraine	*yaranga*	13	yes	Iakovleva 2015
Dobranichevka #3	Ukraine	*yaranga*	13	yes	Iakovleva 2015
Dobranichevka #4	Ukraine	*yaranga*	13	yes	Iakovleva 2015
Mezhirich #3	Ukraine	*yaranga*	20	yes	Iakovleva 2015
Yudinovo #2	Russia	*yaranga*	20		Iakovleva 2015
Gontsy #1	Ukraine	*yaranga*	24	yes	Iakovleva 2015
Mezin #1	Ukraine	*yaranga*	24	yes	Iakovleva 2015
Suponevo #1	Russia	*yaranga*	24	yes	Iakovleva 2015
Mezhirich #2	Ukraine	*yaranga*	27	yes	Iakovleva 2015
Mezhirich #4	Ukraine	*yaranga*	27	yes	Iakovleva 2015
Yudinovo #3	Russia	*yaranga*	35	yes	Iakovleva 2015
Yudinovo #4	Russia	*yaranga*	35		Iakovleva 2015
Mezhirich #1	Ukraine	*yaranga*	39	yes	Iakovleva 2015

		structure type	m²	internal hearth	Reference
Gontsy #5	Ukraine	*yaranga*	50	yes	Iakovleva 2015
Yudinovo #1	Russia	*yaranga*	63		Iakovleva 2015
Contemporary					
Ojibwa	Ojibwa nation (Canada)	tent	7	yes	Morgan 1881
Small hogan	Navajo nation (USA)	hogan	10	yes	Jett and Spencer 1981
Ojibwa	Ojibwa nation (Canada)	tent	18	yes	Morgan 1881
Evenki	Transbaikal (Siberia)	Tent *duytchar*	25	yes	Grøn and Kuznetsov 2003
Medium hogan	Navajo nation (USA)	hogan	26	yes	Jett and Spencer 1981
Large hogan	Navajo nation (USA)	hogan	65	yes	Jett and Spencer 1981

Table 1. Form and dimensions of Lateglacial huts and tents (see also Soffer 1985, Pidoplichko & Allsworth-Jones 1998).

Huts

The appearance during the Late Middle Pleistocene of persistent places, fire, and the evolution of its management (Gowlett 2010, 2016) is not matched by the chronology of artificial shelters. The evidence comes predominantly from Europe. Nothing has changed since Jan Kolen's (1999) seminal review which found neither huts nor houses prior to 32ka . There have been claims for huts of various dimensions and shapes at Terra Amata (de Lumley 1969), where refitting between levels casts doubt on the reconstructions (Villa 1982), and Bilzingsleben (Mania 1991) where the geological context does not support the level of preservation claimed (Kolen 1999: 144). Kolen interprets these patterns as the effect of active human bodies making unintended 'living structures'. These *Centrifugal Living Structures* (CLS) are the result of hominin embodiment (Kolen and Oosterwijk 2017, 215) moving materials 'out' while creating spaces for living 'within'. The implication is that they are an unplanned rather than formal architecture. The principles under which CLS creates patterns that resemble circular 'houses' accounts for the patterning at the locales of Molodova, Ukraine, and the 'cabin' with hearths within Grotte Lazaret, France (Kolen 1999; Kolen and Oosterwijk 2017).

Hearths in huts

After 32ka tents, hearths and pits are common at open locales such as Dolní Věstonice/Pavlov, Czech Republic (Svoboda 2016: 48; 1994) and Kostenki, Russia (Praslov and Rogachev 1982; Soffer and Praslov 1993). At Kostenki 11 three circular accumulations of mammoth bone have been excavated; the most recent, K 11-3 in 2015, is dated to 25ka cal BP (Pryor *et al.* 2020). It is currently the oldest of the three structures and it is also the largest; its external dimensions are 11m x 12m, covering 132m^{2,} and its internal dimensions are 10 × 9m, covering 90m^2 (Alexander Dudin pers. comm). It has an internal combustion deposit, possibly a hearth, consisting of burnt bone, charcoal and rubified loess (op.cit.: 4 and Figure 3). Fine sieving and residue analysis of this and other deposits was undertaken and the results show that occupation was low intensity and of short duration (op.cit.). These support the conclusion from its large dimensions that it is unlikely to be a house.

Its neighbour, K11-1, preserved in-situ following its excavation in the 1960s, is more convincing as a mammoth bone house (Figure 1a and b). It has a diameter of 9m covering 63m^2. It has internal and external hearths as well as pits outside the structure. The waterlogged remains of Natufian brush-huts at Ohalo II, Israel (Nadel 2003; Nadel, Weiss, and Tschauner 2011), extend the sample geographically in the time frame of 20-30 ka (Table 1). Hearths are only found outside the Ohalo huts and there are no pits.

Elsewhere, in Ukraine's Lateglacial at Mezhirich, Mezin and Gontsy (Soffer 1985; Pidoplichko and Allsworth-Jones 1998; Iakovleva 2015), and in Russia at Iudinovo (Abramova 1993), multiple circular mammoth bone structures are widely accepted examples of Palaeolithic houses of *yaranga* form: circular solid walls with a removable tent roof. In her overview of the evidence, Lioudmila Iakovleva

(2015, 328), recognises two sizes of *yaranga* dwellings; standard with a 5m diameter, and large with an 8m diameter. Oval-shaped structures are also present in both small and standard sizes. Structure 5 at Gontsy highlights the difficulties of measuring living space. The circular spread of mammoth bones covers 50m^2 within which the central dwelling was much smaller at 15m^2 (Iakovleva *et al.* 2021: 2).

Comparing the two time periods either side of the LGM (Table 1), median (13m^2: 24m^2), and mean (18.2m^2: 25m^2) values for hut size are greater in the post-LGM sample where *yarangas* dominate. The standard deviations (15.5: 14.8) are, however, similar. Hearths are regularly found in both tent and *yaranga* structures (Table 1) as are external hearths and pits of varying sizes.

Statement of the problem

The evidence is straightforward. The ability to kindle fire in hearths occurred, ~400ka years ago, during the long hominin occupation of *Terra* 2. In Europe, huts and tents are no older than 35 – 32ka, well into *Terra* 3. Hearths within huts and tents become common during the Lateglacial, after 20ka. The problem is how to account for the disconnect as revealed by the time lags in the sequential appearance of hearths, huts, and hearths in huts?

Invoking different abilities for *H.heidelbergensis*, *H.neanderthalensis* and *H.sapiens* does not explain the chronological disconnect. They all kindled fire, but only *H.sapiens*, and 30,000 years *after* arriving in Europe, had the idea to build a hut around a hearth; hardly a ringing endorsement of what it takes to be a modern human. All three species had the ability for speech, but how and what they spoke about may have differed significantly. The evidence suggests *H.heidelbergensis* was less driven by imagined possibilities. Neanderthals, however, buried their dead, indicating a psychological concept of the After-person if not the imagined After-life (Gamble 2014). Now the discovery of Neanderthal feather cloaks and ornaments of raptor body-parts (Finlayson 2019), hand stencils and cave paintings (Hoffmann *et al.* 2018), suggests an imagination like *H.sapiens* that could fly and, as at Bruniquel (Jaubert *et al.* 2016), also travel deep underground. This is unexceptional for a large-brained hominin, who is linguistically adept and equipped with 4**E** cognition. And yet, like their long-time contemporaries *H.sapiens*, they stopped short of building huts.

Neither do climate and environment explain the disconnect. In the last 400,000 years there were repeated interglacials, interstadials, glaciations and concomitant sea level falls and rises as well as floral and faunal successions. The environmental trigger to regularly put hearths inside huts 15,000 years ago remains unclear.

Population as the motor for change through enhanced cultural transmission struggles to make sense of the timing. While some have read the deep sea curves as a reflection of climate-controlled demography (Shennan 2001, Figure 4), the greater number of 'minds' in the Upper Palaeolithic of *Terra 3* did not immediately come up with the idea of putting a hut around a hearth. More convincing is the Lateglacial re-settlement of northern Europe when archaeological evidence shows that population grew almost four-fold during a major geographical expansion (Gamble *et al.* 2004: Table 6). That timing at least matches the chronological evidence (Table 1) for huts with hearths from Ukraine and Southern Russia. But it still doesn't explain why the imagined became material at this time.

Containers

Single accounts for the disconnect either by species, climate or demography misunderstands what needs to be explained. I start with Kolen's (1999) conclusion that the patterns in the settlement data of deep history are to be understood as embodied actions. I would expand his insight by arguing that hearths, CLS, and huts are examples of containers, a universal embodied concept expressed metaphorically both through linguistics (Johnson 2017) and through material objects (Gamble 2023). Containers are the scaffold for our conceptual worlds, so much so that we have stopped appreciating their pervasiveness. This can be called a container-*habitus*. Tim Ingold sums it up succinctly when he states that the

> *'experience of containment influences our thinking about what it means to inhabit a world to an extent that even psychologists and philosophers, who are tasked with the investigation of such matters, are ill prepared to recognise (Ingold 2015: 41).'*

Archaeologists are particularly well placed to both recognise and explain how our 4E cognition came to be so container-based (Gamble 2007). Or put another way, how containers came to do our thinking for us. The problem, therefore, is not explaining the coming together of two separate cultural elements – hearths

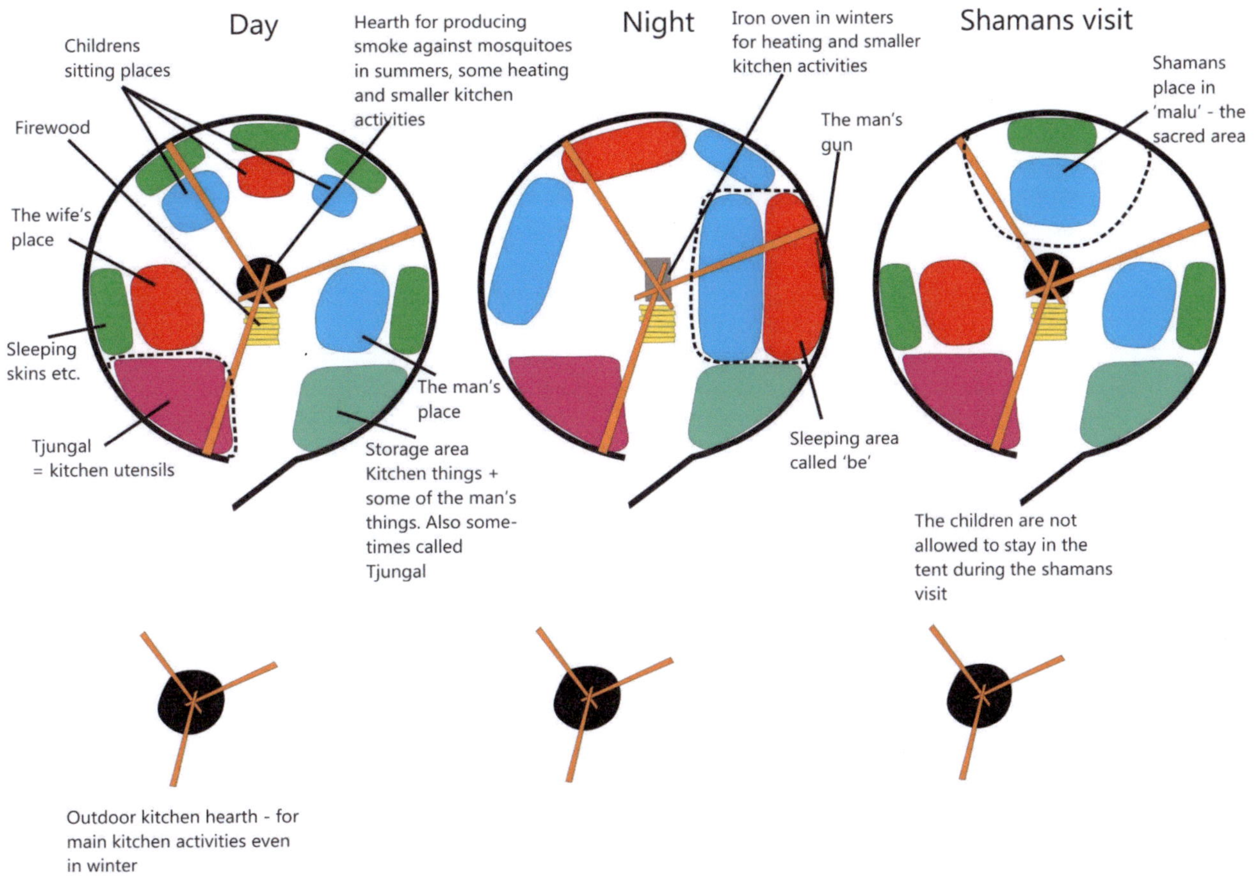

Figure 2. Evenki tent spatial organisation (Grøn 2011, Fig 3). Normally, cooking does not take place within the tent but at the outside hearth, even in winter (Grøn pers.comm.). Food is eaten inside the tent where the main purpose of the fire is to make smoke to repel mosquitos in the warmer months. The door opening, left or right, determines which side of the tent is used by men and women. The clan leader and a high-status visitor, for example a shaman, are seated in the *malu*.
Key: red, female areas; blue, male areas; purple and light green, storage for kitchen utensils and other gear; dark green, sleeping skins. Figure courtesy of Ole Grøn.

in huts – but rather how, when two containers were combined, they referenced a new experience of human embodiment. In the rest of this paper, I will suggest that hospitality, the honouring of strangers and guests within homes and around hearths, was the basis of this understanding.

Round houses and hospitality

The laws of hospitality have been stated by philosophers and anthropologists (Gamble in press). The former emphasise (Levinas 1961 [2002]; Kearney and Semonovitch 2011) the ethical imperative of hospitality and where, as remarked by Jacques Derrida, 'there is no culture or social bond without a principle of hospitality (2005: 6).'

The anthropologists' (Candea and Da Col 2012; Pitt-Rivers 1977 [2017]) starting point is Lewis Henry Morgan's law of hospitality, which he examined through indigenous North American houses. Morgan (1881: 44-5) linked the architecture of these houses with four pan-continental customs that underpinned a universal law of hospitality. These were

- Communism in living
- The ownership of lands in common
- The practice of having one prepared meal each day
- The separation of men and women at this meal with the men eating first.

Morgan examined societies that lived by farming. Most of his examples of houses are either *rectangular* with multiple hearths along the central axis, as in the Seneca-Iroquois long house (Morgan 1881, 126 and Fig. 13), or *modular* with the pueblos of New Mexico. In both cases, the houses are associated with storage pits for agricultural produce upon which the generosity of their hosts was founded.

Morgan also illustrated *circular* houses, notably the large Mandan lodges of the Upper Missouri. For fishers and hunters he described the smaller Ojibwa frame tents from the sub-Arctic (Morgan 1881: 117 and Fig.7). According to Morgan, these tents had a central fire pit and floor areas of between 7m² to 18m² (3m and 4.8m diameters respectively) (Table 1).

In the sub-Arctic of eastern Siberia, the Evenki (Tungus-speaking reindeer herders) compartmentalise the interiors of similar circular, conical tents (Anderson 2006; 2007). Around the central hearth, as recorded by Sergei Shirokogoroff (1929), men use the left and women the right sides of the tent. Directly opposite the entrance is a place of honour (*malu*), while lower-status guests and children are seated on the left of the fire (Anderson 2006: 3). Visiting anthropologists were also seated in the *malu* to get them out of the way while other occupants went about their work (Anderson 2007: 54).

Further detail is provided by Ole Grøn and Oleg Kuznetsov's (2003) ethno-archaeological study of conical larch bark tents (*duytchar*) built by Evenki people in the northern Transbaikal.

In single-family dwellings they found a common pattern where women use the left and men the right-hand sides. The spatial code varies depending on which way the door opens. If it is to the left, as in Figure 2, that is the women's side, which is reversed when the door opens to the right (Grøn and Kuznetsov 2003: 217). The important point is not that men and women follow an absolute spatial rule in these small circular structures, but that irrespective of which side they take up, the genders are invariably spatially demarcated.

The Evenki explain their spatial code as necessary to avoid awkward situations when visiting other Evenki (Grøn 2014: 33). Embarrassment, a powerful social emotion between hosts and guests, works as a constraint to ensure that correct positions in the tent are taken according to status, age, and gender.

Strict spatial codes between men and women are found in other small circular dwellings such as Navajo hogans (Jett and Spencer 1981, Fig 2.6). The Many Sheep hogan that Susan Kent (1984: figure 18) studied in detail was log-built, roofed and measured 5.7m in diameter with a floor area of 25.5m², making it comparable in size to an Evenki *duytchar* (Table 1). Both cooking and eating food took place inside and storage areas were located close by (Kent 1984, Figure 19). In their comprehensive survey of Navajo architecture Stephen Jett and Virginia Spencer (1981, 231) conclude that hogans supported Navajo pragmatic and social values. These resulted in the formal compartmentalization of space by age, gender and status as recorded ethnoarchaeologically by Kent.

The conceptual sectors of Navajo hogans are similar to the Evenki's *duytchars*. For example, the hogan has a place of honour usually reserved for the family's matriarch or patriarch (Jett and Spencer 1981, 23), and is also used by high-status visitors. Kent found that men and women used different areas of the hogan. Food was prepared by the women in the north and east while the south and west was male (1984: 59-61). On entering, women go to the right side (north) of the hogan, and men to the left (south). A hogan is both a domestic and ceremonial space. During ceremonies, those entering must pass south of the hearth and exit to its north, thereby following the daily path of the sun. The entrance is oriented to the east and the rising sun. This is not the case among the Evenki where tent doorways observe wind direction.

Kent found that reconstructing discrete activity areas within hogans was difficult. But this is a case where absence of evidence is not evidence of absence. In their excavation of an abandoned conical tent structure, Grøn and Kuznetsov (2003: 220) found clearly demarcated patterns between two distinctive artefacts, rifle cartridges and net floats. They conclude (2003: 218 and figure 30.1) that the three Evenki families who used this structure between c. 1930 and 1970, were seated in accordance with a persistent set of spatial rules as illustrated diagrammatically in Figure 3. Elsewhere, Grøn (2018: fig. 5) shows that a bipartite pattern is a feature also of circular Mesolithic dwellings in Denmark.

Environment and cosmology

But why are houses built to a circular plan? And why are rectangular structures so rare, even non-existent, before farming? Ease of construction may be a practical answer. Alternatively, Richard Bradley's (2012) study of the round and rectangular houses of Neolithic

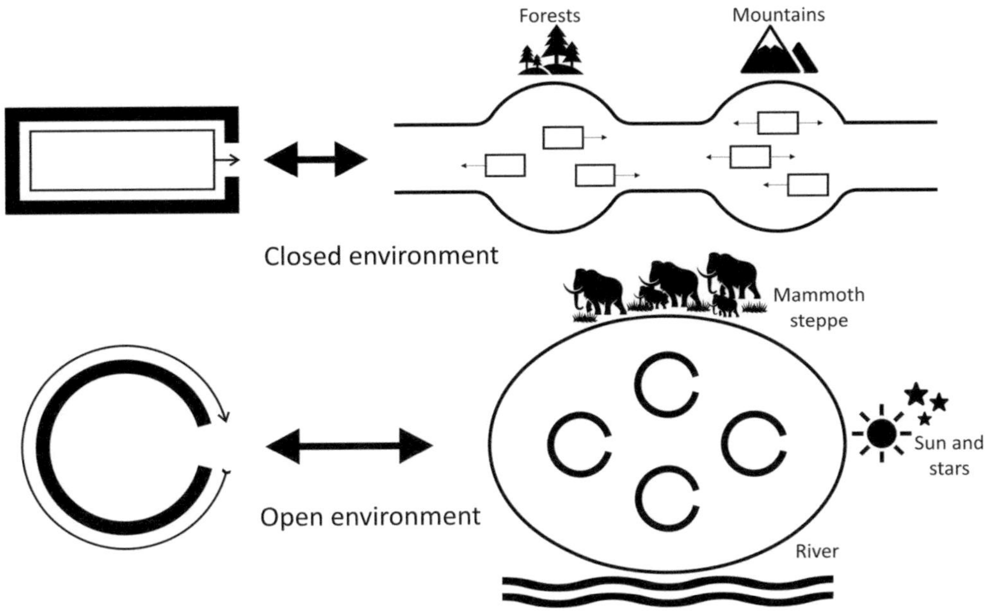

Figure 3. Round and rectangular buildings as microcosms of the local environment which structures their cosmological significance (after Bradley 2012, Figure 15). Illustration Rebecca Ferreira.

Europe makes the distinction between the practice of building round houses in open, and rectangular ones in closed environments. His central point is that living in grasslands (open) or among forests (closed) affected not only the materials available for construction but also the philosophy behind the forms which houses took; a philosophy based on cosmology rather than the pragmatics of immediate, lived experiences. In this way houses provided a commentary for people on the world and their place within it (2012: 42). In Bradley's study (Figure 3) house construction, use, and internal organisation offer interpretations of the ideas on which the natural order depends (2012: 39). In closed forest environments notions of space are dominated by the importance of tracks and directions through the landscape. In open habitats the mobile hunter or herder say they stay 'in the same place' and it is the world that moves around them (Grøn and Kuznetsov 2003, 217), as in the progression of the moon and stars across the sky. An example of this natural order is provided by Kent's (1984: 200-1) study of segmented space. She found that the Navajo's round, sacred hogan is divided into the same male and female areas as the round, sacred cosmos. 'The division of the hogan space by the sex of the user is not a reflection of the Navajo's emphasis on the differences between the sexes or of their division of labour; instead, it is a reflection of the sacred, circular cosmos (ibid).' Similarly, Evenki dwellings and settlements are best *explained* by referencing their cosmology and mythology (Grøn and Kuznetsov 2003: 218-220 their emphasis).

The insights of Bradley, Grøn and Kent can be applied to circular houses in open environments such as the Lateglacial *yarangas* of Ukraine. The form they took can be linked to a cosmology that stems from open habitats rather than solely, as Olga Soffer (1985) argued, to sedentism and mammoths. It is also reasonable to suggest that these circular containers would have been conceptually compartmentalized.

Living in small dwellings, as Grøn (2018, 929) points out using findings from proxemics, needs to restrict individual activities to customary positions. Female: male compartmentalisation is one example. The regular appearance of internal hearths in Lateglacial *yarangas* and tents provides the pivot around which this compartmentalisation revolved, while the walls of the *yarangas* contained and intensified these spatial rules and gendered social roles. I cannot say which side of the *yarangas* the men and women habitually took, but I can be confident that these small internal spaces were organised by gender.

One reason for my confidence in the gendered segmentation of space is that the *yarangas* also embodied, embedded, and enacted a major transformation in the relation between men and women. As discussed by Foley and Gamble (2009), during human evolution there was a break with primate socioecology where females, who principally bear the high energetic costs of reproduction, are distributed to the best resources available so as to minimize those costs. With humans, males have come to control those resources and hence the distribution of females. This socioecological outcome is a hallmark of agriculture where food is stored in well-protected granaries. But it also applies to hunters and gatherers where the circulation of ceremonially important objects and sacred knowledge can be stored and controlled by men in a similar manner, and from which their control over women stems (Gamble 2013: 105-7, 259-260). The abundant art, ceremonial objects and storage pits associated with *yarangas* (Iakovleva 2015) is evidence that a major shift in the power relations between the sexes had, by the Lateglacial, become the basis of society.

Hospitality, welcoming the stranger

So far, I have said little about these circular spaces as sites of hospitality. The pre-eminent virtue of place and dwellings are their affordances for hospitality, that ethic which makes social life possible through the reception of visitors and strangers (Kearney and Semonovitch 2011, 50). Pitt-Rivers (1977 [2017], 178) states that hospitality 'imposes order through an appeal to the sacred, makes the unknown knowable, and replaces conflict by reciprocal honour'. The stranger becomes familiar through rites of incorporation, and as Morgan (see above) pointed out, this means they are fed. Hospitality has its limits. Benjamin Franklin's quip that fish and visitors stink in three days, is matched by the Swahili proverb; feed guests for two days, then give them a hoe (Dikeç *et al.* 2009: 5).

The alternative is aggression towards strangers. Anne Pisor and Martin Surbeck (2019, 218) examine the assumption of evolutionary psychologists that a strong human universal toward parochial altruism exists; defined as in-group favouritism and aggression towards out-groups (Romano *et al.* 2017). They point out, however, that environments where resources vary in temporal and spatial availability require social relationships that span distances. In these circumstances strangers are not a threat but a source to negotiate with in order to access other regions and their resources. This ability to network at inter-regional scales has long been investigated by archaeologists through concepts such as planning depth, alliance networks and connubia (Roebroeks *et al.* 1988; Gamble 1982; Wobst 1976) using the changing scales over which raw materials were distributed and artistic styles shared. The revolution in stable isotopes (Linscott *et al.* 2023) and aDNA (Skov *et al.* 2022) promises to provide even better data on the movement of individuals between populations.

Living apart was a consequence of settling open environments with low population densities and fluctuations in resources. Relationships must be remembered *in-absentia*. Just as men and women know which way to turn on entering a round house, they also need to know how to act towards strangers. Ceremonial rules, as Pitt-Rivers states, are used to make the unknown known, thereby turning a stranger and potential enemy into a guest and ally.

One way this is achieved is through the concept of kinshipping (Shryock *et al.* 2011); an imaginative exercise that binds people to the cosmos, materials and things, as much as to the living. In short, kinshipping navigates those cosmological beliefs on which the natural order depends. Kinshipping establishes relationships through cultural means such as fictive kinship, trading arrangements, myth, and ritual. Kinshipping is a facet of hospitality because it provides the basis of negotiation towards a shared understanding of how people stand in relation to each other.

The archaeological evidence for expansive inter-group relations based on kinshipping points to its appearance in a late phase of hominin evolution. I have argued previously that the standout evidence for such behaviour is the settling of new continents. It occurs at the transition from *Terra* 2 to *Terra* 3 commencing 50,000 years ago when almost 75 per cent of the earth was settled for the first time. Traveling to Australia, the Americas and later the Pacific needed new technology. But while ocean-going canoes and domestic animals were a necessary condition for dispersal, they were not a sufficient one. There had to be social networks that linked people over time and space and mechanisms to keep in touch in widely separated social worlds. There also had to be a mechanism in these new social worlds that structured meetings with strangers, and that was the law of hospitality.

Hearths in huts: a solution and a conclusion

I promised a solution to why hearths commonly appear inside huts during the Lateglacial and rarely before. I have argued that by this time human social worlds had assumed a form familiar to us where males control resources; kinshipping facilitates movement among strangers who might otherwise be hostile; cosmologies explain why these practices are customary; and the ethics of hospitality underpin the strength of social bonds beyond the in-group. All four aspects connect with hearths in huts. I don't have space to examine further the course in deep history of this momentous change, only noting in passing that agriculture did not initiate it, but it did amplify the consequences to such an extent that its beginnings in deep history are veiled.

What happened in the deep history of *Terra* 3, especially during the Lateglacial, was a culmination of embedded engagements with places and materials enacted through containers such as hearths and huts. Dwellings were place-saturated with hospitality and spatial order within them arose through embodied concepts of containment. By placing one container (the house), around another older container (the hearth) the possibilities of formal, embodied spaces were extended and augmented. Segmentation now became the dominant spatial order for social life (Kent 1984, 205).

The customs creating social order – hospitality, spatial segregation by gender, asymmetries of power between men and women and their cosmological explanation – were the outcome of a hominin adept with language and 4E cognition. Earlier in the Late Middle Pleistocene of *Terra* 2, we know that food was shared around hearths (Stiner 2018), and no doubt the enduring topic of conversation was remembering acts of generosity and who to welcome into the circle around the fire. The threshold crossed by those 4E hominins needed hospitality to structure social relations for the local in-group and hearths provided a focus. To paraphrase Pitt-Rivers, there were fewer social unknowns to know and rare strangers to welcome as guests. A subsequent threshold was crossed in *Terra* 3. The incorporation of hearths inside huts, as evidenced by the Ukrainian *yarangas*, did indeed reference a new experience of human embodiment. It was no longer a question of who to include in the circle but rather who to allow a place in the home.

Acknowledgements

I am grateful to Ole Grøn for sharing his field-observations on the Evenki and for Figure 2, and to Richard Bradley, Jan Kolen, John Gowlett, and Alex Pryor for their comments on earlier drafts. Rebecca Ferreira expertly drew the illustrations. The research was generously supported by the Leverhulme Trust's project Seasonality, Mobility and Storage in Palaeolithic Hunting Societies (RPG-2013-318). Thanks also to Olga Soffer and Nikolai Praslov for arranging the visit to Kostenki in 1989, and to Wil who invited me into the hospitality of the European Science Foundation network, The Palaeolithic Occupation of Europe, that was such an enriching experience.

References

Abramova, Z.A.. 1993. Two examples of terminal Palaeolithic adaptations. In: O. Soffer and N. Praslov (eds.), *From Kostenki to Clovis: Upper Paleolithic – Paleo-Indian adaptations*: 85-100. New York: Plenum.

Aiello, L. and R Dunbar. 1993. Neocortex size, group size and the evolution of language. *Current Anthropology* 34: 184-193.

Aiello, L. and P Wheeler. 1995. The expensive-tissue hypothesis: the brain and the digestive system in human and primate evolution. *Current Anthropology* 36: 199-221.

Anderson, D.G.. 2007. Mobile architecture and social life: the case of the conical skin lodge in the Putoran Plateau Region. In: S. Beyries and V. Vaté (eds.), *Les Civilisations du renne d'hier et d'aujourd'hui. Approches ethnohistoriques, archéologiques et anthropologiques:* 43-64. Antibes: Editions APDCA.

Anderson, D.G.. 2006. Dwellings, Storage and Summer Site Structure among Siberian Orochen Evenkis: Hunter-Gatherer Vernacular Architecture under Post-Socialist Conditions. *Norwegian Archaeological Review* 39: 1-26.

Barkai, R, J Rosell, R Blasco and A Gopher. 2018. A land of flint and fallow deer: human persistence at Middle Pleistocene Qesem Cave. In: M Pope, J McNabb and C Gamble (eds.), *Crossing the human threshold: dynamic transformation and persistent places during the Middle Pleistocene*: 60-82. London: Routledge.

Bradley, R.. 2012. *The idea of order: The circular archetype in prehistoric Europe*. Oxford: Oxford University Press.

Candea, M, and G Da Col. 2012. The return to hospitality. *Journal of the Royal Anthropological Institute*:S1-S19.

Cole, J.. 2015. Handaxe symmetry in the Lower and Middle Palaeolithic: implications for the Acheulean gaze. In: F Coward, R. Hosfield, M Pope and F. Wenban-Smith (eds.), *Settlement, society and cognition in human evolution: Landscapes in mind*: 234-257. Cambridge: Cambridge University Press.

de Lumley, H.. 1969. A Paleolithic Camp at Nice. *Scientific American* 220: 42-51.

de Lumley, H.. 2009. *Terra Amata Nice Alpes-Maritimes, France Tome I.* Paris: CNRS Éditions.

Dennell, R, and W Roebroeks. 2005. An Asian perspective on early human dispersal from Africa. *Nature* 438: 1099-1104.

Derrida, J.. 2005. The Principle of Hospitality. *Parallax* 11: 6-9.

Dikeç, M, N Clark, and C Barnett. 2009. Extending hospitality: giving space, taking time. *Paragraph* 32: 1-14.

Dunbar, R.. 2014. How conversations around campfires came to be. *PNAS* 111:14013-14014.

Finlayson, C.. 2019. *The smart Neanderthal: bird catching, cave art and the cognitive revolution.* Oxford: Oxford University Press.

Foley, R. and C Gamble. 2009. The ecology of social transitions in human evolution. *Philosophical Transactions of the Royal Society of London B* 364: 3267-3279.

Gamble, C.S.. 1982. Interaction and alliance in Palaeolithic society. *Man* 17: 92-107.

Gamble, C.S.. 2007. *Origins and revolutions: human identity in earliest prehistory.* New York: Cambridge University Press.

Gamble, C.S.. 2012. When the words dry up: Music and material metaphors half a million years ago. In: N Bannan (ed.), *Music, language and human evolution*: 85-106. Oxford: Oxford University Press.

Gamble, C.S.. 2013. *Settling the Earth: the archaeology of deep human history.* Cambridge: Cambridge University Press.

Gamble, C.S.. 2014. The after-life. In: W Brown and A Fabian (eds.), *Life*: 147-165. Cambridge University Press, Cambridge.

Gamble, C.S.. 2023. Thinking inside the mask. In: O.P Nieuwenhuyse, R Bernbeck and K Berghuijs (eds.), *Containers of Change. Ancient Container Technologies from Eastern to Western Asia*: 19-30. Leiden: Sidestone Press.

Gamble, C.S. in press. The alterity of deep history and the hospitality of things. In: O Moro-Abadia and M Porr (eds.), *Alterity and human evolution: deep time and multispecies perspectives on difference and variation,*. Oxford: Berghahn.

Gamble, C.S, W Davies, P Pettitt and M Richards. 2004. Climate change and evolving human diversity in Europe during the last glacial. *Philosophical Transactions of the Royal Society Biological Sciences* 359: 243-254.

Gaudzinski-Windheuser, S, L Kindler, K MacDonald and W Roebroeks. 2023. Hunting and processing of straight-tusked elephants 125.000 years ago: Implications for Neanderthal behavior. *Science Advances* 9: eadd8186.

Gosden, C.. 2010. The death of the mind. In: L Malafouris and C Renfrew (eds.), *The cognitive life of things: recasting the boundaries of the mind*: 39-46. Cambridge: McDonald Institute for Archaeological Research.

Gowlett, J A J.. 2010. Firing up the social social brain. In: R Dunbar, C Gamble and J A J Gowlett (eds.), *Social brain and distributed mind*: 341-366. Oxford: Oxford University Press.

Gowlett, J A J.. 2016. The discovery of fire by humans: a long and convoluted process. *Philosophical Transactions of the Royal Society B* 371: 20150164.

Grøn, O.. 2011. A 7000 year old Mesolithic dwelling with a brush floor excavated under water. A window into behavioural organization of hunter-gatherer dwellings. In: S. Gaudzinski-Windheuser, O. Jöris, M. Sensburg, M. Street and E. Turner (eds.), *Site-internal spatial organization of hunter-gatherer societies: Case studies from the European Palaeolithic and Mesolithic*: 249-261. Mainz: Verlag des Römisch-Germanischen Zentralmuseums.

Grøn, O.. 2014. Human spatial behaviour in dwellings and social psychology. In M.V. Kristiansen and K. Giles (eds.), *Dwellings, Identities and Homes. European Housing Culture from the Viking Age to the Renaissance*: 29-38. Højbjerg: Jutland Archaeological Society Publications.

Grøn, O.. 2018. Interdisciplinary reflections on repetitive distribution patterns in Scandinavian Mesolithic dwelling spaces. *Journal of Archaeological Science: Reports* 18: 925-935.

Grøn, O. and O Kuznetsov. 2003. Ethno-archaeology among Evenkian forest hunters: Preliminary results and a different approach to reality! In: L Larsson, H Kindgren, K Knitsson, D Loeffler and A Akerlund (eds.), *Mesolithic on the move: papers presented at the sixth international conference on the Mesolithic in Europe, Stockholm 2000*: 197-215. Oxford: Oxbow Books.

Hérisson, D., J-L Locht, P Auguste, and A Tuffreau. 2013. Néandertal et le feu au Paléolithique moyen ancien. Tour d'horizon des traces de son utilisation dans le Nord de la France. *L'Anthropologie* 117: 541-578.

Hoffmann, D. L., C. D. Standish, M. García-Diez, P. B. Pettitt, J. A. Milton, J. Zilhão, J. J. Alcolea-González,

P. Cantalejo-Duarte, H. Collado, R. de Balbín, M. Lorblanchet, J. Ramos-Muñoz, G.-Ch. Weniger and A. W. G. Pike. 2018. U-Th dating of carbonate crusts reveals Neandertal origin of Iberian cave art. *Science* 359: 912-915.

Iakovleva, L.. 2015. The architecture of mammoth bone circular dwellings of the Upper Palaeolithic settlements in Central and Eastern Europe and their socio-symbolic meanings. *Quaternary International* 359-360: 324-334.

Iakovleva, L., F Djindjian and A-M Moigne. 2021. La cabane no 5 du site paléolithique de Gontsy (Ukraine). *L'Anthropologie* 125: 102912.

Ingold, T.. 2015. *Lines: a brief history*. London: Routledge.

Jaubert, J., S Verheyden, D Genty, M Soulier, H Cheng, D Blamart, C Burlet, H Camus, S Delaby, D Deldicque, R. L Edwards, C Ferrier, F Lacrampe-Cuyaubère, F Lévêque, F Maksud, P Mora, X Muth, É Régnier, J-N Rouzaud and F Santos. 2016. Early Neanderthal constructions deep in Bruniquel Cave in southwestern France. *Nature* 534: 111-114.

Jeffares, B.. 2010. The co-evolution of tools and minds: cognition and material culture in the hominin lineage. *Phenomenology and the Cognitive Sciences* 9: 503-520.

Jett, S.C. and V.E. Spencer. 1981. *Navajo architectrure: forms, history, distributions*. Tucson: The University of Arizona Press.

Johnson, M.. 2017. *Embodied mind, meaning, and reason: How our bodies give rise to understanding*. Chicago: University of Chicago Press.

Karkanas, P., R Shahack-Gross, A Ayalon, M Bar-Matthews, R Barkai, A Frumkin, A Gopher and M C Stiner. 2007. Evidence for habitual use of fire at the end of the Lower Palaeolithic: Site formation processes at Qesem Cave, Israel. *Journal of Human Evolution* 53: 197-212.

Kearney, R., and K Semonovitch. 2011. *Phenomenologies of the stranger: between hostility and hospitality*. New York: Fordham University Press.

Kent, S.. 1984. *Analysing activity areas: an ethnoarchaeological study of the use of space*. Albuquerque: University of New Mexico press.

Kolen, J.. 1999. Hominids without homes: on the nature of Middle Palaeolithic settlement in Europe. In: W. Roebroeks and C. Gamble (eds.), *The Middle Palaeolithic occupation of Europe*: 139-175. Leiden: Leiden University Press.

Kolen, J.C.A. and B Oosterwijk. 2017. The first cultural landscapes of Europe: a true enigma. In: S. Kluiving, L. Kootker and R. Hermans (eds.), *Interdisciplinarity between humanities and science: A Festschrift in honour of Prof. Dr. Henk Kars*: 23-33. Leiden: Sidestone Press.

Kuhn, S. L., R Shimelmitz and A.E Clark. 2018. The road to differentiated land use and domestic space in the Middle Pleistocene of southwestern Asia. In: M. Pope, J. McNabb and C. Gamble (eds.), *Crossing the human threshold: dynamic transformation and persistent places during the Middle Pleistocene*: 43-59. London: Routledge.

Kuhn, S.L. and M.C Stiner. 2019. Hearth and home in the Middle Pleistocene. *Journal of Anthropological Research* 75: 305-327.

Levinas, E. 1961 [2002]. *Totality and infinity: an essay on exteriority*. Pittsburgh: Dusquesne University Press.

Linscott, B., A. W. G Pike, D. E Angelucci, M. J Cooper, J. S Milton, H. Matias and J. Zilhão. 2023. Reconstructing Middle and Upper Paleolithic human mobility in Portuguese Estremadura through laser ablation strontium isotope analysis. *PNAS* 120: e2204501120.

MacDonald, K. F Scherjon, E van Veen, K Vaesen and W Roebroeks. 2021. Middle Pleistocene fire use: The first signal of widespread cultural diffusion in human evolution. *PNAS* 118: e2101108118.

Malafouris, L.. 2013. *How things shape the mind: a theory of material engagement*. Cambridge Massachusetts: MIT Press.

Mania, D. 1991. The zonal division of the Lower Palaeolithic open-air site Bilzingsleben. *Anthropologie* 29: 17-24.

McNabb, J., F Binyon and L Hazelwood. 2004. The large cutting tools from the South African Acheulean and the question of social traditions. *Current Anthropology* 45: 653-677.

Menary, R.. 2010. Introduction to the special issue on 4E cognition. *Phenomenology and the Cognitive Sciences* 9: 459-463.

Morgan, L.H.. 1881. *Houses and house-life of the American Aborigines, Volume IV of Contributions to North American Ethnology*. Washington: Government printing Office.

Morley, I.. 2013. *The Prehistory of Music: Evolutionary Origins and Archaeology of Human Musicality*. Oxford: Oxford University Press.

Nadel, D.. 2003. The Ohalo II brush huts and the dwelling structures of the Natufian and PPNA sites in the Jordan Valley. *Archaeology, Ethnology and Anthropology of Eurasia* 1: 34-48.

Nadel, D. E Weiss and H Tschauner. 2011. Gender-specific division of indoor space during the Upper Palaeolithic? A brush-hut floor as a case study. In: S. Gaudzinski-Windheuser, O. Jöris, M. Sensburg, M. Street and E. Turner (eds.), *Site-internal spatial organization of hunter-gatherer societies: Case studies from the European Palaeolithic and Mesolithic*: 263-273. Mainz: Verlag des Römisch-Germanischen Zentralmuseums.

Newen, A., L De Bruin and S Gallagher (eds.) 2018. *The Oxford handbook of 4E cognition*. Oxford: Oxford University Press.

Pettitt, P. 2011. *The Palaeolithic origins of human burial*. London: Routledge.

Pidoplichko, I.G. and P Allsworth-Jones. 1998. *Upper palaeolithic dwellings of mammoth bones in the Ukraine: Kiev-Kirillovskii, Gontsy, Dobranichevka, Mezin and Mezhirich*. Oxford (BAR International Series 712).

Pisor, A.C. and M Surbeck. 2019. The evolution of intergroup tolerance in nonhuman primates and humans. *Evolutionary Anthropology: Issues, News, and Reviews* 28: 210-223.

Pitt-Rivers, J.. 1977 [2017]. The law of hospitality. In: G. daCol and A. Shryock (eds.), *From Hospitality to Grace: A Julian Pitt- Rivers Omnibus*: 163-184. Chicago: Hau Books.

Pope, M, J. McNabb and C.S. Gamble (eds.) 2018. *Crossing the human threshold: dynamic transformation and persistent places during the Middle Pleistocene*. London: Routledge.

Praslov, N.D. and A.N. Rogachev (eds.) 1982. *Palaeolithic of the Kostenki-Borshevo area on the Don river, 1879-1979 (In Russian)*. Leningrad: NAUKA.

Pryor, A. J. E., A. Pullen, D. G. Beresford-Jones, J. A. Svoboda, and C. S. Gamble. 2016. Reflections on Gravettian firewood procurement near the Pavlov Hills, Czech Republic. *Journal of Anthropological Archaeology* 43: 1-12.

Pryor, A.J.E., D.G. Beresford-Jones, A.E Dudin, E.M Ikonnikova, J.F Hoffecker and C.S Gamble. 2020. The chronology and function of a new circular mammoth-bone structure at Kostenki 11. *Antiquity* 94: 323-341.

Roebroeks, W.. 1988. From find scatters to early hominid behaviour: a study of Middle Palaeolithic riverside settlements at Maastricht-Belvédère *(The* Netherlands). *Analecta Praehistorica Leidensia* 21.

Roebroeks, W.. 1996. The English Palaeolithic Record: absence of evidence, evidence of absence and the first occupation of Europe. In: C. Gamble and A.J. Lawson (eds.), *The English Palaeolithic Reviewed*: 57-62. Trust for Wessex Archaeology.

Roebroeks, W.. 2006. The human colonisation of Europe: where are we? *Journal of Quaternary Science* 21: 425-435.

Roebroeks, W. (ed.) 2007. *Guts and brains: An integrative approach to the hominin record*. Leiden: Leiden University Press.

Roebroeks, W. and C.C. Bakels. 2015. 'Forest furniture' or 'forest managers'? On Neanderthal presence in last interglacial environments. In: F. Coward, R. Hosfield, M. Pope and F. Wenban-Smith (eds.), *Settlement, society and cognition in human evolution: Landscapes in mind*: 174-188. Cambridge: Cambridge University Press.

Roebroeks, W., D. De Loecker, P. Hennekens and M. van Ieperen. 1992. A veil of stone: on the interpretation of an early Middle palaeolithic low density scatter at Maastricht Belvédère (The Netherlands). *Analecta Praehistorica Leidensia* 25: 1-16.

Roebroeks, W. and P. Villa. 2011a. On the earliest evidence for habitual use of fire in Europe. *PNAS* 108: 5209-5214.

Roebroeks, W. and P. Villa. 2011b. Reply to Sandgathe *et al.*: Neandertal use of fire. *PNAS* 108: E299-E299.

Roebroeks, W., J. Kolen and E Rensink. 1988. Planning depth, anticipation and the organization of middle Palaeolithic technology: the "archaic natives" meet Eve's descendants." *Helinium* 28:17-34.

Roebroeks, W. and T van Kolfschoten. 1994. The earliest occupation of Europe: A short chronology. *Antiquity* 68: 489-503.

Rolland, N.. 2004. Was the emergence of home bases and domestic fire a punctuated event? A review of the Middle Pleistocene record in Eurasia. *Asian Perspectives* 43: 248-80.

Romano, A., D. Balliet, T. Yamagishi and J.H. Liu. 2017. Parochial trust and cooperation across 17 societies. *PNAS* 114: 12702-12707.

Sandgathe, D.M, H.L. Dibble, P. Goldberg, S.P. McPherron, A. Turq, L. Niven and J. Hodgkins. 2011. On the role of fire in Neandertal adaptations in Western Europe: evidence from Pech de l'Azé IV and Roc de Marsal, France. *PaleoAnthropology* 2011: 216-242.

Scott, B, A. Shaw, K. Scott and M. Pope (eds.). 2023. *Repeopling La Manche: new perspectives on Neanderthal archaeology and landscapes from La Cotte de St Brelade*. Oxford: Oxbow Books (Prehistoric Society Research Paper No. 10).

Shaw, A., M. Bates, C. Conneller, C. Gamble, M.-A. Julien, J. McNabb, M. Pope and B. Scott. 2016. The archaeology of persistent places: the palaeolithic case of La Cotte de St Brelade, Jersey UK. *Antiquity* 90: 1437-1453.

Shennan, S.J. 2001. Demography and cultural innovation: a model and its implications for the emergence of modern human culture. *Cambridge Archaeological Journal* 11: 5-16.

Shirokogoroff, S. M. 1929. *Social Organization of the Northern Tungus.* Shanghai: The Commercial Press Ltd.

Shryock, A. T. Trautmann and C. Gamble. 2011. Imagining the human in deep time. In: A Shryock and D.L. Smail (eds.), *Deep history: the architecture of past and present*: 21-52. Berkeley: University of California Press.

Skov, L. S. Peyrégne, D. Popli, L.N. M Iasi, T. Devièse, V. Slon, E.I. Zavala, M. Hajdinjak, A.P. Sümer, S. Grote, A. Bossoms Mesa, D. López Herráez, B. Nickel, S. Nagel, J. Richter, E. Essel, M. Gansauge, A. Schmidt, P. Korlević, D. Comeskey, A.P. Derevianko, A. Kharevich, S.V. Markin, S. Talamo, K. Douka, M.T. Krajcarz, R.G. Roberts, T. Higham, B. Viola, A.I. Krivoshapkin, K.A. Kolobova, J. Kelso, M. Meyer, S. Pääbo and B.M. Peter. 2022. Genetic insights into the social organization of Neanderthals. *Nature* 610 : 519-525.

Soffer, O.. 1985. *The Upper Palaeolithic of the Central Russian Plain.* New York: Academic Press.

Soffer, O. and N.D. Praslov (eds.). 1993. *From Kostenki to Clovis: Upper palaeolithic – Paleo-Indian adaptations.* New York: Plenum.

Stiner, M. C.. 2002. Carnivory, coevolution, and the geographic spread of the genus *Homo. Journal of Archaeological Research* 10: 1-64.

Stiner, M.C.. 2018. On the co-evolution of hearth and home-making during the Middle Palaeolithic of the Levant. In: M. Pope, J. McNabb and C. Gamble (eds.), *Crossing the human threshold: dynamic transformation and persistent places during the Middle Pleistocene*: 83:-105. London: Routledge.

Svoboda, J. (ed.). 1994. *Pavlov I: excavations 1952-53.* Liège: Université de Liège. (ERAUL 66).

Svoboda, J.(ed). 2016. *Dolni Vestonice II: chronostratigraphy, paleoethology, palaeoanthropology.* Brno: Academy of Sciences of the Czech Republic, Institute of Archaeology.

Villa, P.. 1982. Conjoinable pieces and site formation processes. *American Antiquity* 47: 276-90.

Wobst, H.M.. 1976. Locational relationships in Palaeolithic society. *Journal of Human Evolution* 5: 49-58.

Wrangham, R.W. 2009. *Catching Fire: How Cooking Made Us Human.* London: Profile Books.

Zohar, I., N. Alperson-Afil, N. Goren-Inbar, M. Prévost, T. Tütken, G. Sisma-Ventura, I. Hershkovitz and J. Najorka. 2022. Evidence for the cooking of fish 780,000 years ago at Gesher Benot Ya'aqov, Israel. *Nature Ecology and Evolution* 6: 2016-2028.

In Dusseldorp, G.L., Chu, W., Bakels, C.C. and M.A. Soressi (eds) 2025. Intent on the Paleolithic. Papers in honour of Prof. Dr. Wil Roebroeks. *Analecta Praehistorica Leidensia* 55. Leiden: Sidestone Press, pp. 213-232.

Terra Incognita Revisited: the settlement patterns of early humans in north-west Europe

Nick Ashton, Rob Davis, Simon G. Lewis, Simon Parfitt

Abstract

North-west Europe has long been recognised to be on the margins of the Palaeolithic world, but with its long history of research has been used as a laboratory to explore ideas on human dispersal and range-limits against a backdrop of periodic climate change. This paper re-examines the distribution of Lower and early Middle Palaeolithic sites across the region, and the historical and geological factors that have influenced the record. For the Lower Palaeolithic, the imbalance in sites between Britain and mainland Europe is argued to be largely the result of thicker loess cover over northern France and Belgium, making historic access to fluvial sediments more challenging. This, together with the effects of glaciation in Britain, make it difficult to determine range-limits to human occupation, which are likely to have been considerably beyond known distributions. For the early Middle Palaeolithic there is an impoverished record in Britain, which has been emphasised by new discoveries in adjacent submerged areas of the North Sea Basin. It is suggested that a real boundary to human occupation lies in south-east England, being on the margins of the deeper loess-rich soils which supported the mammoth steppe.

Keywords: Palaeolithic; north-west Europe; population distribution; range-limits

Nick Ashton
Department Britain, Europe and Prehistory, British Museum, Franks House, London, UK.
nashton@britishmuseum.org

Rob Davis
Department Britain, Europe and Prehistory, British Museum, Franks House, London, UK

Simon G. Lewis
School of Geography, Queen Mary University of London, London, UK

Simon Parfitt
Institute of Archaeology, UCL, London, UK; Centre for Human Evolution Research, Natural History Museum, London, UK

Introduction

North-west Europe lies on the margins of the first human expansions out of Africa, through Asia and into southern Europe. Yet, since the 1840s it has a history of research that provides a wealth of information still relevant to questions asked today. The geographical, climatic, biological and technological reasons for the delayed and intermittent occupation of north-west Europe compared to southern Eurasia are important for understanding the broader delimiting factors that affected human expansion. This was the focus of Roebroeks's long-term interest in the region for defining and studying hominin range limits (2006; 2014; Hublin and Roebroeks 2009), and also forming part of Gamble's biotidal zone (2009) and one of the 'sinks' of Dennell and colleagues source and sinks model (2011). Building on this work, our paper further examines how current understanding of the north-west European record has been influenced by differing histories of research and how variations in bedrock and surface geology have affected archaeological visibility. It considers whether changes in site distribution across the region can really inform

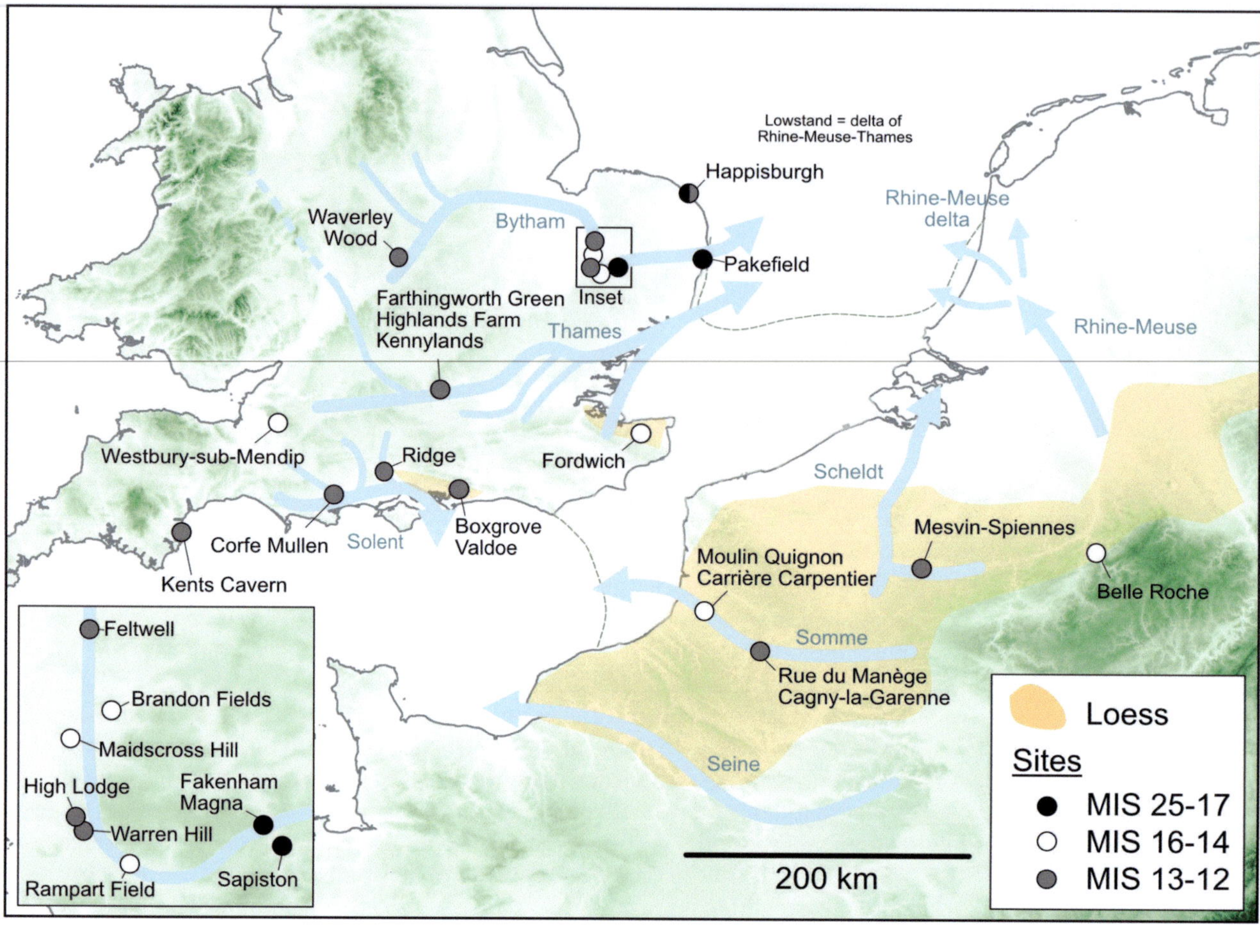

Figure 1. Suggested palaeogeography of north-west Europe from c. 1 Ma to 500 ka. Distribution of loess (>2m depth) is based on Antoine *et al.* 2003 and Lehmkuhl *et al.* 2021.

us about the factors that governed human dispersal at the margins of the Palaeolithic world.

Intensive research across north-west Europe since the mid-19th century (Evans 1861) has produced large numbers of sites and artefacts with accompanying datasets. Bordes' thesis (1954) covered the valley of the Seine, while that of Tuffreau (1987) examined the Somme. The work of Ulrix-Closset (1981) listed many sites for the Meuse Basin in Belgium, with a more expansive update in the thesis of Di Modica (2010). For Britain, the comprehensive gazetteers of Roe (1968) and Wymer (1968; 1985) provided the foundations for The English Rivers Palaeolithic Survey (TERPS) and its resulting datasets (e.g. Wessex Archaeology 1993, 1996a,b). Much of the British record has limitations due to poor contextual information or biases towards the collection of handaxes. However, through acknowledgment of these biases and a more robust chronological framework (Bridgland 1994; Schreve 2001; Penkman *et al.* 2011), the data can still be used to address broad questions about changes in human occupation, distribution and periods of absence, or at least archaeological invisibility (Ashton and Lewis 2002; Ashton *et al.* 2011; Bridgland and White 2014; White *et al.* 2018, 2019; Davis *et al.* 2021a; Lewis *et al.* 2021).

For Britain, a fundamental geographical factor has been its changing peninsula/island status, with the duration and degree of isolation increasing through successive interglacials from MIS 11 (White and Schreve 2000; Ashton and Lewis 2002; Stringer 2005; Ashton *et al.* 2011). Therefore, to understand other constraints to occupation, we need to consider north-west Europe as a whole, and incorporate the North Sea Basin -the *Terra Incognita* of Roebroeks (2014)- as a central zone of occupation, rather than as a transit route to and from Britain.

The geography of the study area during the early Palaeolithic

For this paper, the study area is defined as southern Britain (south of the Mersey-Humber estuaries), the English Channel, the southern North Sea Basin (Southern Bight), Belgium, the Netherlands and the north-west French regions of Hauts de France and Normandy. For Britain, the vast majority of Lower and early Middle Palaeolithic sites are associated with terrace deposits of the major rivers or their tributaries, in particular the now-extinct Bytham (prior to MIS 12), the Solent and the Thames (Figure 1; Wymer 1999). After MIS 12, glacial reorganisation of drainage patterns resulted in new rivers, including the catchment of the Great Ouse, together with the Yare, Waveney and Stour (Figure 2; Wymer 1985). Only a few sites come from coastal deposits or from caves (see below). Although upland sites survive in dolines on the North Downs and Chilterns, there has been limited success in providing age constraints for the contained archaeology (Bradley and Sampson 1978; Bailiff *et al.* 2013).

The surface geology of the north-western parts of Belgium and much of the Netherlands consists of stacked fluvial sediments of the major rivers, including the Scheldt, Meuse (or Maas) and Rhine with all the Middle Pleistocene sediments deeply buried with limited access to Palaeolithic sites (Di Modica 2010; Di Modica and Pirson 2016). Repeated glaciation of the northern Netherlands has further reduced the chances of discovering sites. However, in the Limburg Peninsula of the Netherlands and in central Belgium there are river terraces of the Meuse/Maas and of the Haine (a tributary of the Scheldt), where several late Middle Pleistocene sites have been identified in both fluvial and overlying loess deposits. Despite numerous caves and rock-shelters in the Meuse Basin and the Ardennes, other than Belle-Roche there are no assemblages that can be attributed with confidence to the late Middle Pleistocene or earlier.

For north-west France the river systems of the Scarpe (a tributary of the Scheldt), Somme and Seine have provided important sites either from terrace deposits (occasionally with tufa), or from long sequences of overlying loess with intermittent palaeosols (Tuffreau 1987; Cliquet 2013). On the plateaux there are also a significant number of dolines infilled with similar loess sequences that contribute understanding of the late Middle Pleistocene (Hérisson *et al.* 2016). Several coastal sites have been identified, mostly on raised beaches on the Normandy coast, together with the cave site of La Cotte de St Brelade on Jersey (Scott *et al.* 2023).

There is now better understanding of the changes in configuration of both the southern North Sea Basin and the English Channel (Gibbard 1995; Gupta *et al.* 2007; Hijma *et al.* 2012; Tizzard *et al.* 2014). The Solent, Somme and Seine rivers have always flowed into the Channel region and at times of low sea-level into the Channel River. Prior to MIS 12, the Thames, Bytham, Scheldt, Meuse and Rhine flowed north into the southern North Sea Basin, and at times were probably confluent (Figure 1). After MIS 12 and the formation of the Strait of Dover, there is a progressive shift over several marine isotope stages with these rivers eventually flowing in a southerly direction into the Channel River at times of lower sea-level (Figures 2 and 3).

Site distributions in north-west Europe

The survey below includes most of the principal Lower and early Middle Palaeolithic sites in north-west Europe that have some age constraint to at least a specific glacial-interglacial cycle. The time considered is up to, but not including MIS 5e, when Britain was an island throughout the interglacial (Gibbard 1995; Lewis *et al.* 2011). Little information is given about artefact typology or technology as the main concern is human presence and distribution.

900-700 ka (MIS 25-17)

The North Sea Basin was a major region of low-lying land that linked northern France with Britain. There is evidence of sparse human occupation with the earliest signs at the coastal site of Happisburgh 3, where a small flake assemblage comes from fluvial sediments of the proto-Thames, dating to either MIS 25 or 21 (Parfitt *et al.* 2010; Bynoe *et al.* 2021). Some 30 km to the south, the small flake assemblage from Pakefield comes from alluvial sediments of either the Bytham or Thames and dates to MIS 17 (Parfitt *et al.* 2005). Other hints of human occupation come from upstream terrace remnants of the Bytham River with a few flakes from Fakenham Magna, attributed to MIS 19 and Sapiston to MIS 17 (Davis *et al.* 2021a; Lewis *et al.* 2021). There is no unequivocal evidence from the rest of the study area, despite the existence of high terraces of a broadly similar age in the Thames, Solent, Seine, Somme and Meuse.

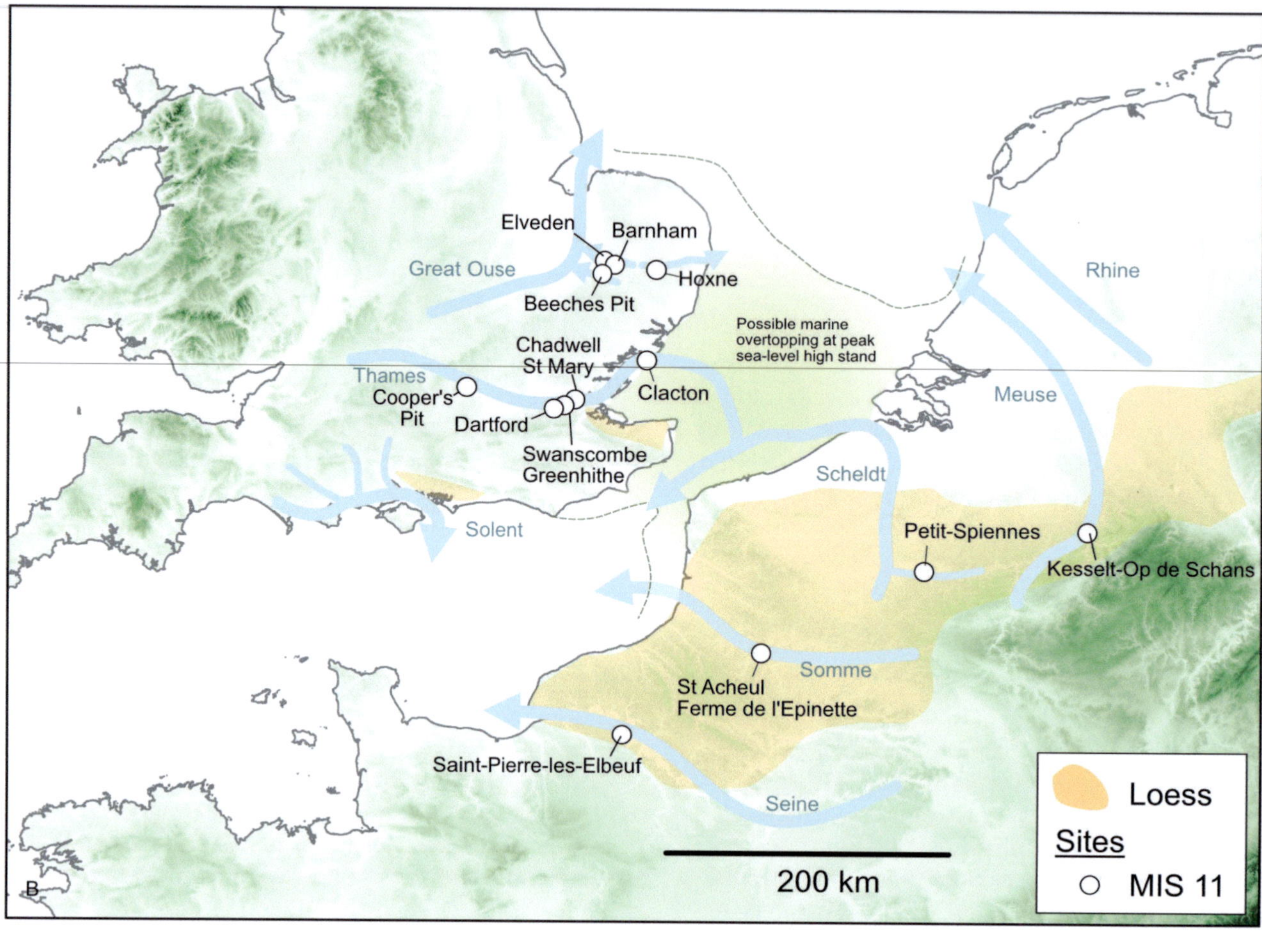

Figure 2. Suggested palaeogeography of north-west Europe during MIS 11. Distribution of loess (>2m depth) is based on Antoine *et al.* 2003 and Lehmkuhl *et al.* 2021.

650-550 ka (MIS 16-15-14)

Although there were undoubtedly changes in palaeogeography, the structure of the region probably remained essentially the same, with the North Sea Basin retaining its central position in linking the surrounding areas. In Britain, Bytham River deposits at Brandon Fields, Maidscross Hill and Rampart Field are attributed to MIS 14, and the associated rolled handaxes are suggested to be MIS 15 in age (Davis *et al.* 2021a; Lewis *et al.* 2021). Handaxes from a high terrace of the Stour at Fordwich are probably a similar age (Bridgland *et al.* 1998; Key *et al.* 2022). In the Solent and Thames catchments, there are handaxes from locations interpreted as pre-MIS 13 terraces, but there are no examples with unequivocal contexts. The cave deposits of Westbury-sub-Mendip have yielded degraded flint artefacts and cut-marked bone. Mammalian biostratigraphy suggests that they are older than Boxgrove, probably dating to MIS 15, or possibly earlier in MIS 13 (Andrews *et al.* 1999; Parfitt and Preece 2022).

In Belgium, the cave deposits of La Belle-Roche, south of Liège, are biostratigraphically the same age as Westbury, and have produced cut-marked bone, but disputed artefacts (Cordy *et al.* 1993). The earliest site in north-west France, is Moulin Quignon in Abbeville where handaxes have been recovered from fluvial gravels towards the base of the Alluvial Formation VII of the Somme, attributed to early MIS 16 (Antoine *et al.* 2019). At an adjacent quarry – Carrière Carpentier – handaxes were also recovered from the base of slope deposits immediately above the same alluvial formation, and attributed to early MIS 14 (Antoine *et al.* 2016).

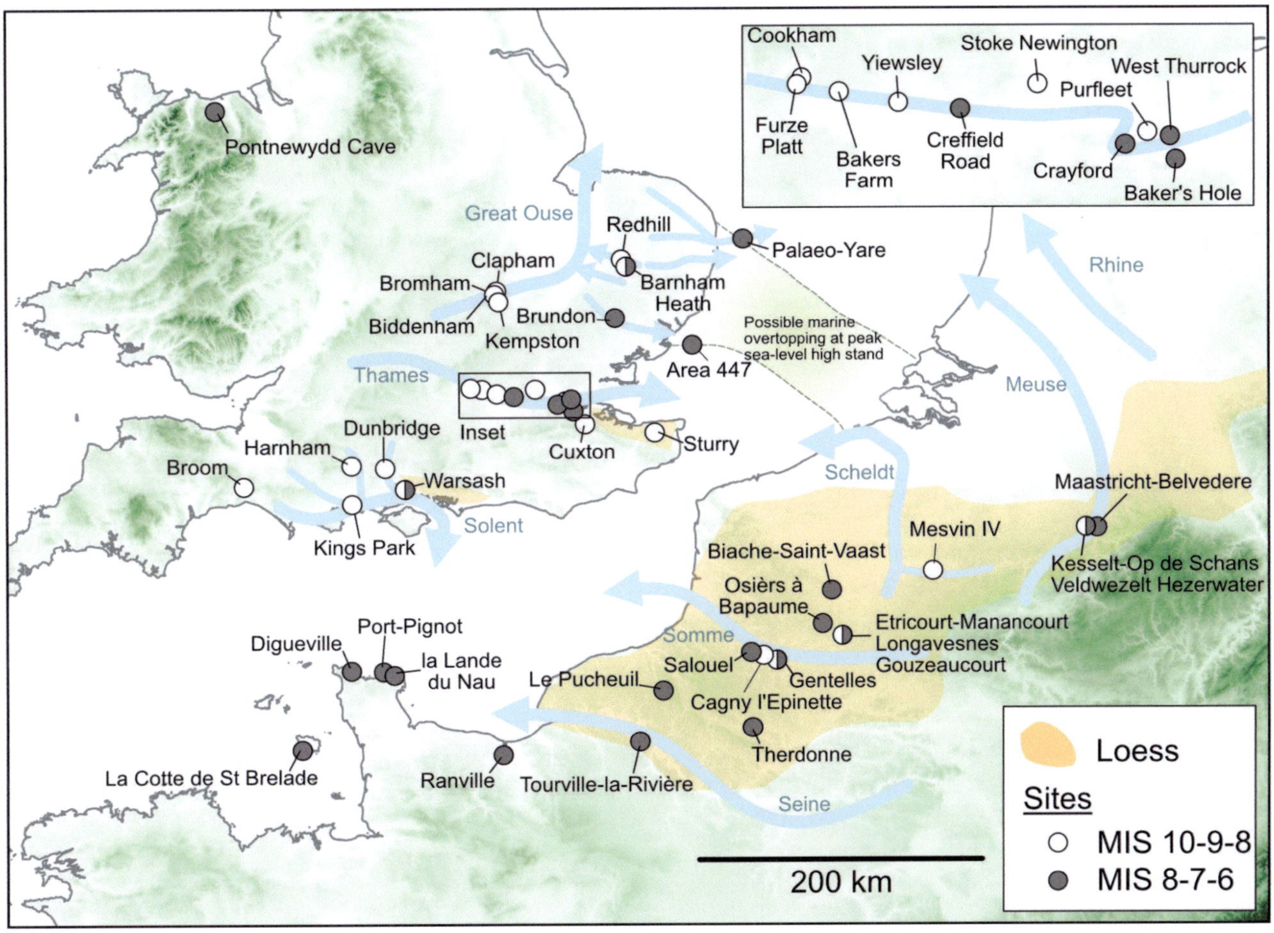

Figure 3. Suggested palaeogeography of north-west Europe from MIS 10 to MIS 6. Distribution of loess (>2m depth) is based on Antoine *et al.* 2003 and Lehmkuhl *et al.* 2021.

500 ka (MIS 13-early 12)

The geographic structure of the region was essentially unchanged in MIS 13, but from this period there is a marked increase in the number of sites in Britain. On the Bytham River, there are large assemblages from Warren Hill and High Lodge (Ashton *et al.* 1992; Davis *et al.* 2021a; Lewis *et al.* 2021), and smaller assemblages from Waverley Wood and Feltwell (Shotton *et al.* 1993; Keen *et al.* 2006; Lewis *et al.* 2021).

On the coast, Happisburgh Site 1 comes from channel deposits of an unknown river and dates to MIS 13 (Lewis *et al.* 2019), while c. 2 km to the north-west at Ostend, a concentration of 30 handaxes was recovered from coarse gravel on the beach after a storm surge in 2018 (Davis *et al.* 2023). The gravel and handaxes probably derive from underlying or nearby fluvial sediments that became exposed through erosion of the cliffs. The cliffs consist of glacial sediments, attributed to MIS 12, and the assemblage is therefore likely to be MIS 13 or earlier, but more work is needed to test this interpretation.

In the Middle Thames, there are several sites in the Black Park Terrace, attributed to MIS 12 with significant handaxe assemblages from Highlands Farm, Farthingworth Green and Kennylands Gravel Pit which likely derive from MIS 13 landsurfaces (Wymer 1968, 1999; Bridgland 1994). There are numerous locations from the tributaries of the Solent that probably date to MIS 13 (Ashton and Hosfield 2010; Davis 2013), the largest sites being Corfe Mullen on the Stour (McNabb *et al.* 2012) and Ridge Gravel Pit on the Test (Davis *et al.* 2021b). To the east of the Solent, there is the coastal site of Boxgrove and its lesser-known neighbour Valdoe, both dating to MIS 13 (Roberts and Parfitt 1999; Pope *et al.* 2009, 2020; Preece and Parfitt 2022). Finally, the small handaxe assemblage from Kent's Cavern is also likely to be MIS 13 in age (Cook and Jacobi 1998).

Outside Britain, there are very few sites that can be attributed to this period. In the Mesvin-Spiennes area of Belgium, there are several flakes from the Pa d'la l'iau terrace of the River Haine, attributed to MIS 12 (Pirson *et al.* 2009). In France, a small flake assemblage has been excavated from Rue du Manège in Amiens, at the top of Alluvial Formation VI of the Somme, dated to late MIS 14 or early MIS 13 (Antoine *et al.* 2015). Nearby, the handaxe assemblages of Cagny-la-Garenne come from the lower part of Alluvial Formation V and are attributed to early MIS 12.

400 ka (MIS 11)

Glaciation during MIS 12 had a major impact on the palaeogeography of the region. Ice sheets reached southern England, destroying the Bytham River and diverting the Thames into its current valley. At the end of the glaciation, the pro-glacial lake that had formed in the southern North Sea Basin over-topped the Kent-Artois Anticline, forcing a passage through the Chalk to form the Strait of Dover (Smith 1985; Gibbard 1995; Gupta *et al.* 2007). The Thames and Scheldt now flowed through the Strait into the Channel River, while the Rhine continued to flow north. This left an interfluve between these drainage systems, composed of tills and Palaeogene sediments, between East Anglia and mainland Europe (Hijma *et al.* 2012; Figure 2). Strong evidence of a land-connection across the southern North Sea during the peak of MIS 11c is provided by the arrival of the molluscan 'Rhenish fauna' in the Thames, indicating a fluvial link with the River Scheldt (Kerney 1971; White *et al.* 2013) with terrestrial molluscs providing further evidence of a land connection (Preece *et al.* 2007; Limondin-Lozouet *et al.* 2015). However, marine molluscan evidence suggests occasional overtopping of the interfluve (Wesselingh *et al.* 2022).

Several Palaeolithic sites in Britain are dated to MIS 11c, including Swanscombe (Phases 1 and 2), Clacton, Barnham, Elveden and Beeches Pit (Wenban-Smith 2013; Ashton 2016; Davis and Ashton 2019 and references therein), while those attributed to MIS 11a include the upper sequence at Hoxne, Greenhithe, Dartford Heath and Swanscombe (Phase 3; White *et al.* 2019). There are also numerous assemblages and findspots from Thames terrace deposits that formed during the glacial-interglacial cycle of MIS 12-11-10, such as Chadwell St Mary and Cooper's Pit (Wymer 1968; Bridgland 1994). Although there may be assemblages from other river systems attributable to this cycle, the taphonomy of the assemblages is poorly understood and the dating less secure.

In Belgium the Petit-Spiennes terrace of the River Haine is dated to MIS 10 with the contained handaxes attributed to MIS 11 (Di Modica and Pirson 2016). A few simple cores and flakes have also been found in the lowest palaeosol (level 4) in the long loess sequence at Kesselt-Op de Schans on the Meuse, and suggested to date to MIS 11 (Di Modica and Pirson 2016).

In France, the eponymous site of St Acheul in Amiens has been dated to MIS 11, with handaxes coming from within Alluvial Formation V of the Somme and from overlying tufa (Commont 1908; Bahain *et al.* 2007). Nearby, Ferme de l'Epinette contains handaxes from a palaeosol within slope deposits attributed to MIS 11/10 (Lamotte *et al.* 2001). At St-Pierre-lès-Elbeuf on the Seine, handaxe assemblages lie towards the base of an 18 m sequence of palaeosols and loess, again attributed to MIS 11 (Cliquet *et al.* 2009; Moncel *et al.* 2015). Although there are references to other sites in the Somme and the Seine that might relate to this period, few if any have secure contexts or age control (Bordes 1954; Tuffreau 1987).

370-250 ka (MIS 10-8)

It is likely that the land connection between Belgium, the Netherlands and Britain persisted during the high sea-stand of MIS 9 (Hijma *et al.* 2012), although subsidence of the floor of the North Sea Basin would have required increasingly lower sea-levels for wider land connections to be maintained. For Britain, there are few, if any, primary context sites for MIS 9, but an abundance of locations and findspots from river terrace deposits attributed to MIS 10-9-8. For the Thames the major sites include Furze Platt, Bakers Farm, Cookham, the Hillingdon-Yiewsley-Acton areas of west London, Stoke Newington and Purfleet (Wymer 1968). South bank tributary sites are Cuxton on the Medway, and Sturry on the Stour (Wenban Smith 2004; Bridgland *et al.* 1998). The Great Ouse catchment probably includes the Bedford sites of Biddenham, Kempston, Bromham and Clapham, as well as Barnham Heath and Redhill near Thetford (Harding *et al.* 1991; Gibbard *et al.* 2008; Davis *et al.* 2024). The former Solent River and its tributaries again have numerous locations and findspots on terrace deposits that have been attributed to this glacial-interglacial cycle, the largest being Warsash, the fresher assemblage from Dunbridge and the Ensbury-King's Park area of Bournemouth (Ashton and Hosfield 2010; Davis 2013, Davis *et al.* 2021b). Harnham on the Hampshire Avon and Broom on the River Axe have assemblages dated to MIS 8 (Bates *et al.* 2014; Hosfield *et al.* 2013).

In Belgium, Kesselt-Op de Schans has further assemblages from the loess/palaeosol sequence in late MIS 9 and into MIS 8 (levels 3 and 2; Van Baelen *et al.* 2017), while in the Haine valley, the assemblages from Mesvin IV on the Mesvin Terrace have been dated to MIS 8 (Ryssaert 2006). In neighbouring north-west France, the lower assemblages in the doline at Etricourt-Manancourt are dated to MIS 9 (Hérisson *et al.* 2016b). Cagny l'Épinette has handaxe assemblages from fine-grained sediments within Alluvial Formation IV of the Somme and suggested to date to early MIS 9 (Tuffreau 2001). More broadly attributed to this period are the assemblages from cycle II of the doline at Gentelles, on a plateau east of the Somme (Tuffreau *et al.* 2001, 2008). There are records of other assemblages that might date to this period, but currently lacking age control or firm contexts (Bordes 1954; Tuffreau 1987).

250-180 ka (late MIS 8-early MIS 6)

The continued subsidence of the North Sea Basin meant that, during higher sea-level stands, land connections across this area were probably reduced to the narrow interfluve between the Rhine and the Thames/Scheldt (Hijma *et al.* 2012). However, erosion of the interfluve may have led to occasional marine flooding, as suggested by molluscan evidence (Meijer and Cleveringa 2009). This created a 'bottleneck' that appears to have had an impact on the occupation of Britain, as beyond the lower reaches of the Thames there are few notable sites compared to mainland north-west Europe, perhaps reflecting shorter periods of occupation (Ashton *et al.* 2018). In the Hillingdon-Yiewsley-Acton area, Levallois material comes from sediments immediately overlying Thames terrace deposits attributed to MIS 10-9-8. The assemblages probably date to late MIS 8 or early MIS 7 with Creffield Road being the best-recorded site (Scott 2011). In the Lower Thames, the sites of Baker's Hole/Ebbsfleet and West Thurrock, Lion Pit Tramway Cutting, have assemblages from the lower parts of terrace sequences, attributed to early MIS 7, while the assemblage from Crayford is higher in the terrace sequence and might date to later in MIS 7 (Schreve *et al.* 2006; Scott 2011; Wenban Smith *et al.* 2020).

Beyond London, the record is generally very poor with just a few small Levallois assemblages that are probably MIS 7 in age. These include Warsash in the Solent (Davis *et al.* 2016; Hatch *et al.* 2017) and the East Anglian sites of Barnham Heath (Wymer 1985; Davis *et al.* 2024) and Brundon (Wymer 1985). The only major site outside these areas is Pontnewydd Cave in north Wales (Aldhouse-Green *et al.* 2012).

In the southern North Sea, there are three recently discovered assemblages from dredged deposits, which, based on survey, dating work and lithic analysis, are probably early Middle Palaeolithic in age. The first two assemblages comprise over 50 handaxes and 100 Levallois artefacts, recovered initially from aggregate landed on Dutch wharfs from dredging license Area 240 (Tizzard *et al.* 2014; Shaw *et al.* 2023), and subsequently from sand originating from adjacent dredging areas used for beach replenishment at Bacton and Walcott, Norfolk (Davis *et al.* 2023). The dredged deposits are from the Palaeo-Yare, c. 10 km off the East Anglian coast and have been dated by OSL to MIS 7-6. The third assemblage at Clacton consists of over 200 Levallois artefacts from beach replenishment sediments that originate from Area 447, 18 km offshore (Bynoe *et al.* 2022) and may correspond with a submerged River Stour. The Levallois record from these two areas of the North Sea Basin now far surpasses that of the rest of Britain, beyond the Thames.

In the Netherlands, the Levallois site of Maastricht-Belvedere lies in alluvial interglacial sediments of the Maas and probably dates to MIS 7 (Roebroeks 1988; De Loecker and Roebroeks 2012), although attribution to MIS 9 has also been suggested (Meijer and Cleveringa 2009). In Belgium, the Levallois site at Veldwezelt Hezerwater lies within a loess sequence and dates to MIS 7 (Bringmans 2006).

There are a substantial number of sites in north-west France that date to this period from a variety of depositional contexts (Hérisson *et al.* 2016a). In the valley or surrounding plateaux of the River Scarpe, the main fluvial site is in a low-lying terrace at Biache-Saint-Vaast (Tuffreau 1988; Hérisson 2012), while most other sites come from loess deposits within dolines, such as Longavesnes (Amerloot-Van der Heijden 1993), Osièrs à Bapaume (series B; Koehler 2008), Gouzeaucourt (Tuffreau and Bouchet 1985; Marcy 1989) and Etricourt-Manancourt (Hérisson *et al.* 2016b).

In the Somme, the assemblage from Salouel was recovered from fluvial sediments of a tributary, the

		Britain				NW France, Belgium, Netherlands			
Period	**Discovery**	**Fluvial**	**Loess/ Doline**	**Cave**	**Raised beach**	**Fluvial**	**Loess/ Doline**	**Cave**	**Raised beach**
>MIS 14	pre-1900	3				2			
	1900-1950	1							
	Post-1950	4							
MIS14-12	pre-1900	2		1		1			
	1900-1950	4			2	1			
	Post-1950	5		1		1		1	
MIS12-10	pre-1900	8				2	1		
	1900-1950	5							
	Post-1950						2		
MIS10-8	pre-1900	10				1			
	1900-1950	4							
	Post-1950	2				1	1		
MIS 8-6	pre-1900	5		1				1	1
	1900-1950	3							
	Post-1950					4	7	1	2

Table 1. Sites in north-west Europe by age, depositional environment and discovery date.

Selle (Amerloot-Van der Heijden *et al.* 1996), while a further three assemblages come from Cycle III of the doline at Gentelles (Tuffreau *et al.* 2001, 2008). In the Seine, the only fluvial site is at Tourville-la-Rivière (Cliquet 2013), other sites being in plateau dolines, such as Le Pucheuil (Delagnes and Ropars 1996; Delagnes 1993), or valley-edge at Therdonne (Locht *et al.* 2010; Hérisson 2012).

In the west of Normandy, Ranville (assemblage B) lies in a karstic fissure above the estuary of the River Orne (Cliquet dir. 2008; Cliquet and Lautridou 2009), and several sites lie on Saalian raised beaches on the Cotentin, such as at la Lande du Nau (Gouberville), Port-Pignot (Fermanville) and Digueville (Coutard and Cliquet 2005; Cliquet 2013). Finally, the site of La Cotte de St Brelade lies in a coastal fissure on Jersey (Callow and Cornford 1986; Pope *et al.* 2023).

Discussion

It is immediately obvious that there is an uneven distribution across the region with a greater number of sites in Britain until c. MIS 8, with comparatively few sites in mainland north-west Europe, but the converse after this date, a pattern that has been previously recognised (Scott and Ashton 2011; Roebroeks 2014; Table 1).

But what are the factors behind these patterns, other than a reflection of human occupational history? Despite having the appearance of a coherent study area, connected by the North Sea Basin, there are important disconnects between Britain and mainland north-west Europe, which concern surface geology, the resulting systems of geological interpretation, and the history of research. Better understanding of these factors should enable improved comparison between the two records and contribute to studies attempting to identify marginal areas of human occupation.

Although there are similarities in the pre-Quaternary geology between southern Britain and mainland north-west Europe, there is a fundamental difference in the surface geology – the mantling of loess over large areas of mainland Europe from the Seine northwards to central Belgium and across the Limburg Peninsula of the Netherlands (Antoine *et al.* 2003; Roebroeks 2014). The loess both contained and protected Palaeolithic archaeology, but also made access to underlying fluvial sediments far more difficult. In the Seine valley, the loess is up to 18 m thick, and is typically over 5 m in the valleys of the Somme, Haine and Meuse/Maas (Bordes 1954; Antoine *et al.* 2007; Di Modica and Pirson 2016). By contrast, in the Thames, Solent and Great Ouse River basins, the loess is a minor component of slope deposits that are typically under 2 m thick above terrace sediments (Wymer 1968; Bridgland 1994).

One consequence has been that the geological systems of interpretation have developed in different ways. In north-west Europe, particularly the Somme, the geological framework is not just based on terrace stratigraphy, but also on the recognition of overlying sequences of loess interspersed with palaeosols and 'cailloutis' (thin seams of gravel) that can be attributed to subsequent glacial-interglacial cycles (Antoine *et al.* 2003, 2007; Bahain *et al.* 2007). This overlying sedimentation has also acted as a protection for interglacial sediments, including tufas, that sometimes survive at the top of a fluvial sequence and provide biostratigraphically significant and datable materials (Antoine *et al.* 2006; Bahain *et al.* 2007).

By contrast, in southern and eastern England, the geological framework is predominantly based on terrace stratigraphy, while the thin veneer of overlying non-fluvial deposits contributes little information. In most terrace systems there are few if any surviving interglacial sediments, including classic areas such as the Middle Thames. The Lower Thames is far more complex due to the effects of the 'coastal prism' (Lewis *et al.* 2004), whereby periods of sea-level rise have led to much thicker sequences of interglacial sediments, such as at Swanscombe, Thurrock, Purfleet, Crayford and Trafalgar Square (Schreve *et al.* 2001). Importantly, these interglacial sediments have provided datable and biostratigraphic materials to distinguish terrace deposits, independent of altitudinal considerations (Schreve 2001; Bridgland and Schreve 2004; Penkman *et al.* 2011).

The surface geology has had a profound effect on the recovery of Palaeolithic archaeology. After the ratification of the antiquity of handaxes in the Somme, the 'call to arms' by Evans (1860) led to a widespread search of other river systems in north-west Europe (e.g. Arnould *et al.* 1868; D'Acy 1875; Laville 1896; Gamble and Kruszynski 2009). Initially, the greatest success was in the Great Ouse catchment where terrace gravels, readily exploited in small, hand-dug workings, soon provided large quantities of handaxes in the Bedford area (Wyatt 1861) and in the Breckland (Flower 1867). This is exemplified by Warren Hill, where over 2000 handaxes were collected from small pits exploiting Bytham River gravel (Lewis *et al.* 2021). Further success was met in the Thames and Solent catchments in the growing conurbations of London, Reading, Southampton and Bournemouth, through quarries, road and railway construction, and foundation trenches (Evans 1872, 1896). Despite concerted efforts in the Somme, Seine, Meuse and Haine there were far fewer accessible gravel deposits for easy handaxe discoveries. The exceptions were the pits around Abbeville and Amiens where loess cover was unusually shallow (Antoine *et al.* 2007). Efforts elsewhere naturally waned, and attention was drawn to the findings in the caves of the Ardennes and southern France. Although occasional discoveries were made in the overlying loess, mainly associated with cailloutis, the aggregating power of river systems inevitably led to higher concentrations of artefacts in the fluvial gravels, albeit in secondary context. These were being discovered in their thousands in Britain (Roe 1968).

So how can understanding of these differences in collection history aid in the interpretation of the imbalance in distribution patterns in the archaeological records up to c. MIS 8 (Lower Palaeolithic) and from c. MIS 8 (early Middle Palaeolithic)? And can we still address questions of human dispersal and range limits for these periods?

Lower Palaeolithic: 0.9 Ma – 250 ka

For the Lower Palaeolithic, the combination of geological and historical circumstances, rather than occupation distribution or intensity, has caused the uneven distribution of sites between Britain and mainland north-west Europe. This is due to historically easier access to terrace deposits in Britain. Likewise, for the earliest sites of Happisburgh and Pakefield, discovery was through access to coastal exposures of the Cromer Forest-bed Formation, compared to more deeply buried deposits of a similar time-range elsewhere in north-west Europe. It is also clear that the northern and western boundaries of occupation cannot be defined, in part due to the burial or destruction of earlier landscapes by glacial sediments, but also because of the difficulty of identifying artefacts made on poor-quality flint, or other materials, beyond the rivers that drain the Cretaceous bedrock of southern and eastern England, northern France and central Belgium. Rare exceptions in Britain are discoveries in the Trent to the north

(Bridgland *et al.* 2014) and Warwickshire Avon to the west (Maddy *et al.* 1999; Shaw *et al.* 2016).

It seems likely that the pattern in England should be mirrored in mainland north-west Europe. Equally, the high concentration of sites that have been discovered in south-east England and East Anglia from MIS 13, is not the pattern that would be expected from a marginal zone, so occupation should be envisaged considerably to the north and west of the known distribution. In other words, although north-west Europe was an area of ebb and flow driven by climate, we cannot detect the margins of that zone.

Early Middle Palaeolithic: 250 ka – 180 ka

The pattern for the early Middle Palaeolithic sites is the converse, with only nine main sites from Britain and over 16 from mainland north-west Europe (Table 1). The imbalance would be far more marked if the large number of Middle Palaeolithic sites listed by Bordes (1954) were added, with most not included here due to problems of accessibility and consequently any recent dating work. Instead, knowledge of this period is based on fieldwork over the last 40 years, particularly in France, in response to building and infrastructure schemes (Table 1). A combination of better dating methods and understanding of the loess/palaeosol succession across the region has provided a wealth of well-documented archaeological sites.

The question remains as to whether there is a similar wealth of undiscovered sites in Britain? The paucity of protective loess on the interfluves must be a factor, and where artefacts occasionally occur within the infill of dolines, the deposits have been difficult to date (Bradley and Sampson 1978; Bailiff *et al.* 2013). However, lithic artefacts are difficult to destroy, so on eroded interfluves they should either survive as sub-surface artefact scatters or, through slope processes, be transported into lower terrace deposits. Despite extensive fieldwalking programmes across large areas of Britain, very few Middle Palaeolithic discoveries have been made on the interfluves (e.g. Halliwell and Parfitt 1993; Graf 2004; Ashton *et al.* 2018). Equally, large-scale collecting in lower terraces since the 1860s suggests that artefacts are genuinely scarce in these deposits (Ashton and Lewis 2002). There is also a paucity of artefacts in raised beach deposits, unlike north-west France. We conclude that there is a genuine low signature of early Middle Palaeolithic presence in Britain, other than in the lower reaches of the Thames. This interpretation is supported by the recent discovery of concentrations of early Middle Palaeolithic artefacts in the submerged river systems of the southern North Sea, where over 300 Levallois artefacts have been recovered, compared to only 35 from the whole of East Anglia (Davis *et al.* 2023). In combination, the evidence suggests that East Anglia and south-east England were indeed marginal areas for early Middle Palaeolithic populations, with more substantive occupation of the riverine landscapes now submerged beneath the North Sea, the lower reaches of the Thames and the rest of north-west Europe.

The reason why most of Britain may have been a marginal area for early Middle Palaeolithic occupation has been attributed to the reduction in land connections to mainland Europe (Ashton and Lewis 2002; Roebroeks 2014; Ashton *et al.* 2019). But a further reason might lie in the surface geology. Loess provides a rich source of nutrients for vegetation with large areas of the mammoth steppe thriving on soils developed on this substrate (Zimov *et al.* 2012). These authors have argued that maintenance of the steppe was dependent on persistent trampling from large herbivores, which prevented the growth of larger shrubs and trees. The expansion of the mammoth steppe into western Europe is well documented from MIS 7 (Kahlke 1999; 2014) with the suggestion that Neanderthals were better equipped to exploit the richness of this biome through more efficient hunting, better planning and extended social networks (Schreve 2001; Ashton and Lewis 2002; Scott 2011; Hérisson *et al.* 2016a). Any reduction in loess cover has been argued to result in less productive vegetational environments with lower biomass (Zimov *et al.* 2012). In Britain, the thinner loess cover has been redistributed through erosion to form a mosaic of poorer and richer soils (Catt 1978), a process that would have resulted in a fragmented mammoth steppe and more dispersed herds of herbivores. For humans, it can be suggested that Britain formed a zone where social networks would have to be over-extended to successfully exploit the landscape and its resources, and therefore only occasional forays into the interior are recorded.

Conclusion

Examination of the geological and historical factors that have influenced the discovery of earlier Palaeolithic sites in north-west Europe suggests that a northern limit to Lower Palaeolithic occupation cannot be identified due to glaciation of potential areas of occupation and the low visibility of artefacts in north-west France, Belgium and the Netherlands. Indeed, the high density of sites in south-east England and East Anglia from MIS 13, suggests that the occupational boundaries were considerably to the north and west.

By contrast, a western boundary in south-east England can be suggested for early Middle Palaeolithic populations, which was dependent on the productivity of the mammoth-steppe. We suggest that a key limiting factor on Neanderthal expansion into Britain was the difficulty of maintaining more extended social networks in marginal areas with dispersed resources. Despite land access to Britain during most of the late Middle Pleistocene there is still a paucity of archaeological data compared to the rest of north-west Europe. Therefore, this limiting factor may also apply to later Neanderthal and early Modern Human populations. The potential importance of the archaeological record from the North Sea Basin for the understanding of spatial distributions patterns was emphasised by Roebroeks (2014). Here we suggest that through the richness of the recent discoveries in the Palaeo-Yare and adjacent river systems we can identify a real boundary to the Middle Palaeolithic world.

Acknowledgements

We thank the Calleva Foundation for support of this work through the Pathways to Ancient Britain Project, and to Mark White for discussion of the ideas. We would also like to pay tribute to Wil Roebroeks, who has inspired this paper and many others, and whose work continues to be a catalyst for many of the current agendas in Palaeolithic studies.

References

Aldhouse-Green, S., R. Peterson and E.A. Walker (eds.). 2012. *Neanderthals in Wales: Pontnewydd and the Elwy Valley Caves*. Oxford: Oxbow Books.

Ameloot-Van der Heijden, N.. 1993. L'ensemble lithique du gisement de Longavesnes (Somme): illustration d'un problème de reconnaissance du débitage levallois dans une industrie à bifaces de la phase ancienne du Paléolithique moyen. *Bulletin de la Société Préhistorique Française* 90: 257-264

Ameloot-Van der Heijden, N., C. Dupuis, A.V. Munaut and J.J. Puissegur. 1996. Le gisement Paléolithique Moyen de Salouel (Somme, France). *L'Anthropologie* 100: 555-573.

Andrews, P., J. Cook, A. Currant and C.B. Stringer. 1999. *Westbury Cave: The Natural History Museum Excavations 1976 – 1984*. Bristol: CHERUB.

Antoine, P., P. Auguste, J-J. Bahain, P. Coudret, P. Depaepe, J-P. Fagnart, C. Falguères, M. Fontugne, M. Frechen, C. Hatté, A. Lamotte, M. Laurent, N. Limondin-Lozouet, J-L. Locht, N. Mercier, A-M. Moigne, A.V. Munaut, P. Ponel and D.D. Rousseau. 2003. Paléoenvironnements Pléistocènes et peuplements Paléolithiques dans le bassin de la Somme (Nord de la France). *Bulletin de la Société Préhistorique Française* 100: 5-28.

Antoine, P., N. Limondin-Lozouet, P. Auguste, J-L. Locht, B. Galehb, J.L. Reyss, E. Escudé, P. Carbonel, N. Mercier, J-J. Bahain, C. Falguères and P. Voinchet. 2006. Le tuf de Caours (Somme, France): mise en évidence d'une séquence éemienne et d'un site Paléolithique associé. *Quaternaire* 17: 281-320.

Antoine, P., N. Limondin-Lozouet, C. Chaussé, J-P. Lautridou, J-F. Pastre, P. Auguste, J-J. Bahain, C. Falguères and B. Galehb. 2007. Pleistocene fluvial terraces from northern France (Seine, Yonne, Somme): synthesis and new results. *Quaternary Science Reviews* 26: 2701-2723.

Antoine, P., M-H. Moncel, N. Limondin-Lozouet, J-L. Locht, J-J. Bahain, D. Moreno, P. Voinchet, P. Auguste, E. Stoetzel, J. Dabkowski, S. Bello, S. Parfitt, O. Tombret and B. Hardy. 2016. Palaeoenvironment and dating of the Early Acheulian from the type area of the (River) Somme basin (Northern France): new discoveries from the high terrace at Abbeville-Carrière Carpentier. *Quaternary Science Reviews* 149: 338-371.

Antoine, P., M-H. Moncel, J-L. Locht, J-J. Bahain, P. Voinchet, D. Hérisson and A. Hurel. 2019. The earliest record of Acheulean Human occupation from Northern Europe and the rediscovery of the Moulin Quignon site. *Nature Scientific Reports*. https://www.nature.com/articles/s41598-019-49400-w.

Antoine P., M-H. Moncel, J-L. Locht, N. Limondin-Lozouet, P. Auguste, E. Stoetzel, J. Dabkowski, P. Voinchet, J-J. Bahain and C. Falguères. 2015. Dating the earliest human occupation of Western Europe: New evidence from the fluvial terrace system of the Somme basin (Northern France). *Quaternary International* 370: 77-99.

Arnould, G., A. Briart, F-L. Cornet, A. Houzeau de Lehaie, N.L. Michot and A. Wesmael. 1868. Rapport sur les découvertes géologiques et archéologiques faites à Spiennes en 1867. *Mémoires et publications de la Société des Sciences, des Arts et des Lettres du Hainaut* 3e série, t. II (1866-1867): 356-392.

Ashton, N.M.. 2016. The human occupation of Britain during the Hoxnian Interglacial. *Quaternary International* 409: 41-53.

Ashton, N.M., J. Cook, S.G. Lewis and J. Rose (eds.). 1992. *High Lodge. Excavations by G. de G. Sieveking 1962-68 and J. Cook 1988*. London: British Museum Press.

Ashton, N.M., C.R.E. Harris and S.G. Lewis. 2018. Frontiers and route-ways from Europe: the Early Middle Palaeolithic of Britain. *Journal of Quaternary Science* 33: 194-211.

Ashton, N.M. and R.T. Hosfield. 2010. Mapping the human record in the British early Palaeolithic: evidence from the Solent River system. *Journal of Quaternary Science* 25: 737-753.

Ashton, N.M. and S.G. Lewis. 2002. Deserted Britain: Declining Populations in the British Late Middle Pleistocene. *Antiquity* 76: 388-396

Ashton, N.M., S.G. Lewis and R.T. Hosfiel.d 2011. Mapping the human record: population change during the later Middle Pleistocene. In: N.M. Ashton, S.G. Lewis and C.B. Stringer (eds.), *The Ancient Human Occupation of Britain*: 39-52. Amsterdam: Elsevier.

Ashton, N.M., S.G. Lewis and S.A. Parfitt 1998. *Excavations at the Lower Palaeolithic site at East Farm, Barnham, Suffolk 1989-94*. London: British Museum Press.

Bahain, J-J., C. Falgueres, M. Laurent, P. Voinchet, J-M. Dolo, P. Antoine and A. Tuffreau. 2007. ESR chronology of the Somme River Terraces system and of the first human settlements of northern France. *Quaternary Geochronology* 2: 356-362.

Bailiff, I.K., S.G. Lewis, H.C. Drinkall and M.J. White. 2013. Luminescence dating of sediments from a Palaeolithic site associated with a solution feature on the North Downs of Kent, UK. *Quaternary Geochronology* 18: 135-148.

Bates, M.R., F.F. Wenban-Smith, S.M. Bello, D.R. Bridgland, L.T. Buck, M.J. Collins, D.H. Keen, J. Leary, S.A. Parfitt, K.E.H. Penkman, E. Rhodes, C. Ryssaert and J.E. Whittaker. 2014. Late persistence of the Acheulian in southern Britain in an MIS 8 interstadial: evidence from Harnham, Wiltshire. *Quaternary Science Reviews* 101: 159-176.

Bordes, F. 1954.. *Les Limons Quaternaire du Bassin de la Seine. Stratigraphie et Archéologie Paléolithique*. Paris: Masson. (Archives de l'Institut de Paléontologie Humaine Memoire 26).

Bradley, B. and C.G. Sampson. 1978. Artefacts from the Cottages Site. In: C.G. Sampson (ed.), *Paleoecology and Archaeology of an Acheulian Site at Caddington, England*: 83-137. Dallas: Department of Anthropology, Southern Methodist University.

Bridgland, D.R.. 1994. *Quaternary of the Thames*. London: Chapman and Hall.

Bridgland, D.R., A. Howard, M.J. White and T.S. White. (eds.) 2014. *Quaternary of the Trent*. Oxford: Oxbow Books.

Bridgland, D.R., D.H. Keen, D.C. Schreve and M.J. White. 1998. Quaternary drainage of the Kentish Stour. In: J.B. Murton, C.A. Whiteman, M.R. Bates, D.R. Bridgland, A.J. Long, M.B. Roberts and M.P. Waller (eds.), *The Quaternary of Kent and Sussex*: 39-54. London: Quaternary Research Association.

Bridgland, D.R. and D.C. Schreve. 2004. Quaternary lithostratigraphy and mammalian biostratigraphy of the Lower Thames terrace system, south-east England. *Quaternaire* 15: 29-40.

Bridgland, D.R. and M.J. White. 2014. Fluvial archives as a framework for the Lower and Middle Palaeolithic: patterns of British artefact distribution and potential chronological implications. *Boreas* 43: 543-555.

Bringmans, P.M.M.A.. 2006. Seven well-preserved in-situ Middle Palaeolithic open-air sites in a loess-soil "climate-calender" sequence at Veldwezelt-Hezerwater, Belgium. *Notae Praehistoricae* 26: 9-18.

Bynoe, R., N.M. Ashton, T. Grimmer, P. Hoare, J. Leonard, S.G. Lewis, D. Nicholas and S.A. Parfitt. 2021. Coastal curios? An analysis of ex situ beach finds for mapping new Palaeolithic sites at Happisburgh, UK. *Journal of Quaternary Science* 36: 191-210.

Bynoe, R., M. Grant and J. Dix. 2022. *Strategic Support for Marine Development Management: Palaeolithic Archaeology and Landscape Reconstruction Offshore*. London: Historic England.

Callow, P. and J.M. Cornford. 1986. *La Cotte de St Brelade (Jersey) 1961-1978. Excavations by C.B.M. McBurney*. Norwich: Geobooks.

Catt, J.A.. 1978. The contribution of loess to soils in lowland Britain. In: S. Limbrey and J.G. Evans (eds.), *The Effect of Man on the Landscape: the Lowland Zone*. CBA Research Report 21: 12-20. York: CBA.

Cliquet, D. (ed.). 2008. *Le Site Pléistocène Moyen Recent de Ranville (Calvados – France) dans son Contexte Environnemental. Analyse du Fonctionnement d'une Aire de Boucherie Soutirée par un Réseau Karstique*. Liege: Université de Liège (ERAUL 119).

Cliquet, D., J-P. Lautridou, P. Antoine, A. Lamotte, M. Leroyer, N. Limondin-Lozouet and N. Mercier.

2009. La séquence loessique de Saint-Pierre-lès-Elbeuf (Normandie, France): nouvelles données archéologiques, géochronologiques et paléontologiques. *Quaternaire* 20: 321-343.

Cliquet, D.. 2013. Les occupations Paléolithiques en Normandie dans leur context chronostratigraphique: "bribes archéologiques". *Quaternaire* 24: 315-358.

Cliquet, D. and J-P. Lautridou. 2009. Les occupations humaines du Pléistocène moyen de Normandie dans leur cadre environnemental. *Quaternaire* 20: 303-320.

Commont, V.. 1908. Les industries de l'ancien Saint-Acheul. *L'Anthropologie* 19: 527-572.

Conway, B., J. McNabb and N.M. Ashton (eds.). 1996. *Excavations at Barnfield Pit, Swanscombe, 1968-1972*. London: British Museum.

Cook, J. and R.M. Jacobi. 1998. Observations on the artefacts from the Breccia at Kent's Cavern. In: N.M. Ashton, F. Healy and P. Pettitt (eds.), *Stone Age Archaeology. Essays in Honour of John Wymer*: 77-89. Oxford. (Lithic Studies Society Occasional Paper 6).

Cordy, J.-M., B. Bastin, M. Demaret-Fairon, C. Ek, R. Geeraerts, M-C. Groessens-Van Dyck, A. Ozer, R. Peuchot, Y. Quinif, J. Thorez and M. Ulrix-Closset. 1993. La grotte de la Belle-Roche (Sprimont, Province de Liège): un gisement paléontologique et archéologique d'exception au Benelux. *Bulletin de la Classe des Sciences. Académie Royale de Belgique*. 6e série IV: 165-186.

Coutard, S. and D. Cliquet. 2005. Chronostratigraphie des formations Pléistocènes et peuplement Paléolithique en contexte littoral: le Val de Saire (Normandie). *Bulletin de la Société Préhistorique Française* 102: 477-499.

D'Acy, E.. 1875. Quelques observations sur la succession chronologique des types appelés généralement type de Saint-Acheul et type du Moustier. *Matériaux pour l'Histoire Positive et Philosophique de l'Homme, II* Séries 6: 281-287.

Davis, R.J.. 2013. *The Palaeolithic Archaeology of the Solent River: Human Settlement History and Technology*. Unpublished PhD Thesis, University of Reading.

Davis, R.J. and N.M. Ashton. 2019. Landscapes, environments and societies: The development of culture in Lower Palaeolithic Europe. *Journal of Anthropological Archaeology* 56: 101107.

Davis, R.J, N.M. Ashton, R. Bynoe, J. Craven, R. Ferguson, C.E.R. Harris, M. Hatch, C. Johnson, J. Leonard, S.G. Lewis, D. Nicholas and M. Stevens. 2023. Middle Palaeolithic occupation of the southern North Sea Basin: evidence from the sandscaping sediments emplaced on the beach between Bacton and Walcott, Norfolk, UK. *Journal of Quaternary Science* 38: 866-890.

Davis, R.J., N.M. Ashton, M. Hatch, P.G. Hoare and S.G. Lewis. 2021a. Palaeolithic archaeology of the Bytham River: human occupation of Britain during the early Middle Pleistocene and its European context. *Journal of Quaternary Science* 36: 526-546.

Davis, R.J., N.M. Ashton, M. Hatch, R.T. Hosfield and S.G. Lewis. 2021b. Lower and early Middle Palaeolithic of Southern England: the evidence from the River Test. *Journal of Paleolithic Archaeology* 4: 23.

Davis, R.J., M. Hatch, N.M. Ashton, R.T. Hosfield and S.G. Lewis. 2016. The Palaeolithic record of Warsash, Hampshire, UK: implications for late Lower and early Middle Palaeolithic occupation history of Southern Britain. *Proceedings of the Geologists' Association* 127: 558-574.

Davis, R.J., S.G. Lewis, M. Hatch, N.M. Ashton, P. Voinchet, J-J. Bahain, L. Dale, F. Foulds, A. Rawlinson and M.J. White. 2024. A revised terrace stratigraphy and chronology for the Little Ouse River as a framework for interpreting the late Lower and early Middle Palaeolithic of central East Anglia, UK. *Quaternary Environments and Humans* 2, 100045.

Delagnes, A.. 1993. Un mode de production inédit au Paléolithique Moyen dans l'industrie du niveau 6e du Pucheuil (Seine-Maritime). *Paléo* 1993: 111-120.

Delagnes, A. and A. Ropars. 1996. *Paléolithique Moyen en Pays de Caux (Haute-Normandie). Le Pucheuil, Ettouteville: Deux Gisements de Plein Air en Milieu Loessique*. Documents d'Archéologie Française.

De Loecker, D. and W. Roebroeks. 2012. Beyond '15-minutes': revisiting the late Middle Pleistocene archaeological record of Maastricht-Belvédère (The Netherlands). *Analecta Praehistorica Leidensia* 43/44: 349-367.

Dennell, R.W., M. Martinón-Torres and J.M. Bermúdez de Castro. 2011. Hominin variability, climatic instability and population demography in Middle Pleistocene Europe. *Quaternary Science Reviews* 30: 1511-1524.

Di Modica, K.. 2010. *Les productions Lithiques du Paléolithique Moyen de Belgique: Variabilité des Systèmes d'Acquisition et des Technologies en Réponse à une Mosaïque d'Environnements Contrastés*. PhD Département de Préhistoire. Liège: Université de Liège.

Di Modica, K. and S. Pirson. 2016. The Lower to Middle Palaeolithic transition and the onset of prepared-core technologies in Belgium. *Quaternary International* 411 B: 95-107.

Di Modica, K., M. Toussaint, G. Abrams and S. Pirson. 2016. The Middle Palaeolithic from Belgium: Chronostratigraphy, territorial management and culture on a mosaic of contrasting environments. *Quaternary International* 411 A: 77-106.

Dupont, E.. 1866. Etude sur les fouilles scientifiques exécutées pendant l'hiver de 1865-1866 dans les cavernes des bords de la Lesse. *Bulletins de l'Académie Royale des Sciences des Lettres et des Beaux-Arts de Belgique*, 2e série, t. XXI: 31-54.

Evans, J.. 1861. On the occurrence of flint implements in undisturbed beds of gravel, sand and clay. *Archaeologia* 38: 280-307.

Evans, J.. 1872. *Ancient Stone Implements, Weapons and Ornaments of Great Britain*. London: Longmans and Co..

Evans, J.. 1896. *The Ancient Stone Implements, Weapons and Ornaments of Great Britain*. London: Longmans.

Flower, J.W.. 1867. On some flint implements lately found in the valley of the Little Ouse River, at Thetford, Norfolk. *Quarterly Journal of the Geological Society of London* 22: 45-56.

Gamble, C.S.. 2009. Human display and dispersal: A case study from biotidal Britain in the Middle and Upper Pleistocene. *Evolutionary Anthropology* 18: 144-156.

Gamble, C.S. and R. Kruszynski. 2009. John Evans, Joseph Prestwich, and the stone that shattered the time barrier. *Antiquity* 83: 461-75.

Gibbard, P.L.. 1995. The formation of the Strait of Dover. In: R.C. Preece (ed.), *Island Britain: a Quaternary Perspective*: 15-26. London. (Geological Society Special Publication 96).

Gibbard, P.L., R.G. West, A. Pasanen, J.J. Wymer, S. Boreham, K.M. Cohen and C. Rolfe. 2008. Pleistocene geology of the Palaeolithic sequence at Redhill, Thetford, Norfolk, England. *Proceedings of the Geologists' Association* 119: 175-192.

Graf, A.. 2004. 'A cause for wonder': preliminary observations on further Palaeolithic surface finds by Ron Waite from around Nuneaton, Warwickshire. *Lithics* 25: 78-105.

Gupta, S., J.S. Collier, A. Palmer-Felgate and G. Potter. 2007. Catastrophic flooding origin of the shelf valley systems of the English Channel. *Nature* 448: 342-345.

Halliwell, G. and K. Parfitt. 1993. Non-river gravel Lower and Middle Palaeolithic discoveries in east Kent. *Kent Archaeological Review* 114: 80-89.

Harding, P., D.R. Bridgland, D.H. Keen and R. Rogerson. 1991. A Palaeolithic site rediscovered at Biddenham, Bedfordshire. *Bedfordshire Archaeology* 19: 87-90.

Hatch, M., R.J. Davis, S.G. Lewis, N.M. Ashton, R.M. Briant and S. Lukas. 2017. The stratigraphy and chronology of the fluvial sediments at Warsash, UK: Implications for the Palaeolithic archaeology of the River Test. *Proceedings of the Geologists' Association* 128: 198-221.

Hérisson, D.. 2012. *Etude des Comportements des Premiers Néandertaliens du Nord de la France: les occupations Saaliennes des Gisements de Biache-Saint-Vaast et de Therdonne*. Thèse de doctorat. Université de Lille, Villeneuve-d'Ascq.

Hérisson, D., M. Brenet, D. Cliquet, M-H. Moncel, J. Richter, B. Scott, A. Van Baelen, K. Di Modica, D. De Loecker, N. Ashton, L. Bourguignon, A. Delagnes, J-P. Faivre, M. Folgado-Lopez, J-L. Locht, M. Pope, J-P. Raynal, W. Roebroeks, C. Santagata, A. Turq and P. Van Peer. 2016a. The emergence of the Middle Palaeolithic in north-western Europe and its southern fringes. *Quaternary International* 411: 233-283.

Hérisson, D., S. Coutard, E. Goval, J-L. Locht, P. Antoine, Y. Chantreau and N. Debenham. 2016b. A new key-site for the end of Lower Palaeolithic and the onset of Middle Palaeolithic at Etricourt-Manancourt (Somme, France). *Quaternary International* 409: 73-91.

Hijma, M.P., K.M. Cohen, W. Roebroeks, W.E. Westerhoff and F.S. Busschers. 2012. Pleistocene Rhine – Thames landscapes: geological background for hominin occupation of the southern North Sea region. *Journal of Quaternary Science* 27: 17-39.

Hosfield, R.T. and C.P. Green. 2013. *Quaternary History and Palaeolithic Archaeology in the Axe Valley at Broom, South West England*. Oxford: Oxbow Books.

Hublin, J.-J. and W. Roebroeks. 2009. Ebb and flow or regional extinctions? On the character of Neandertal occupation of northern environments. *Comptes rendus – Palevol* 8: 503-509.

Kahlke, R.-D.. 1999. *The History of the Origin, Evolution and Dispersal of the Late Pleistocene* Mammuthus-Coelodonta *Faunal Complex in Eurasia (Large Mammals)*. Fenske Companies, Rapid City.

Kahlke, R-D.. 2014. The origin of Eurasian Mammoth Faunas (*Mammuthus – Coelodonta* Faunal Complex). *Quaternary Science Reviews* 96: 32-49.

Keen, D.H., T. Hardaker and A.O.T. Lang. 2006. A Lower Palaeolithic industry from the Cromerian (MIS 13) Baginton Formation of Waverley Wood and Wood Farm Pits, Bubbenhall, Warwickshire, UK. *Journal of Quaternary Science* 21: 457-470.

Kerney, M.P.. 1971. Interglacial deposits at Barnfield Pit, Swanscombe and their molluscan fauna. *Journal of the Geological Society of London* 127: 69-86.

Key, A., T. Lauer, M.M. Skinner, M. Pope, D.R. Bridgland, L. Nobel and T. Proffitt. 2022. On the earliest Acheulean in Britain: first dates and in-situ artefacts from the MIS 15 site of Fordwich, Kent (UK). *Royal Society Open Science* 9: 211904.

Koehler, H.. 2008. L'apport du gisement des Osiers à Bapaume (Pas-de-Calais) au débat sur l'émergence du Paléolithique moyen dans le Nord de la France. *Bulletin de la Société Préhistorique Française* 105: 709-735.

Lamotte, A., A. Tuffreau and J-L. Marcy. 2001. La série MS du gisement Acheuléen de La Ferme de l'Epinette á Cagny (Somme, France). In: A. Tuffreau (ed.), *L'Acheuléen dans la Vallée de la Somme: Données Récent*: 137-148. Lille: Université des Sciences et Technologies de Lille. (Publications du CERP 6).

Laville, A.. 1896. Découverte d'un gisement de silex taillés moustériens à Paris. *L'Anthropologie* 7: 683.

Lehmkuhl, F., J.J. Nett, S. Pötter, P. Schulte, T. Sprafke, C. Jary, P. Antoine, L. Wacha, D. Wolf, A. Zerboni, J. Hošeki, S.B. Marković, I. Obreht, P. Sümegi, D. Veres, C. Zeeden, B. Boemke, V. Schauber, J. Viehweger and U. Hambach. 2021. Loess landscapes of Europe- Mapping, geomorphology, and zonal differentiation. *Earth Science Reviews* 215: 103496.

Lewis, S.G., N.M. Ashton, M. Field, P.G. Hoare, H. Kamermans, M. Knul, H.J. Mücher, S.A. Parfitt, W. Roebroeks and M.J. Sier. 2019. Human occupation of northern Europe in MIS 13: Happisburgh Site 1 (Norfolk, UK) and its European context. *Quaternary Science Reviews* 211: 34-58.

Lewis, S.G., N.M. Ashton and R.M. Jacobi. 2011. Testing human presence in the last interglacial (MIS 5e): A review of the British evidence. In: N.M. Ashton, S.G. Lewis and C.B. Stringer (eds.), *The Ancient Human Occupation of Britain*: 125-164. Amsterdam: Elsevier.

Lewis, S.G., D. Maddy and S. Glenday. 2004. The Thames Valley Sediment Conveyor: Fluvial system development over the last two interglacial-glacial cycles. *Quaternaire* 15: 17-28.

Lewis, S.G., N.M. Ashton, R.J. Davis, M. Hatch, P.G. Hoare, P. Voinchet and J-J. Bahain. 2021. A revised terrace stratigraphy and new ESR geochronology of the early Middle Pleistocene Bytham River in the Breckland of East Anglia, UK. *Quaternary Science Reviews* 269: 107113.

Limondin-Lozouet, N., J. Dabkowski and P. Antoine. 2020. Palaeoenvironmental dynamics of the MIS 11 interglacial in north-western Europe based on the malacological succession from La Celle (Seine Valley, France): Relationship with glacial refugia and palaeobiodiversity. *Palaeogeography, Palaeoclimatology, Palaeoecology* 560: 110044.

Locht, J-L., P. Antoine, D. Hérisson, G. Gadebois and N. Debenham. 2010. Une occupation de la phase ancienne du Paléolithique Moyen à Therdonne (Oise). Chronostratigraphie, production de pointes Levallois et réduction des nucléus. *Gallia Préhistoire* 52: 1-32.

Maddy, D., S.G. Lewis and D.H. Keen. 1999. Pleistocene palaeomeanders of the River Avon, Warwickshire. *Proceedings of the Geologists' Association* 110: 163-172.

Marcy, J.-L.. 1989. L'outillage sur éclat du gisement acheuléen de la vallée du Muid à Gouzeaucourt (Nord): Premiers résultats. In: A. Tuffreau (ed.), *Paléolithique et Mésolithique Du Nord de La France. Nouvelles Recherches*: 31-41. Lille: Université des Sciences et Technologies de Lille. (Publications du CERP 1).

McNabb, J., R.T. Hosfield, K. Dearling, D. Barker, K. Strutt, J. Cole, M. Bates and P. Toms. 2012. Recent Work at the Lower Palaeolithic Site of Corfe Mullen, Dorset, England. *Proceedings of the Prehistoric Society* 78: 35-50.

Meijer, T. and P. Cleveringa. 2009. Aminostratigraphy of Middle and Late Pleistocene deposits in The Netherlands and the southern part of the North Sea Basin. *Global and Planetary Change* 68: 326-345.

Moncel, M-H., N.M. Ashton, A. Lamotte, A. Tuffreau, D. Cliquet and J. Despriée. 2015. The early Acheulean of north-west Europe. *Journal of Anthropological Archaeology* 40: 302-331.

Parfitt, S.A., R.W. Barendregt, M. Breda, I. Candy, M.J. Collins, G.R. Coope, P. Durbridge, M.H. Field, J.R. Lee, A.M. Lister, R. Mutch, K.E.H. Penkman, R.C. Preece, J. Rose, C.B. Stringer, R. Symmons, J.E. Whittaker, J.J. Wymer and A.J. Stuart. 2005. The earliest record of human activity in northern Europe. *Nature* 438: 1008-1012.

Parfitt, S.A., N.M. Ashton, S.G. Lewis, R.L. Abel, G.R. Coope, M.H. Field, M.H., R. Gale, P.G. Hoare, N.R. Larkin, M.D. Lewis, V. Karloukovski, B.A. Maher, S.M. Peglar, R.C. Preece, J.E. Whittaker and C.B. Stringer. 2010. Early Pleistocene human occupation at the edge of the boreal zone in northwest Europe. *Nature* 466: 229-233.

Parfitt, S.A. and R.C. Preece. 2022. New palaeontological evidence suggests an early Middle Pleistocene age for the lower levels of Sun Hole Cave, Cheddar, Somerset, UK. *Proceedings of the Geologists' Association* 133: 162-175.

Penkman, K.E.H., R.C. Preece, D. R. Bridgland, D.H. Keen, T. Meijer, S. A. Parfitt, T. S. White and M. J. Collins. 2011. A chronological framework for the British Quaternary based on *Bithynia* opercula. *Nature* 476: 446-449.

Pirson, S., P. Haesaerts and K. Di Modica. 2009. Cadre chronostratigraphique des principaux gisements du Paléolithique moyen du bassin de la Haine: un état de la question. In: K. Di Modica and C. Jungels (eds.), *Paléolithique moyen en Wallonie. La collection Louis Eloy*: 58-77. Bruxelles: Service du Patrimoine Culturel de la Communauté Française de Belgique.

Pope, M., M.B. Roberts, A. Maxted and P. Jones. 2009. The Valdoe: Archaeology of a locality within the Boxgrove palaeolandscape, West Sussex. *Proceedings of the Prehistoric Society* 75: 239-263.

Pope M., S.A. Parfitt and M.B. Roberts. 2020. *The Horse Butchery Site: A high-resolution record of Lower Palaeolithic hominin behaviour at Boxgrove, UK.* Portslade: SpoilHeap Publications.

Preece, R.C. and S.A. Parfitt. 2022. Environmental heterogeneity of the Lower Palaeolithic land surface on the Goodwood-Slindon Raised Beach: comparisons of the records from Boxgrove and Valdoe, Sussex, UK. *Journal of Quaternary Science* 37: 572-592.

Preece, R.C., S.A. Parfitt, D.R. Bridgland, S.G. Lewis, P.J. Rowe, T.C. Atkinson, I. Candy, N.C. Debenham, K.E.H. Penkman, E.J. Rhodes, J.-L. Schwenninger, H.I. Griffiths, J.E. Whittaker and C. Gleed-Owen. 2007. Terrestrial environments during MIS 11: evidence from the Palaeolithic site at West Stow, Suffolk, UK. *Quaternary Science Reviews* 26: 1236-1300.

Roberts, M.B. and S.A. Parfitt (eds.). 1999. *Boxgrove: A Middle Palaeolithic Pleistocene Hominid Site at Eartham Quarry, Boxgrove, West Sussex*. London: English Heritage.

Roe, D.A.. 1968. *A Gazetteer of British Lower and Middle Palaeolithic Sites*. York (CBA Research Report No. 8).

Roebroeks, W.. 1988. From find scatters to early hominid behaviour. A study of Middle Paleolithic riverside settlements at Maastricht-Belvédère (The Netherlands). *Analecta Praehistorica Leidensia* 21.

Roebroeks, W. 2006. The human colonisation of Europe: where are we? *Journal of Quaternary Science* 21: 425-436.

Roebroeks, W.. 2014. Terra incognita: The Palaeolithic record of northwest Europe and the information potential of the southern North Sea. *Netherlands Journal of Geosciences* 93: 43-53.

Ryssaert, C.. 2006. Some new insights in an old collection: lithic technology at Mesvin IV. *Notae Praehistoricae* 26: 91-99.

Schreve, D.C.. 2001. Differentiation of the British late Middle Pleistocene interglacials: the evidence from mammalian biostratigraphy. *Quaternary Science Reviews* 20: 1693-1705.

Schreve, D.C., D.R. Bridgland, P. Allen, J.J. Blackford, C.P. Gleed-Owen, H.I. Griffiths, D.H. Keen and M.J. White. 2002. Sedimentology, palaeontology and archaeology of late Middle Pleistocene River Thames terrace deposits at Purfleet, Essex, UK. *Quaternary Science Reviews* 21: 1423-1464.

Schreve, D.C., P. Harding, M.J. White, D.R. Bridgland, P. Allen, F. Clayton, D.H. Keen and K.E.H. Penkman. 2006. A Levallois knapping site at West Thurrock, Lower Thames, UK: Its Quaternary context, environment and age. *Proceedings of the Prehistoric Society* 72: 21-52.

Scott, B.. 2011. *Becoming Neanderthal: The Earlier British Middle Palaeolithic*. Oxford: Oxbow Books.

Scott, B. and N.M. Ashton. 2011. The early Middle Palaeolithic: The European context. In: N.M. Ashton, S.G. Lewis and C.B. Stringer (eds.), *The Ancient Human Occupation of Britain*: 91-112. Amsterdam: Elsevier.

Scott, B. A.D. Shaw, K. Scott and M. Pope (eds.). 2023. *Repeopling La Manche New Perspectives on Neanderthal Archaeology and Landscapes from La Cotte de St Brelade*. Prehistoric Society Research Paper No. 10, Oxford: Oxbow.

Shaw, A.D., N. Daffern and O. Russell. 2016. The Palaeolithic in Worcestershire, UK A.D. *Lithics* 36, 41-54.

Shaw, A., D. Young and H. Hawkins. 2023. The submerged Palaeo-Yare: new Middle Palaeolithic archaeological finds from the southern North Sea. *Proceedings of the Prehistoric Society* 89: 273-297.

Shotton, F.W., D.H. Keen, G.R. Coope, A.P. Currant, P.L. Gibbard, M. Aalto, S.M. Peglar and J.E. Robinson. 1993. The Middle Pleistocene deposits of Waverly Wood pit, Warwickshire, England. *Journal of Quaternary Science* 8: 293-325.

Smith, A.J.. 1985. A catastrophic origin for the palaeovalley system of the eastern English Channel. *Marine Geology* 64: 65-75.

Stringer, C.B. 2005. *Homo Britannicus. The Incredible Story of Human Life in Britain*. London: Penguin.

Tizzard, L., A.R. Bicket, J. Benjamin and D. De Loecker. 2014. A Middle Palaeolithic site in the southern North Sea: investigating the archaeology and palaeogeography of Area 240. *Journal of Quaternary Science* 29: 698-710.

Tuffreau, A.. 1987. *Le Paléolithique inferieur et moyen du Nord de la France (Nord, Pas-de-Calais, Picardie) dans son Cadre Stratigraphique*. Thèse de Doctorat d'Etat, Université de Sciences et Technologie Lille.

Tuffreau, A.. 2001. Le gisement de Cagny-l'Epinette: Introduction. In: A. Tuffreau (ed.) *L'Acheuléen dans la Vallée de la Somme et le Paléolithique Moyen dans le Nord de la France: Données Récentes*: 109-112. Lille (Publications du CERP 6).

Tuffreau, A.. 1988. Historique des fouilles effectuées à Biache-Saint-Vaast. In: A. Tuffreau and J. Sommé (eds.), *Le Gisement Paléolithique Moyen de Biache-Saint-Vaast (Pas-de-Calais). Stratigraphie, Environnement, Études Archéologiques (1ère Partie)*: 15-24. Paris: Société Préhistorique Française.

Tuffreau, A., P. Antoine, J-L. Marcy and N. Segard. 2001. Les industries Paléolithiques à nombreux bifaces du Mont de l'Evangile à Gentelles (Somme). In: D. Cliquet (ed.), *Les Industries à Outils Bifaciaux du Paléolithique Moyen d'Europe Occidentale*: 29-41 Liège: Université de Liège. (ERAUL 98).

Tuffreau, A. and J-P. Bouchet. 1985. Le gisement Acheuléen de la Vallée du Muid à Gouzeaucourt (Nord). *Bulletin de la Société Préhistorique Française* 82: 291-306.

Tuffreau, A., A. Lamotte and E. Goval. 2008. Les industries Acheuléennes de la France septentrionale. *Anthropologie* 112: 104-139.

Ulrix-Closset, M.. 1981. *Le Paléolithique Moyen dans le Bassin Mosan en Belgique*. Liège: Bibliothèque de la Faculté de Philosophie et Lettres de l'Université de Liège.

Van Baelen, A., P. Van Peer, E.P.M. Meijs, J.J. Hus, F. Lehmkuhl, P. Schulte and J. Zens. 2017. *The Lower to Middle Palaeolithic Transition in Northwestern Europe: Evidence from Kesselt-Op de Schans*. Leuven: Leuven University Press.

Wenban-Smith, F.F. 2004. Handaxe typology and Lower Palaeolithic cultural development: ficrons, cleavers and two giant handaxes from Cuxton. *Lithics* 25: 11-21.

Wenban-Smith, F.F. (ed.). 2013. *The Ebbsfleet Elephant: excavations at Southfleet Road, Swanscombe in advance of High Speed 1, 2003-4.* Oxford: Oxford Archaeology.

Wenban-Smith, F.F., E. Stafford, M. Bates and S.A. Parfitt. 2020. *Prehistoric Ebbsfleet: Excavations and research in advance of High Speed 1 and South Thameside Development Route 4, 1989-2003*. Oxford. (Wessex Archaeology Monograph 7).

Wesselingh, F.P., T. Meijer, R. Harting, M. Bakker and F.S. Busschers. 2022. New marine warm-temperate molluscan assemblage demonstrates warm conditions during the Middle Pleistocene of the North Sea Basin. *Netherlands Journal of Geosciences 102*, e3.

Wessex Archaeology. 1993. *The Southern Rivers Palaeolithic Project Report No. 1. 1991-1992, The Upper Thames Valley, the Kennet Valley and the Solent Drainage System*. Salisbury: Wessex Archaeology.

Wessex Archaeology. 1996a. *Regions 7 (Thames) and 10 (Warwickshire Avon). English Rivers Palaeolithic Project Report*. Salisbury: Wessex Archaeology.

Wessex Archaeology. 1996b. *The Great Ouse Drainage and the Yorkshire and Lincolnshire Wolds. The English Rivers Palaeolithic Project Report 2*. Salisbury: Wessex Archaeology.

White, M.J. and D.C. Schreve. 2000. Island Britain – Peninsula Britain: Palaeogeography, colonization, and the Lower Palaeolithic settlement of the British Isles. *Proceedings of the Prehistoric Society* 66: 1-28.

White, T.S., R.C. Preece and J.E. Whittaker. 2013. Molluscan and ostracod successions from Dierden's Pit, Swanscombe: insights into the fluvial history, sea-level record and human occupation of the Hoxnian Thames. *Quaternary Science Reviews* 70: 73-90.

White, M.J., N.M. Ashton and D.R. Bridgland. 2019. Twisted handaxes in Middle Pleistocene Britain and their implications for regional-scale cultural variation and the deep history of Acheulean hominin groups. *Proceedings of the Prehistoric Society* 85: 61-81.

White, M.J., D.R. Bridgland, D.C. Schreve, T.S. White and K.E.H. Penkman. 2018. Well-dated fluvial sequences as templates for patterns of handaxe distribution:

Understanding the record of Acheulean activity in the Thames and its correlatives. *Quaternary International* 480: 118-131.

Wyatt, J. 1861. Flint implements in the drift of Bedfordshire. *Notes of the Bedfordshire Architectural and Archaeological Society* 10: 146-158.

Wymer, J.J.. 1968. *The Lower Palaeolithic of Britain: as represented by the Thames Valley*. London: John Baker.

Wymer, J.J.. 1985. *Palaeolithic Sites of East Anglia*. Norwich: Geo Books.

Wymer, J. J.. 1999. *The Lower Palaeolithic Occupation of Britain*. Salisbury: Wessex Archaeology and English Heritage.

Zimov, S.A., N.S. Zimov, A.N. Tikhonov and F.S. Chapin III. 2012. Mammoth steppe: a high-productivity phenomenon. *Quaternary Science Reviews* 57: 26-45.